Green RFID Systems

Combining cutting-edge technologies and techniques with existing approaches, this book equips you with the tools and knowledge needed to develop new energy-efficient and environmentally friendly RFID (radio frequency identification) systems.

As well as covering RFID basics, a wide range of new technologies is discussed, including biodegradable and recyclable material use, energy scavenging, passive and chipless architectures, RFID passive sensors, networked RFID and RFID sensors, organic electronic devices, textile electronics, and distributed and wide area electronics. Providing a clear description of how RFID technology can enable the evolution of the Internet of Things, the book guides you down the path to facing new challenges as we move towards ubiquitous sensing for smart environments and a networked society. This is an ideal guide for researchers in academia and industry, technical managers, and graduate students in RF and wireless communications.

Luca Roselli is a Professor of Applied Electronics at the University of Perugia, Italy, where he has been working since 1991. His research interests focus on the design of high frequency electronic circuits and systems, including development of numerical methods for electronic circuit analyses and new material electronics. He has published more than 200 papers and founded two companies (WiS Srl and DiES Srl) in this field. He is strongly involved in MTT-S activity, participating in several MTT Technical Committees (RFID – past chair, RF nanotechnologies, Wireless Power Transfer), organizing the first Wireless Power Transfer Conference (WPTC 2013) and serving as a reviewer for several journals and peer reviewed conferences. He is also a member of the ERC panel.

EuMA High Frequency Technologies Series

Series Editor
Peter Russer, Technical University of Munich

Homayoun Nikookar, *Wavelet Radio*
Thomas Zwick, Werner Wiesbeck, Jens Timmermann, and Grzegorz Adamiuk (Eds.), *Ultra-wideband RF System Engineering*
Er-Ping Li and Hong-Son Chu, *Plasmonic Nanoelectronics and Sensing*
Luca Roselli (Ed.), *Green RFID Systems*

Forthcoming

Peter Russer, Johannes Russer, Uwe Siart, and Andreas Cangellaris, *Interference and Noise in Electromagnetics*
Maurizio Bozzi, Apostolos Georgiadis, and Ke Wu, *Substrate Integrated Waveguides*
George Deligeorgis, *Graphene Device Engineering*
Luca Pierantoni and Fabio Coccetti, *Radiofrequency Nanoelectronics Engineering*
Alexander Yarovoy, *Introduction to UWB Wireless Technology and Applications*
Natalia Nikolova, *Introduction to Microwave Imaging*
Vesna Crnojevic-Bengin, *Advances in Multi-band Microstrip Filters*
Philippe Ferrari, Rolf Jakoby, Onur Karabey, and Gustavo Rehder, *Reconfigurable Circuits and Technologies for Smart Millimeter-Wave Systems*

Green RFID Systems

Edited by

LUCA ROSELLI

University of Perugia, Italy

CAMBRIDGE
UNIVERSITY PRESS

CAMBRIDGE
UNIVERSITY PRESS

University Printing House, Cambridge CB2 8BS, United Kingdom

Cambridge University Press is part of the University of Cambridge.

It furthers the University's mission by disseminating knowledge in the pursuit of education, learning, and research at the highest international levels of excellence.

www.cambridge.org
Information on this title: www.cambridge.org/9781107030404

First published 2014

Printed in the United Kingdom by Clays, St Ives plc

A catalog record for this publication is available from the British Library

ISBN 978-1-107-03040-4 Hardback

Contents

Contributors

Federico Alimenti
Università degli studi di Perugia

Alírio Boaventura
Universidade de Aveiro

Nuno Borges Carvalho
Universidade de Aveiro

Ana Collado
Centre Tecnològic de Telecomunicacions de Catalunya (CTTC)

Benjamin S. Cook
Georgia Institute of Technology

Alessandra Costanzo
University of Bologna

Ricardo Fernandes
Universidade de Aveiro

Apostolos Georgiadis
Centre Tecnològic de Telecomunicacions de Catalunya (CTTC)

Amelie Hagelauer
University of Erlangen-Nuremberg

Sangkil Kim
Georgia Institute of Technology

Vasileios Lakafosis
Georgia Institute of Technology

Taoran Le
Georgia Institute of Technology

Paolo Lugli
Technische Universität München

Chiara Mariotti
Università degli studi di Perugia

Diego Masotti
University of Bologna

João Nuno Matos
Universidade de Aveiro

Giulia Orecchini
Università Degli Studi di Perugia

Luca Pierantoni
Università Politecnica delle Marche

Hendrik Rogier
Ghent University

Aldo Romani
University of Bologna

Marco Tartagni
University of Bologna

Manos M. Tentzeris
Georgia Institute of Technology

Anya Traille
Georgia Institute of Technology

Thomas Ussmueller
University of Erlangen-Nuremberg

Luigi Vallozzi
Ghent University

Patrick Van Torre
Ghent University

Rushi Vyas
Georgia Institute of Technology

Robert Weigel
University of Erlangen-Nuremberg

Introduction

Since a beginning in the late forties, the development of solid state electronics has been characterized by relentless progress toward miniaturization and concentration of functionalities, mainly computational operations, in an ever smaller volume. This behavior was first expressed theoretically by Gordon Moore in 1964 when he formulated his greatly renowned "Moore's Law."

In the last decade we have experienced a fairly new scenario: on the one hand, technology development has introduced new concepts and materials. Beyond conventional semiconductors (group IV and III-V elements), carbon-based materials such as carbon nanotubes (CNT) and graphene, along with organic semiconductors, have been investigated. Focusing on substrates and supports, conventional materials (ceramics, Teflon-based, glass fiber, and so on) come alongside others that are new, low cost, easily producible in large areas, and eco-friendly. Material science, to this extent, is in its infancy, nevertheless some materials can be cited already: paper, bioplastics, PET, and likely many others in the near future. On the other hand, traditional electronics, still developing according to Moore's Law in the miniaturization direction, often referred to as "More Moore axis," is experiencing this inherent saturation. Incidentally, Moore himself, in a famous interview in 2004, on the occasion of the celebration of the forty years since the Moore's Law formulation, said, referring to it: "It can't continue forever. The nature of exponentials is that you push them out and eventually disaster happens." He was the only one having the credibility to say that, at that time. I wish to expand a bit on this: it is a matter of fact that the ICT policy of the industrialized world, in supporting societal development and steering investments, is addressing new challenges. From the technological side, this trend can be summarized by the well known expression "More than Moore." This emphatic expression actually means that beside investments to foster miniaturization (More Moore), technology developments have to be directed towards adding more functionalities to electronic systems. It is worth noting that this new direction is not an alternative to the previous one, instead it stimulates new investigations and new ways to exploit all the technologies developed and being developed, ultimately enabling new solutions for societal challenges and needs.

At this point I wish to take this concept to the extreme: we are assisting, combining More Moore and More than Moore approaches, in a large development of multi-functional, heterogeneous, highly miniaturized electronic systems: why not extend these systems to distributed architectures? Once we have multi-functional miniaturized systems, why don't we incorporate them in everyday objects? We are clearly entering the world of Internet of Things (IoT); a world where objects in general (things, animals, humans) can host electronic devices, collect information, and process and react to this information doing something by themselves or just send information to the internet for further decision taking.

Is this a precise vision of the future? We cannot say "precise," but certainly evolution is also in this direction.

Now the question is: how can scientists and technicians foster this evolving scenario? If we refer to what can be seen as a stabilized "networked society," where information is automatically collected by "objects" that can react by themselves (smart objects) or transfer information to humans via the internet, we have to face mainly two big challenges: first, it is impossible to connect all the objects to the grid or put batteries that last for life in them; second, we have to avoid producing long lasting apparatuses that outlive their host objects, thus causing pollution. Internet of Things is a very multidisciplinary evolution area, but whatever the developments towards the Internet of Things, it will need technological platforms that must be eco-friendly and energetically autonomous – in one word "green."

Keeping on thinking about the future networked society and understanding that the adopted technologies must be "green," the next question is: given that the technological platform must be green, what are the architectures and protocols suitable to support new distributed functionalities?

Without lack of generality, a fascinating possibility enabled by the availability of distributed smart objects consists of collecting information in a distributed instead of a concentrated way. Mapping of a parameter over a mesh, the granularity of which depends on the spatial distribution of smart objects, is straightforward. The problem is that these objects have to collect information and transfer it to the network autonomously. One of the most suitable means to satisfy these requirements is the RFID concept. RFID, in fact, is a way to get information from a tag by means of an interrogation protocol between a tag and a reader.

It is not the intention to extend the RFID description in this introduction, but it is just worth underlining that, in principle, this communication approach allows information to be gathered and transferred to a network, without either physical contact to the tagged object or power supplying it via a connection to the grid: thus objects become smart, but remain autonomous.

The main objective of this book is just to give an overview of a basic technology, foreseeable as of now, to face the development of distributed apparatuses for applications in the Internet of Things in line with the vision of a networked society.

To meet this objective the book is structured as follows:

First a historical overview of the development of RFID systems as well as a recap of RFID working principles is given in Chapters 1 and 2.

Second, a description of the main challenges posed by tagging objects in the context of distributed, RFID-based solutions, is given in Chapters 3 and 4, namely: energy harvesting, required to provide objects' autonomy, and RFID sensor development to provide the functionality of getting information from the environment.

Third, the book enters the area of unconventional RFID systems devoted to new distributed scenarios. Chapters 5 and 6 refer to unconventional RFID applications and how to connect RFID systems to a network, respectively.

Fourth, materials and related technologies for the realization of green RFID systems to be used within these new massively distributed applications are described in Chapters 7 and 8.

Fifth, in Chapter 9, chipless architecture has been introduced to provide a means to push to the extreme the possibility of realizing ultra-low cost, low power tags for sensing objects.

Sixth, eventually in Chapter 10, examples of new materials, specifically textile ones, are given to provide new solutions in all those applications directly involving humans.

1 Context analysis

Apostolos Georgiadis and Ana Collado

1.1 Introduction

This chapter addresses the present challenges facing radio frequency identification (RFID) systems and aims to highlight the perspective and potential of this technology which, due to attractive properties such as low power, low cost, and ability to integrate sensing functionality, is enjoying widespread application and a growing market potential. The simplicity and minimalist approach in the design of RFID tag circuits has led to their widespread utilization in supply chain and logistics applications. IDTechEX reported that the total RFID hardware market value was 5.56 billion US$ in 2009 and is expected to increase above 25 billion US$ by 2019 [1]. In 2005 the large volume electronic product code (EPC) tag cost was US$0.13 [2], while the recent volume pricing of UHF RFID transponders is approximately US$0.10 (for example [3]). A further price reduction below US$0.05 is desired in order to achieve additional market penetration [4]. The World Wide Research Forum (WWRF) estimates that 7 trillion wireless devices will be serving 7 billion people by 2017 [5]. RFID systems with sensing functionality are a fundamental technology for realizing a network of interconnected devices, which represents the vision of ubiquitous sensing and communication.

This chapter aims to set out the background to the exciting technological advances and properties that RFID systems are required to accomplish, demonstrating how RFID technology is capable of providing a solution towards ubiquitous sensing smart environments. It begins with a brief historical perspective, showing the major milestones in the evolution and commercialization of RFID technology. The remaining paragraphs highlight existing performance challenges and problems, from a brief description of networking and security issues to more detailed circuit, system, material, and computer aided design (CAD) topics, which will be addressed in more detail in the rest of this book.

1.2 Historical perspective of RFID

RFID systems operate based on the principle of *back-scatter communication* which has its roots in radar technology [6, 7]. The principle of radar is to transmit a radio signal which is then reflected or scattered by objects present in its trajectory, and subsequently received by a receiver placed at the same or different location from the transmitter [8].

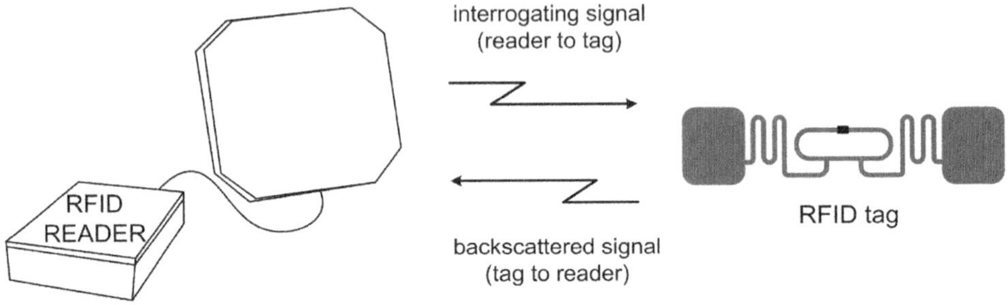

Figure 1.1 Back-scatter communication.

Signal processing is then applied in order to identify the nature of the objects which have interfered with the originally transmitted signal.

Radar technology developed during World War II was widely utilized in identifying incoming planes. The challenge for the technology was to determine whether it was a friendly or an enemy plane approaching, in other words to create an identify friend or foe (IFF) system. As an example, the German military utilized the fact that if the pilots rolled their planes when approaching their base, the reflected radar signal could be distinguished and therefore they could be identified. This is reported as a first implementation of a passive RFID system [6]. Alternatively, the British planes utilized a transmitter which, when interrogated by an incoming radar signal would transmit back a special signal used to identify the plane. This, in turn, represents an implementation of an active RFID system [6].

The 1948 paper by Stockman [9] is, however, arguably the first proposal to utilize the back-scattered signal for communication purposes, by modulating the reflected signal with some selected time-varying signal, and this represents to many the foundation of RFID communication technology. Following the work of Stockman, in the next two decades, more fundamental publications appeared studying back-scatter communication systems [8], notably the 1964 paper by Harrington on the "Theory of loaded scatterers" [10], as well as a number of patents such as the 1960 filing by Harris "Radio transmission systems with modulatable passive transponder" [11] exploring and demonstrating the potential of the technology.

The principle of back-scatter communications is shown in Figure 1.1. The interrogating transmitter – *reader* – sends a signal which is captured by all RFID *tag* antennas in the vicinity of the reader. The scattered signal from the tag antenna has a particular signature in the frequency or time domain (or both), which allows the reader to identify the tag.

Essential to the development of RFID technology has been the invention of the integrated circuit, originally envisioned by Dummer in 1952 [12]. Subsequently, Jack Kilby at Texas Instruments filed the first patent in 1959 [13] and was awarded the Nobel Prize in Physics in 2000 [14]. Independently from Kilby, and only six months later, Robert Noyce of Fairchild Semiconductor filed a patent for a "semiconductor-device-and-lead structure" [15]. The realization of integrated circuits or chips in silicon and other substrates has enabled the *low cost fabrication* and *mass production* of RFID tag

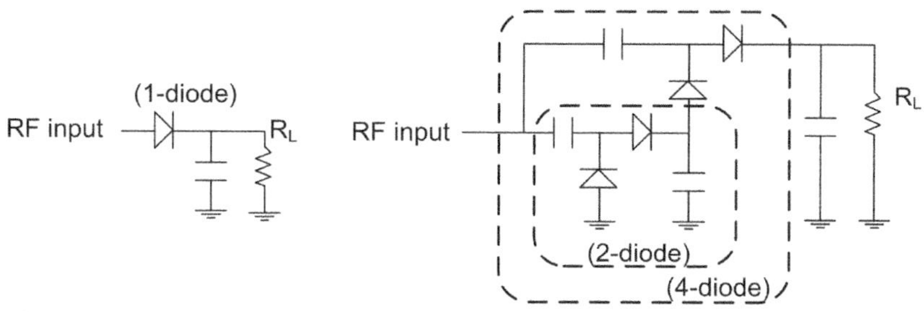

Figure 1.2 Rectifier circuits used in RFID tags.

chips, which ensured their widespread application. The tag chip is a core element of RFID technology, which has led to the introduction of a programmable memory and subsequently identification features, and modulation and detection circuits, as well as the capability of sensing through analog-to-digital (ADC) and digital-to-analog (DAC) conversion circuitry.

RFID technology owes much of its popularity and great potential for emerging applications to the simplicity of the tag circuit, especially when looking towards green technologies that place emphasis on limiting environmental pollution by not requiring the use of batteries and their associated waste treatment. Passive RFID tags extract the DC power required to activate their circuitry from the transmitted reader signal. Key to RFID technology is therefore the concept of wireless power transmission (WPT), which is attributed to the original work of Hertz and Tesla [16]. A fundamental component of the radio front-end of RFID tag chips is the rectenna, an active antenna consisting of a radiating structure and a rectifier circuit, which in its simplest form consists of a single diode, as shown in Figure 1.2. The rectenna was patented by Brown in 1969 [17], and is responsible for powering the circuitry of passive tags. A great challenge in modern RFID circuits is to come up with improved diode devices and original rectenna circuit topologies in order to maximize the efficiency in converting microwave power to DC power sufficient to activate the RFID tag circuits.

Electronic article surveillance (EAS) systems were the first commercial application of RFID technology, which began in the 60s with companies such as Sensormatic, Checkpoint, and Knogo [8]. In the 70s further developments of RFID systems were reported as the number of research laboratories such as Los Alamos Scientific Laboratory and companies such as Raytheon working on RFID systems increased significantly. The paper by Landt [8] presents an excellent historical overview of the technology. Notable publications include the first patent proposing an active RFID system with memory by Cardullo and Parks in 1973 [18], and the 1975 paper by Koelle, Depp, and Freyman from Los Alamos Scientific Laboratory, demonstrating a passive RFID system with temperature sensing capability in addition to identification functionality [19]. RFID was considered in applications such as animal tracking, vehicle identification, and factory automation.

In the 80s and 90s RFID commercialization grew significantly especially in terms of transportation applications. A characteristic example is the first open highway electronic

Table 1.1 RFID technology timeline.

Time	Milestone
1940s	radar technology [6, 7], first publication on back-scattered signals for communication [9]
1950s	integrated circuit vision [12]
1960s	patent on back-scattered communication systems [11], rectenna patent [17]
1970s	active RFID system patent [18], passive RFID system with temperature sensing capability [19]
1980s/1990s	first open highway electronic toll system in Oklahoma
2000s	GS1 launching [21], Auto-ID Labs [22], EPCglobal development [23]

toll system installed in Oklahoma in 1991. The development of the personal computer (PC) further assisted in the widespread utilization of the technology as it provided an efficient way to manage data collection from RFID systems. In 1997, the Uniform Code Council (UCC) and European Article Number (EAN) International initiated discussions about global standardization activities [20]. In 1999, the Massachusetts Institute of Technology (MIT) Auto-ID Center was founded with support from Procter & Gamble, Gillette, and UCC, in order to develop a universal RFID system, the Electronic Product Code (EPC). In 2003 EAN International underwent a significant reorganization and its management board agreed to change its name to GS1 [21]. GS1 was eventually launched in 2005 and UCC became GS1 US. The Auto-ID Center evolved into Auto-ID Labs in 2003 [22], responsible for research activities, and EPCglobal [23] responsible for standardization activities, related to Electronic Product Code technology. Presently Auto-ID Labs has seven laboratories in USA, UK, Australia, China, Republic of Korea, Japan, and Switzerland [22]. Standardization has helped the widespread application of the technology, which is now used by some of the biggest organizations in the world, one of the most notable examples being Wal-Mart, which in 2003 announced that its top 100 suppliers would be required to use EPC compatible RFID tags on all their pallets and cases of goods beginning January 2005 [24]. A summary of the timeline of the major milestones that have led to the mainstream deployment of RFID systems is shown in Table 1.1.

1.3 RFID towards a networked society

Technology advances towards increased processing power together with low cost, low profile, and low power electronics are making possible notions such as ambient intelligence, pervasive computing and communication, and Internet of Things. These concepts aim to improve the quality of life through a network of sensing devices communicating with each other and providing useful information to the network user [25]. Ubiquitous communications and sensing are becoming more and more part of everyday life by embedding wireless transceivers into an increasing array of appliances.

Sensing capability, jointly with communication and networking, leads to an array of smart objects ranging from clothes to cars, to buildings and to cities, with applications from health monitoring and assisted living to energy efficient operation, minimizing environmental pollution. All of these concepts imply everyday all-day communication, *anytime*, but also available connectivity *anywhere*, and effectively by *anything*, by introducing an increasing number of interconnected devices [25].

A characteristic example of the increasing number of communication links among smart objects, is the deployment of near-field communication (NFC) [26] and specifically its introduction in smart phones. In contrast to the traditional trend of communication networks towards high data rate and large range, a significant interest in *low data rate* and *short range* communications has been established, while placing emphasis on a low energy requirement and energy efficient operation. Networks of smart objects, based on back-scatter communication, are envisioned, placing emphasis on minimizing tag circuit complexity and energy requirements, based on a continuous, unidirectional communication from a reader to a large number of nearby tags, with a very low bit rate operation of a few bits per sec [27]. RFID technology is widely considered as the enabling technology for the implementation of such networks of smart objects, due to attributes such as low cost, low complexity circuit architecture, and low power operation [28]. RFID tags are able to provide much more than a simple identification, they can track location, store new information, and provide environmental sensing information [29].

RFID is also a key component in the concepts of augmented reality (AR) [30], where real world scenarios are overlaid with computerized graphics, sounds, videos, or 3D images to create a more informative or entertaining reality. In AR the RFID tags are used to activate these additional layers containing animations or information. AR finds application in a wide range of scenarios, such as urban, educational, cultural, architecture, design, and logistics fields. Further applications include autonomous logistics [31], where RFID technology can be used to achieve unmanned and autonomous control of goods being transported. The use of RFID tags can be considered not only to identify but also to locate and provide additional information about the goods being transported.

More recent solutions include networked RFID (NRFID), where RFID tags identify themselves and provide information to their associated readers that forward it to a central node to create a centralized database [27, 32]. NRFID makes it possible for data obtained by a certain reader to be reached by a network of authorized users that have access to the central database. This type of scheme can be used to improve the efficiency of the supply chain cycles.

In a standard application scenario, a large number of RFID tags are situated within the range of a reader. As a result, multiple tags receive and respond simultaneously to a reader interrogation transmission and, consequently, the reader is receiving simultaneously a large number of interfering responses, which lead to what is known as tag collisions. The result of tag collisions is a delay in the response, as well as wasted bandwidth and power resources. A challenge for RFID systems is that of developing protocols and system architectures which minimize the collisions between the responses of the various tags [33]. Furthermore, the widespread use of RFID technology is resulting in a more general scenario where multiple readers are also simultaneously operating. As a result, one should also consider the reader to reader collision problem [34], where the interrogation signal of one reader interferes with the operation of another reader, and also the reader-to-tag interference scenario, where a single tag is being detected by two readers [33].

In addition to the collision problem, an important challenge for RFID systems is that of security and privacy [35, 36]. Juels [35] identifies two main privacy concerns in covert tracking and inventorying. Directly because of the added functionality of RFID systems, user privacy can be compromised when identification is combined with personal information. Characteristic examples, where user privacy is of concern, include vehicle tolls, libraries, passports, and human implants [35]. Stanford [29] reports a case where user concerns that RFID clothing tags, being able to identify the purchaser and included in databases, have led a well-known commercial retailer to withdraw its plan for deploying RFID systems. In addition to the issue of privacy, which relates to potentially undesired interrogation or utilization of tag information by a reader, a second important issue is that of authentication, which deals with the problem of identifying counterfeit tags. With alarming reports estimating that roughly 8% of world trade yearly is in counterfeit goods [37], authentification is an important problem that RFID systems need to address, as replicating low-end RFID tags does not pose a difficult challenge [35]. While RFID technology already finds application in scenarios where authentification is required [36], there are several challenges that need to be addressed in terms of security and privacy, and RFID technology can profit from RF-based techniques in addition to cryptography and encryption algorithms. As an example, RF fingerprinting, generating certificates of authenticity based on the RF transmission or reflection characteristics of quasi-randomly printed or generated conductive objects, has been proposed to support identification [37].

1.4 Standardization

The International Organization for Standardization (ISO) is a global network of national standards bodies with members such as the American National Standards Institute (ANSI) from USA, the UK British Standards Institution (BSI), the Standardization Administration of the People's Republic of China (SAC), and the Asociación Española de Normalización y Certificación (AENOR) in Spain to name a few [38]. ISO has developed a number of different air protocol standards pertaining to RFID systems based on different applications, as a result of the widespread utilization of the technology [39]. As noted in Section 1.2, in addition to ISO, EPCglobal has defined a set of standards for RFID systems operating at 13.56 MHz and at UHF (860MHz – 960MHz) frequency bands. Table 1.2 includes a non-exhaustive list of existing standards. The present framework of standards facilitates the widespread utilization of RFID technology in an increasing number of applications and consequently the implementation of an Internet of Things.

Table 1.2 RFID standards [24, 39].

Standard	Frequency	Application	Communication
ISO/IEC 11784/5 ISO/IEC 14223 ISO/IEC 18000–2	125/134 kHz	animal identification	passive
ISO/IEC 10536 ISO/IEC 14443 ISO/IEC 15693 ISO/IEC 18000–3 EPC Class 1 [23]	4.9152 MHz 13.56 MHz	contactless smart cards	passive
ISO/IEC 18092	13.56 MHz	near-field communication	passive and active [40]
ISO/IEC 18000–7	433 MHz	item management	active
ISO/IEC 18000–6 EPC Class 0,1 [23]	860–960 MHz	item management	passive
ISO/IEC 18000–4	2.45 GHz	item management	various
IEEE 1451.7–2011	various	smart transducers	various [39]

Nonetheless, the integration of sensing functionality and networking between RFID tags remains a challenge in terms of standardization of the technology. The Institute of Electrical and Electronics Engineers Standards Association (IEEE-SA) has a formal agreement with ISO to cooperate in the development of international standards [41]. Specifically, the Instrumentation and Measurement Society TC9 Sensor Technology (IM/ST) Committee has introduced the IEEE 1451.7 standard for a smart transducer interface, which defines the communication interface between different transducers and RFID systems supporting a number of ISO/IEC air interface specifications such as ISO/IEC 18000–2,3,4,6,7 included in Table 1.2 The standard supports a number of sensors from accelerometers to humidity and temperature sensors, and can be considered as a preliminary step towards standardization of the technology [42].

1.5 Circuit challenges for RFID systems

In order to implement these concepts, issues such as miniaturization, conformal and flexible materials, low power, low cost, mass production fabrication, and environmentally friendly, passive or battery-less operation are fundamental, and lead to critical technological as well as circuit and system design challenges. Such attributes frequently lead to incompatible specifications; however, RFID technology comes with a plurality of categories, each one able to address a suitable subset of the sought for applications and specifications. RFID tags can be loosely classified based on their market objectives into low-end and high-end tags, ranging from chipless tags with no local memory all the way to battery powered transceivers with microcontrollers, multiple sensors, and complete radio front-ends [29]. More accurately, one can classify RFID systems based on the type of electromagnetic interaction into near-field and far-field systems [39], or otherwise based on their operating frequency and corresponding operating standard (Table 1.2).

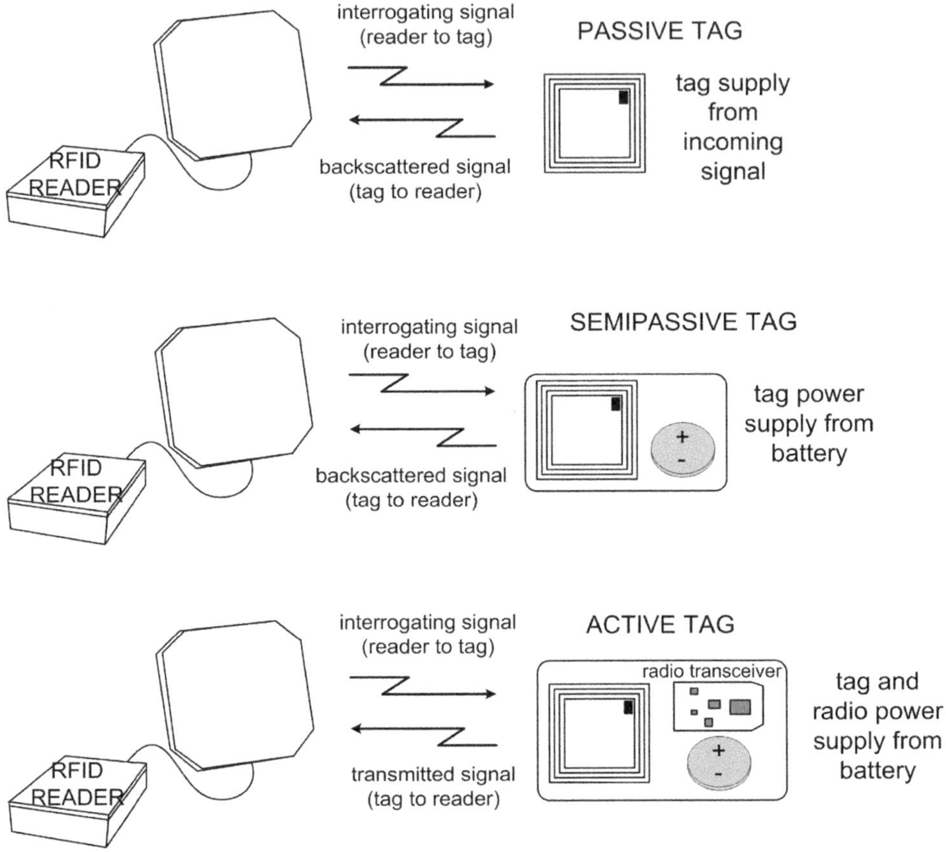

Figure 1.3 Classification of RFID systems.

An alternative classification of RFID systems is based on their functionality, and divides them into passive, semi-passive, and active tags (Figure 1.3) [1, 28, 39]. Passive tags rely on the transmitted power from the reader to both power their circuitry and communicate a response back to the reader. As a result, the operating range of such systems is limited by the minimum required power to operate the tag circuitry, rather than the sensitivity of the reader. The main challenge therefore becomes that of minimizing the dissipated power of the tag circuitry as well as that of maximizing the efficiency of converting the electromagnetic power transmitted by the reader into electrical DC power used by the tag [24]. Furthermore, the inclusion of sensing capability may further reduce the operating range of passive RFID systems, due to the additional power dissipated by the sensor circuits. Consequently, passive sensor design is receiving significant scientific interest, which is further boosted by advances in nanotechnology, with examples such as gas sensors exploring the dependence of the electrical properties of carbon nanotube thin films on the presence of ammonia gas [43]. Chipless RFID systems are a very promising technology, where tags do not require any (silicon) chip [1]. Chipless tags consist of purely passive circuitry or utilize printed thin film transistors and are thus able to minimize cost and energy requirements, leading to an increased recent scientific and market interest [1].

Semi-passive tags utilize an additional power source, a battery, to power-up their circuitry, however, communication is still done passively, by reflecting or scattering the reader signal which arrives at the tag terminals. The use of a power supply maximizes the operating range to the limit set by the reader sensitivity, and therefore the challenge then becomes that of improving the corresponding sensitivity specification of the reader. Furthermore, sensing functionality is additionally enhanced by the inclusion of an auxiliary power source. While the extra power source in its simplest form is a battery, there is a significant environmental concern associated with the disposal of spent batteries, corresponding to an exceptionally large number of sensor tags. As a result, energy harvesting technologies can be utilized as auxiliary power sources.

Finally, fully active tags utilize an auxiliary power source to both power their circuitry and amplify and transmit back the incoming signals from the reader. Such tags combine multiple functionalities with a large operating range at the expense of added circuit complexity, and higher cost. They can also benefit from energy harvesting technologies, but their power requirements are higher than semi-passive, or passive tags.

Researchers from the University of Washington have recently proposed utilizing solar energy as an auxiliary source to power the RFID tag circuitry and therefore extend its operating range [44]. DC power generated from solar cells can be provided to commercial tags utilizing an auxiliary DC terminal and corresponding power management circuitry. Commercially available tag chips with an external supply mode are already on the market [45]. Alternatively, the solar cell output DC electrical power can be efficiently converted to RF power, and the RF power can be directly fed into the RF port of any commercial tag. This was proposed in [46], where a high efficiency class-E oscillator was used to perform the DC–RF power conversion. A suitable matching circuit can be printed on the tag antenna substrate in order to maximize the RF power into the tag at the operating frequency of the oscillator. The oscillator frequency can be chosen so that it does not interfere with the reader signal. The possibility of utilizing RF beacon signals as auxiliary power sources for commercial RFID tags presents yet another new possibility for deploying wireless power nodes separated from the reader in an RFID network scenario. Solar, thermal, vibrational, or even ambient electromagnetic energy harvesters, can be utilized to provide additional power to the sensor tag [47].

The challenge of maximizing the operating range of RFID systems has inspired several circuit and system improvements. The efficiency of a rectifier circuit can be significantly improved by taking advantage of its nonlinear nature. It was found that there is a strong dependence of the rectifier efficiency on the peak-to-average power ratio of the input RF signal. Consequently, an RF signal with a time-varying envelope can result in a higher RF-to-DC conversion efficiency, compared to a simple RF tone signal. RF signals with strongly varying envelope, such as multi-tone signals [48, 49], or chaotic waveforms [50], can be used to increase the range of RFID systems by utilizing the time-varying nature of the RF signal envelope to maximize the RF-to-DC conversion efficiency of the rectifying circuitry of the RFID tags. Trotter *et al.* reported measured gains in the read range of 24% by using four subcarriers compared to a single tone carrier signal [48].

Similarly, Boaventura and Carvalho reported a 3 dB improvement in the conversion gain of a rectifier circuit by utilizing a four tone signal, and a 5 dB improvement using 16 tones [49]. Utilizing a very low complexity circuit consisting of a single transistor Colpitts-based oscillator to generate a chaotic signal, resulted in a 20% efficiency improvement in the rectifier efficiency over a single tone carrier [50].

Another significant challenge consists of maximizing the range of RFID tags when placed on metal or highly conductive surfaces. Placing an electrical dipole very close to a large conductive surface results in a significant reduction of its gain [51]. This is explained by applying image theory, where the electromagnetic problem of an electric dipole near a large conductive surface is equivalent to one where the conductive surface is removed but a second dipole, an image of the original dipole, excited with opposite phase, is placed symmetrically behind the conductive surface and at the same distance from it as the electrical dipole. The total field radiated from the two dipoles tends to zero as their separation is minimized [51]. Consequently, it is customary to use a thick substrate of spacer material such as foam in order to mitigate the effect of the conductive surface on the tag antenna performance [24]. An electromagnetic bandgap (EBG) surface can alternatively be used to eliminate the image phase inversion and, as a result, the antenna gain degradation due to the effect of the metallic ground plane [52]. EBG surfaces have been proposed to minimize the required spacer material thickness in situations where RFID dipole type tags are placed on top of metallic objects [53], as well as to mitigate the effect of the metallic surface size on the performance of patch antenna type tags [54].

Finally, an important challenge is that associated with optimizing the range of RFID systems with respect to the relative alignment of the tag and the reader, and in the presence of multipath propagation. Multiple tag antennas can be used to provide diversity and mitigate the effect of multipath propagation [55]. Tag chips with two antenna terminals are also available commercially [56]. The use of retro-directive array principles, such as the Van-Atta topology, can be used to maximize the signal back-scattered by the tag in the direction of the reader and has been shown to improve the read range by making it less sensitive to the tag alignment with respect to the reader [57].

1.6 Materials and technology

Typically, low cost materials such as polyethylene terephthalate (PET) or paper are used as substrates to host a silicon RFIC chip and its radiating tag antenna circuit. Recently, additional materials have been investigated as suitable candidates for RFID tags such as textiles and conductive threads [58]. The development of organic RFID (ORFID) is also being explored with a view to minimizing the cost of RFID tags by utilizing organic semiconductors for the synthesis of the RFIC chip instead of silicon [59].

In addition to investigating low cost, widely available substrate materials, an important challenge consists of exploring the possibility of using fabrication technologies which permit a low cost, large volume electronic circuit production. Inkjet printing is emerging as a popular alternative to traditional circuit board fabrication techniques, such as chemical etching and milling, finding increased application in the electronics and sensors industries [43]. It has a printing resolution of approximately 50 μm, which makes it suitable for fabricating passive microwave circuits and antennas [60]. Furthermore, it does not require a clean room or use of corrosive solvents, which minimizes the need for specialized waste treatment. Finally, the capability for large volume production using substrate materials such as textiles, PET, and paper have led to a natural application of inkjet printing in radio frequency identification (RFID), and several implementations of passive microwave circuit components such as inductors, transformers, and antennas have appeared in the literature [60]. The possibility of inkjet printing of a complete system-on-substrate (SoS), based on multilayer flexible substrate modules and inkjet deposition of active devices, remains a challenge, especially when it comes to operate at microwave frequencies; however, important steps have been taken in this direction by applying inkjet printing of semiconducting polymers to develop organic thin film transistors (OTFTs) [61].

Integration of off-the-shelf electronic components onto flexible substrates, and development of active circuits and wireless transceivers utilizing inkjet printing to develop passive interconnects, transmission lines, and antennas, has been successfully demonstrated [43, 62]. A further challenge in enabling complete system integration consists of inkjet printing of thin film energy storage elements such as capacitors and batteries [63]. Extending the operational autonomy of wireless sensors and transceivers has spurred significant efforts related to energy harvesting technologies, as well as wireless power transmission [64]. Towards this goal the integration of active circuits on flexible substrates such as PET, and inkjet printing on paper, autonomously powered by amorphous silicon a-Si solar cells [65] or piezoelectric transducers [66], has also been recently demonstrated. Finally, the capability of inkjet deposition of carbon nanotubes has additionally opened the way to gas sensing applications, and furthermore towards the development of RFID systems with sensing capability [43]. A further challenge in system integration is component miniaturization. Antennas and inductors typically require a large substrate area in order to achieve the necessary gain and inductance values respectively. Utilization of high permittivity [67] or high permeability substrate materials permits size reduction and low profile implementation. Recently, inkjet-printed ferromagnetic thin films based on cobalt nanoparticles have been used to design low profile inductors printed on a paper substrate, further demonstrating the potential of the technology [68].

1.7 Computer aided design (CAD) and optimization

A variety of computer aided design (CAD) and simulation techniques are used to accurately model and simulate the performance of RFID systems, but there is still work to be done especially in terms of the co-simulation of the complete reader, channel, and RFID tag system.

Electromagnetic simulation (EM), in addition to linear and nonlinear circuit simulation, is typically used in the design of an RFID system [24]. EM simulation is employed for the antenna and high frequency passive circuit design, and scattering simulation. EM simulation is frequently combined with linear circuit simulation to optimize the design of passive microwave circuits. Nonlinear circuit simulation is necessary for the active circuitry that forms the reader and tag chip. Time domain methods are traditionally used in integrated circuit (IC) design, while harmonic balance (HB) simulation is widely used in the design of nonlinear microwave circuits due to its efficient representation of distributed circuit components such as transmission lines [69].

EM and nonlinear circuit co-simulation allow one to jointly optimize the antenna and active RFID circuit. In [70, 71] the antenna and rectifier circuits are jointly optimized. In [72] combined EM and nonlinear simulations are used to investigate the interactions between passive RFID tags in multiple adjacent tag scenarios. Finally, co-simulation with system simulators can be used to model the complete RFID system performance [73]. There are many challenges associated with system simulation in order to allow the designer to optimize the RFID system performance under real world application scenarios, such as those including interference between tags and readers, modulated signals, and complicated propagation environments.

1.8 Conclusions

There are several milestone developments that have led to the mainstream application of RFID systems. Features such as circuit and architecture simplicity, as well as battery-less operation capability, allow for low cost and mass production making RFID systems an attractive solution for numerous applications, from logistics to monitoring. Nonetheless, there are many challenging problems waiting to be addressed, ranging from circuit and system design, to materials and fabrication techniques. RFID technology combined with sensing capability, and taking advantage of energy harvesting technologies as well as advances in nanotechnology, presents itself as an exciting candidate towards achieving a vision for ubiquitous sensing.

Acknowledgment

The work of A. Georgiadis and A. Collado was supported by Marie Curie project Symbiotic Wireless Autonomous Powered System (SWAP), FP7-PEOPLE-2009-IAPP 251557 (http://www.fp7-swap.eu/).

References

[1] N. C. Karmakar, *Handbook of Smart Antennas for RFID Systems*, Wiley, 2010.

[2] A. Juels, "RFID security and privacy: A research survey," *IEEE J. Selected Areas in Comm.*, **24**, 381–394, (Feb.) 2006.

[3] NXP SL3S1203FTB0 RFID IC tape and reel packaging, Digi-Key authorized distributor of electronic components, 2013, http://www.digikey.com/scripts/dksearch/dksus.dll?pv7=1&k=SL3S1203.

[4] S. E. Sarma, S. A. Weis, and D. W. Engels, "RFID systems, security and privacy implications," AutoID Center, MIT, Cambridge, MA, Tech. Rep. MIT, AUTOID-WH-014, 2002.

[5] M. A. Uusitalo, "Global vision for the future wireless world from the WWRF," *IEEE Vehicular Technology Magazine*, **1**, (*2*), 4–8, 2006.

[6] M. Skolnik, *Introduction to Radar Systems*, 3rd Edn., McGraw-Hill, 2002.

[7] "The history of RFID technology," *RFID Journal*, 2005, http://www.rfidjournal.com/article/articleview/1338/1/129.

[8] J. Landt, "The history of RFID," *IEEE Potentials*, 8–11, (Oct.–Nov.), 2005.

[9] H. Stockman, "Communication by means of reflected power," *Proc. of the IEEE*, 1196–1204, (Oct.), 1948.

[10] R. F. Harrington, "Theory of loaded scatterers," *Proc. of the IEE*, **111**, (*4*), 617–623, 1964.

[11] D. B. Harris, "Radio transmission systems with modulatable passive transponder," US2927321, 1960.

[12] J. Kilby, "Invention of the integrated circuit," *IEEE Transactions on Electron Devices*, **23**, (*7*), 648–654, 1976.

[13] J. Kilby, "Miniaturized electronic circuits," US3138743, 1964.

[14] http://www.nobelprize.org/nobel_prizes/physics/laureates/2000/press.html.

[15] R. N. Noyce, "Semiconductor device-and-lead structure," US2981877, 1961.

[16] W. C. Brown, "The history of power transmission by radio waves," *IEEE Transactions on Microwave Theory and Techniques*, **32**, (*9*), 1230–1242, 1984.

[17] W. C. Brown, "Microwave to DC converter," US3434678, 1969.

[18] Mario Cardullo and William L. Parks, III, "Transponder apparatus and system," US3713148, 1973.

[19] A. R. Koelle, S. W. Depp, and R. W. Freyman, "Short-range radio-telemetry for electronic identification, using modulated RF backscatter," *Proceedings of the IEEE*, **63**, (*8*), 1260–1261, 1975.

[20] http://www.gs1.org/about/media_centre/timeline, 2013.

[21] http://www.gs1.org, 2013.

[22] http://www.autoidlabs.org, 2013.

[23] http://www.gs1.org/epcglobal, 2013.

[24] D. Dobkin, *The RF in RFID*, Newnes, Elsevier, 2008.

[25] "ITU internet reports: The Internet of Things," ITU-T, 2005.

[26] http://www.nfc-forum.org/, 2013.

[27] G. Vannucci, A. Bletsas, and D. Leigh, "A software-defined radio system for backscatter sensor networks," *IEEE Transactions on Wireless Communications*, **7**, (*6*), 2170–2179, 2008.

[28] A. Rida, L. Yang, and M. Tentzeris, *RFID-Enabled Sensor Design and Applications*, Artech House, 2010.

[29] V. Stanford, "Pervasive computing goes the last hundred feet with RFID systems," *IEEE Pervasive Computing*, **2**, (*2*), 9–14, (Apr.) 2003.

[30] S. Kunkel, R. Bieber, Ming-Shih Huang, and M. Vossiek, "A concept for infrastructure independent localization and augmented reality visualization of RFID tags," *Wireless Sensing, Local Positioning, and RFID, 2009. IMWS 2009. IEEE MTT-S International Microwave Workshop*, pp. 1–4, 24–25 Sept. 2009.

[31] V. Palazzari, F. Alimenti, G. Orecchini, P. Mezzanotte, and L. Roselli, "A Ku-band RF self identification (RFSID) system for autonomous logistics," *Antennas and Propagation (EUCAP), Proceedings of the 5th European Conference*, pp. 2828–2830, 11–15 April 2011.

[32] G. Vannucci, D. Leigh, and J. Katz, "Wide-area long-range unidirectional sensor (WALRUS) network," in *Proc. IEEE International Conf. on Mobile Ad hoc and Sensor Systems*, Nov. 2005.

[33] D.-H. Shih, P.-L. Sun, D. C. Yen, and S.-M. Huang, "Taxonomy and survey of RFID anti-collision protocols," *Computer Communications*, **29**, (*11*), 2150–2166, (July) 2006.

[34] D. W. Engels and S. E. Sarma, "The reader collision problem," in *Proc. Intl. Conf. on Systems, Man and Cybernetics*, **3**, 6–9 Oct. 2002.

[35] A. Juels, "RFID security and privacy: A research survey," *IEEE J. Selected Areas in Comm.*, **24**, 381–394, (Feb.) 2006.

[36] S. E. Sarma, S. A. Weis, and D. W. Engels, "RFID systems, security and privacy implications," AutoID Center, MIT, Cambridge, MA, Tech. Rep. MIT, AUTOID-WH-014, 2002.

[37] V. Lakafosis, A. Traille, Hoseon Lee, *et al.*, "RF fingerprinting physical objects for anticounterfeiting applications," *IEEE Transactions on Microwave Theory and Techniques*, **59**, (*2*), 504–514, 2011.

[38] http://www.iso.org/iso/home.html, 2013.

[39] K. Finkenzeller, *RFID Handbook Fundamentals and Applications in Contactless Smart Cards, Radio Frequency Identification and Near-Field Communication*, 3rd Edn., Wiley, 2010.

[40] ISO/IEC 18092:2004, "Information technology – Telecommunications and information exchange between systems – Near field communication – Interface and protocol (NFCIP-1)."

[41] http://standards.ieee.org/develop/intl/iso.html, 2013.

[42] IEEE 1451.7–2011, IEEE Standard for "Smart transducer interface for sensors and actuators – Part 7: Transducers to radio frequency identification (RFID) systems communication protocols and transducer electronic data sheet (TEDS) formats, 2011."

[43] R. Vyas, V. Lakafosis, H. Lee, *et al.*, "Inkjet printed, self powered, wireless sensors for environmental, gas, and authentification-based sensing," *IEEE Sensors Journal*, **11**, (*12*), 3139–3152, 2011.

[44] A. P. Sample, J. Braun, A. Parks, and J. R. Smith, "Photovoltaic enhanced UHF RFID tag antennas for dual purpose energy harvesting," *2011 IEEE International Conference on RFID (RFID)*, pp. 146–153, 12–14 April 2011.

[45] http://www.nxp.com/products/identification_and_security/smart_label_and_tag_ics/ucode/series/SL3S1203_1213.html, 2013.

[46] A. Georgiadis and A. Collado, "Improving range of passive RFID tags utilizing energy harvesting and high efficiency class-E oscillators," *6th European Conference on Antennas and Propagation (EUCAP)*, pp. 3455–3458, 26–30 March 2012.

[47] S. Beeby and N. S. White, *Energy Harvesting for Autonomous Systems*, Artech House, 2010.

[48] M. S. Trotter, J. D. Griffin, and G. D. Durgin, "Power-optimized waveforms for improving the range and reliability of RFID systems," *2009 IEEE International Conference on RFID*, pp. 80–87, 27–28 April 2009.

[49] A. S. Boaventura and N. B. Carvalho, "Maximizing DC power in energy harvesting circuits using multisine excitation," *2011 IEEE MTT-S International Microwave Symposium Digest (MTT)*, pp. 1–4, 5–10 June 2011.

[50] A. Collado and A. Georgiadis, "Improving wireless power transmission efficiency using chaotic waveforms," in *Proc. IEEE MTT-S IMS 2012*, Montreal, 17–22 June 2012.

[51] C. A. Balanis, *Antenna Theory*, 3rd Edn., Wiley, 2005.

[52] D. Sievenpiper, Z. Lijun, R. F. J. Broas, N. G. Alexopoulos, and E. Yablonovitch, "High-impedance electromagnetic surfaces with a forbidden frequency band," *IEEE Transactions on Microwave Theory and Techniques*, **47**, (*11*), 2059–2074, 1999.

[53] H. Lee, S. Kim, D. De Donno, and M. M. Tentzeris, "A novel 'Universal' inkjet-printed EBG-backed flexible RFID for rugged on-body and metal mounted applications," in *Proc. 2012 IEEE MTT-S International Microwave Symposium Digest (IMS)*, pp. 1–3, 17–22 June 2012.

[54] L. Ukkonen, L. Sydanheimo, and M. Kivikoski, "Effects of metallic plate size on the performance of microstrip patch-type tag antennas for passive RFID," *IEEE Antennas and Wireless Propagation Letters*, **4**, 410–413, 2005.

[55] P. V. Nikitin and K. V. S. Rao, "Performance of RFID tags with multiple RF ports," in *Proc. 2007 IEEE Antennas and Propagation Society International Symposium*, pp. 5459–5462, 9–15 June 2007.

[56] http://www.impinj.com/True3DAntennaTechnology.aspx, 2013.

[57] J. Cespedes, F. Giuppi, A. Collado, and A. Georgiadis, "A retro-directive UHF RFID tag on paper substrate," in *Proc. 2012 IEEE RFID-TA*, Nice, 5–7 Nov. 2012.

[58] L. Ukkonen, L. Sydanheimo, and Y. Rahmat-Samii, "Sewed textile RFID tag and sensor antennas for on-body use," in *Proc. 2012 6th European Conference on Antennas and Propagation (EUCAP)*, pp. 3450–3454, 26–30 March 2012.

[59] E. Cantatore, T. Geuns, G. H. Gelinck, *et al.*, "A 13.56-MHz RFID system based on organic transponders," *IEEE Journal of Solid-State Circuits*, **42**, (*1*), 84–92, 2007.

[60] L. Yang, A. Rida, R. Vyas, and M. M. Tentzeris, "RFID tag and RF structures on a paper substrate using inkjet-printing technology," *IEEE Transactions on Microwave Theory and Techniques*, **55**, (*12*), 2894–2901, 2007.

[61] J. Vaillancount, H. Zhang, P. Vasinajindakaw, *et al.*, "All ink-jet-printed carbon nanotube thin-film transistor on a polyimide substrate with an ultrahigh operating frequency over 5 GHz," *Appl. Phys. Lett.*, **93**, 243301, 2008.

[62] A. Rida, L. Yang, R. Vyas, and M. M. Tentzeris, "Conductive inkjet-printed antennas on flexible low-cost paper-based substrates for RFID and WSN applications," *IEEE Antennas and Propagation Magazine*, **51**, (*3*), 13–23, (June) 2009.

[63] D. Pech, M. Brunet, P.-L. Taberna, *et al.*, "Elaboration of a microstructured inkjet-printed carbon electrochemical capacitor," *Journal of Power Sources*, **195**, (*4*), 1266–1269, (Feb.) 2010.

[64] N. Shinohara, "Power without wires," *IEEE Microwave Magazine*, **12**, (*7*), S64–S73, (Dec.) 2011.

[65] S. Kim, A. Georgiadis, A. Collado, and M. M. Tentzeris, "An inkjet-printed solar-powered wireless beacon on paper for identification and wireless power transmission applications," *IEEE Transactions on Microwave Theory and Techniques*, Dec. 2012.

[66] G. Orecchini, L. Yang, M.M. Tentzeris, and L. Roselli, "Wearable battery-free active paper-printed RFID tag with human-energy scavenger," in *Proc. 2011 IEEE MTT-S International Microwave Symposium (IMS)*, Baltimore, p. 1, 5–10 June 2011.

[67] A. A. Babar, V. A. Bhagavati, L. Ukkonen, *et al.*, "Performance of high-permittivity ceramic-polymer composite as a substrate for UHF RFID tag antennas," *International Journal of Antennas and Propagation*, Article ID 905409, 2012.

[68] H. Lee, M. M. Tentzeris, Y. Kawahara, and A. Georgiadis, "Novel inkjet-printed ferromagnetic-based solutions for miniaturized wireless power transfer (WPT) inductors and antennas," in *Proc. ISAP*, Nagoya, 29 Oct.–2 Nov. 2012.

[69] V. Rizzoli and A. Neri, "State of the art and present trends in nonlinear microwave CAD techniques," *IEEE Trans. Microw. Theory Tech.*, **36**, (*2*), 343–356, 1988.

[70] V. Rizzoli, G. Bichicchi, A. Costanzo, F. Donzelli, and D. Masotti, "CAD of multi-resonator rectenna for micro-power generation," in *Proc. 2009 European Microwave Conference (EuMC)*, pp. 1684–1687, 29 Sept.–1 Oct. 2009.

[71] A. Georgiadis, G. Andia-Vera, and A. Collado "Rectenna design and optimization using reciprocity theory and harmonic balance analysis for electromagnetic (EM) energy harvesting," *IEEE Antennas and Wireless Propagation Letters*, **9**, 444–446, 2010.

[72] V. Rizzoli, A. Costanzo, M. Rubini, and D. Masotti, "Rigorous investigation of interactions between passive RFID tags by means of nonlinear/electromagnetic co-simulation," *36th European Microwave Conference*, pp. 722–725, 10–15 Sept. 2006.

[73] F. Fuschini, C. Piersanti, F. Paolazzi, and G. Falciasecca, "Electromagnetic and system level co-simulation for RFID radio link modeling in real environment," *Antennas and Propagation, EuCAP 2007, The Second European Conference*, pp. 1–8, 11–16 Nov. 2007.

2 RFID background

Giulia Orecchini and Luca Roselli

2.1 RFID system architecture

RFID (Radio Frequency IDentification) indicates the capability of identifying by means of radio frequency transmissions. The identification involves assigning a unique identity to an object that is distinguishable in an unambiguous way.

In this original form RFIDs have the same functionality as a barcode. Regarding its evolution, the main purpose of this technology, beyond barcodes, is to obtain identified information about objects, animals, or persons by means of small apparatuses working at radio frequency.

The intake of information is achieved by means of searching operations, identification, selection, spatial localization, and tracking.

Identifier and identified communicate using radio frequency signals, hence no physical contact (unlike, for example, use of magnetic stripe cards) is needed.

The predecessor of the RFID system is commonly recognized as being "Identification Friend or Foe (IFF)," developed in England during World War II (1940) [1–3]. The equipment on board allied aircraft had the functionality of answering if questioned, thus identifying allied planes and distinguishing them from enemy aircraft.

The technology has then evolved differently into systems for following the route of railway wagons, for the automation of processes in the automotive industry, for the location of livestock and wildlife, for anti-theft in the retail trade, for keys and electronic documents, and in agriculture and nature reserves, etc. [4–6].

Real-life RFID deployments use a wide variety of physically distributed RFID readers, access gateways, management interfaces, and databases. The goal of radio frequency identification is mainly to allow computers, and ultimately people, to monitor, decide, and consequently take actions.

2.1.1 RFID system general frame

RFID technology consists of three basic elements (Figure 2.1):

- Tag: a radio frequency transponder of small dimensions consisting of an integrated circuit (chip), connected to an antenna. The tag allows for the transmission of data at short range without physical contact.

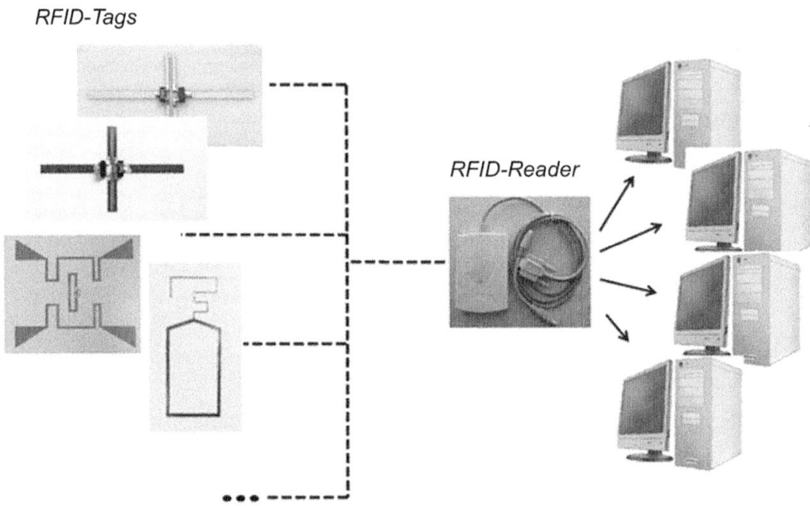

Figure 2.1 RFID system general framework.

- Reader: a transceiver controlled by a microprocessor and used to query and receive information in response from one or more tags.
- Management system: an information system that, when it exists, is connected to a network with the readers. This system allows, starting from the identification codes coming from the tags, all available information associated with the objects to be obtained and managed.

The tags can be categorized first on the basis of the management of the energy sources. The tag, in fact, can be:

- *Passive*: it gets the energy from the RFID reader signal; it does not have a real transmitter, but it circulates, modulates, and irradiates with its antenna, the signal transmitted by the reader. The distances at which it can operate are, at most, of the order of a few meters.
- *Active*: it is powered by batteries. It incorporates a receiver and a transmitter. It usually has a large memory, often rewritable and may contain sensors or other devices. The coverage range depends on the transmitter and batteries and typically is, at most, of the order of 200 meters. For active tags, in addition to the greater amount of memory and the function of rewriting, technological development has made possible the addition of functions that far exceed pure identification. For example, some additional functionalities are radiolocation (RTLS – real-time location system – position identification of objects hosting the RFID) or the measurement of environmental parameters through sensors (temperature, movement, etc.).
- *Battery-assisted passive (BAP) tag*; it uses an energy source to power only certain components of the tag. BAP tags are also classified as semi-passive or semi-active.

BAP tags are equipped with a battery only to power the microchip or auxiliary apparatuses (sensors for instance) and modulator, but not to supply power to the RF transmitter electronics since, for the transmission, they behave as passive tags just circulating the signal coming from the reader. The distances at which they can operate are, at most, of the order of a few tens of meters.

To save energy, the tag is normally deactivated. The activation is achieved through a receiver that operates like the one in the passive tag. Passive tags are typically low cost devices and have small dimensions allowing the field of application to be very wide. When the tag passes through the electromagnetic (EM) field generated by a reader it reacts and transmits its information.

Typically a passive tag that receives a signal from a reader uses the same energy of the signal to power its internal circuitry and, therefore, "wake up" its functions. Once the tag has decoded correctly the signal from the reader, it answers reflecting back the signal through its antenna and adding modulation and thus information.

Often each application is tied to particular dimensional characteristics of the tag itself. In fact, the passive tag, consisting only of an antenna (typically printed) and an integrated circuit, generally miniaturized, can have a thickness of a few hundreds of microns. It can therefore be placed on credit cards, labels, stickers, buttons, and other small items made of plastic or paper, thus generating real "talking" objects.

In the past, tags had primarily read-only memory because the step of writing requires the availability of a high amount of energy, which is hard to obtain from the received signal, and because rewritable memories have a relatively high cost.

2.1.2 RFID regulation

The reader is the element that, in RFID systems, is able to capture the information contained in the tag. It is a real transceiver, governed by a system control and often networked with computerized management systems in order to get information from objects and make it available for any use.

As mentioned before, for the success of an application, the tag and the reader have to communicate easily without disturbing other radio services. Therefore, as always happens in applications that require interoperability between parties, the communication standards are essential.

The two main objectives that the standards must meet are:

- interoperability between reader and tag made by different manufacturers;
- no interference with other radio devices' electrical operations.

The dialogue between reader and tag requires standardization of certain technical parameters:

- Parameters of the two radio frequency communication links, reader-tag and tag-reader (uplink and downlink or forward link and return link);
- Communication protocols used in the same link (definition of procedures and data format for the dialogue);
- Type of signal modulation;
- Coding and packaging of data;
- Transmission of data (bit rate) and bit transmission order;
- Technique of interrogation of the reader ("frequency hopping," "listen before talk");

- Anti-collision procedures (ability to detect and classify the greatest possible number of tags at the same time in the reading range).

In addition to those listed above, specifications and standards are required to deal with the following topics:

- The format of the data contained in the tag (the way in which the data are organized or formatted);
- The compliance (the way in which the products meet a standard);
- "Middleware" protocols (specify instructions and how data are processed).

The standardization process is usually managed both at global and regional level, by two classes of organization:

- Those involved in the management of the radio spectrum, which provides essentially the principle of non-interference. To this category belong the ITU (International Telecommunications Union), the CEPT (European Conference of Postal and Telecommunications Administration), and the ERC (European Radio communications Committee);
- Those that deal with communication interfaces, to ensure interoperability between reader and tag with different manufacturers (this category includes ISO and EPCglobal).

The main use of RFID tags so far is the identification of objects and more generally in logistics (identification of packaging, pallet, container, and so on down the chain of distribution). The standards for these applications belong to two organizations already mentioned, whose activities are converging:

- EPCglobal, set up and acting as a private association;
- ISO (and bodies connected to it), which is the world body of legislation in almost all fields of technology.

For example, the standards for the radio interface are proposed by both ISO [7] and EPCglobal [8] (the former oriented on all fields, the latter mostly polarized on the specific needs of the supply chain). EPCglobal also covers the application layer, while the standard ISO/IEC 18000 [9] is limited to the radio interface.

The frequencies used for the communication between reader and tag depend on both the nature of the tag and the intended application. They are adjusted (to control emissions of power and to prevent interference) by the international and national organizations. The regulation, however, is divided in geographical regions with different standards from region to region.

The portions of frequency bands used for RFID are shown in Figure 2.2; the most commonly used are:

- LF band (low frequencies) and in particular the sub-band 120 to 145 kHz. It is located in the lower part of the RF spectrum; it is historically the first frequency band used for automatic identification and, even today, continues to have a significant presence in the market.

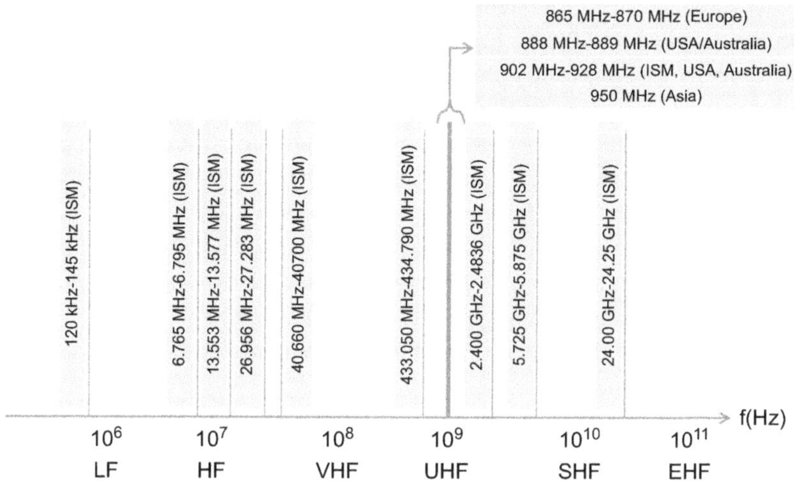

Figure 2.2 Available worldwide frequency bands for RFID.

- HF band (high frequencies) and in particular the sub-band centered on 13.56 MHz. It is considered the "universal" frequency band, usable throughout the world, therefore the most popular band to date.
- UHF band (ultra-high frequencies), and in particular the sub-band 865–870 MHz in Europe, 902–928 MHz in USA, and 950 MHz in Asia. It is the "new band" for RFID for logistics and for the management of individual objects with operating distances significantly higher than the ones allowed by LF and HF bands. Unfortunately, the bandwidth is not uniformly allocated in all countries.
- Microwave frequency band, centered on 2.4 GHz. It has similar characteristics to the previous UHF band, but allows for a further miniaturization of the tag. It is, however, a very crowded band because of other technologies (WiFi, Bluetooth, ZigBee), with which it is necessary to coexist.

There are also other used frequencies such as 433–435 MHz in the UHF low-band and 5.8 GHz SHF (super-high frequencies).

To date (2014), some frequency bands (usually in the LF or HF) are accepted throughout the world. An example is the band of 13.56 MHz, used by many passive tags incorporated mainly in smart cards for access control, identification, and payments, but also into the labels associated with objects, such as luggage checking, laundries, libraries, etc.

The choice of the working frequency influences the operational distance (range) of the system, interference with other radio systems, the transfer speed of data, and the size of the antenna. Systems that use lower frequencies are often based on passive tag structures and are capable of transmitting data at maximum distances of the order of feet (30–120 cm). Higher frequencies are mainly used for active tags resulting in a higher working distance. In these cases the speed of data transfer is generally increased while the size of the antennas is reduced; this allows smaller tags to be built.

The sub-band of LF frequencies, located in the lower part of the RF spectrum (125–145 kHz), is historically the first frequency band used for automatic identification and still remains a significant presence in the market. The coupling between the reader and the tag takes place by induction, with the same physical principle as electrical transformers. In the case of passive tags, the operating distance is approximately equal to the diameter of the reader antenna and varies from 30 cm to a meter; beyond this the field is reduced very quickly, due to the $1/d^3$ effect, while the energy picked up by the tag varies according to $1/d^6$. For this reason, the distance that enables the possible operation of writing into the memory, which has a higher energy consumption, is normally less than the reading range, being typically less than 30 to 50%.

It should be noted that within the LF band there are actually two frequencies that are used most:

- 125.5 kHz mainly in the automotive sector;
- 134.2 kHz in animal traceability.

The frequency of the 125.5 kHz carrier is relatively low and allows a maximum speed of data transmission of the order of thousands of bits per second. The tags at a frequency of 134.2 kHz are used primarily in animal traceability because of the very slight effect that water and fabrics have on transmission.

The sub-band of HF frequencies located at 13.56 MHz is recognized by all regulatory organizations worldwide and this has made it the most popular band up to now. The coupling between the reader and the tag is made by induction, as in LF tags. The shapes and the type of packaging are available in the most disparate forms. The typical configuration provides an antenna formed by a winding made normally of copper, but aluminum, etched from a thin sheet of metal or inkjet printed with conductive ink is also used. The size and the number of turns determine the sensitivity and the operating distance. The costs are lower than those of LF tags but strictly dependent on the support type and size. The HF band is currently the most widely used for the so-called "smart labels" (smart tags) used in logistics and object management, even if, for the latter application, it is anticipated that, in the long term, systems in the UHF band will prevail. In this frequency the contactless smart cards that are the industry's most developed field of application operate too.

The technological evolution of semiconductors, which led to the realization of very-low-power chips, has allowed RFID tags operating at 860–950 MHz with an operating distance much larger than in LF and HF bands. The coupling between the reader and the tag is made in an electromagnetic way, as in conventional radio systems. A working distance of three meters is now common, and extensions to five meters and more are possible. Thanks to this, the UHF band is certainly destined to be confirmed as the "queen band" in logistics and, above all, in the management of objects. However, some issues, have slowed its introduction:

- Operating frequencies: USA, Europe, and Asia have to account for different frequencies: frequencies already occupied by cell phones and then not usable for RFID applications. Tags are thus often built to enhance the ability to work in a "broadband," although performances are usually worsened. For active tags, instead, the transceivers, if necessary, are more easily calibrated to operate on multiple frequencies at the price of an increment of production costs.

- Transmitted power: in USA and Europe there are different limits for the maximum power emitted and different widths of the UHF frequency band dedicated; therefore there is a great number of channels on which the reader can operate.
- At these frequencies (UHF) there are issues not observed at lower frequencies:
  - Reflections: metal structures near the antenna can reflect electromagnetic waves: these reflections can generate areas in which the electromagnetic field is null. The tag in this area would be unreadable.
  - Liquids: electromagnetic absorption by water becomes more consistent. The reading efficiency in particularly humid environments or of tagged objects containing liquids can be difficult.

The transmission speed is higher than that of systems operating at lower frequencies. The systems, moreover, are able to manage multiple readings (more than 100 tags per second). The characteristics (technological and of the anti-collision algorithms) of tags conform to the EPC Class1/Gen2 [10] which should allow, in theory, the reading of 600 (in Europe) and 1500 (USA) tags entering at the same time into the area illuminated by the reader.

In the 2.4 GHz band wireless networks (WLAN, Bluetooth, ZigBee) are already operating. Similarly to HF, these bands are recognized worldwide as frequencies also dedicated to RFID systems. The tags operating at these frequencies have very similar characteristics and behavior to UHF tags and, thanks to the relation between antenna size and wavelength, the size of the antenna tag is in general smaller.

2.1.3 RFID technology

RFID systems have two ways to exchange information, inductive coupling and electromagnetic back-scattered. The inductive coupling between passive tag and reader is realized by coupled coils often referred to generally as "antennas," to supply the tag with energy needed for the transmission. The two antennas are, from the electrical point of view, tuned LC (inductance–capacitance) circuits and they act as an electrical transformer. The circuit can be summarized as shown in Figure 2.3. The equations (2.1–2.4) characterize the tag resonant frequency, the voltage, and the current induced in the same tag:

$$f_o = \frac{1}{2\pi\sqrt{L_{\mathrm{TAG}}C_{\mathrm{TAG}}}}, \tag{2.1}$$

$$V_{\mathrm{TAG}} = Z_{\mathrm{TAG}} I_{\mathrm{TAG}}, \tag{2.2}$$

$$I_{\mathrm{TAG}} = \frac{\mathrm{j}\omega M I_R}{R_{\mathrm{TAG}} + Z_{\mathrm{TAG}} + \mathrm{j}\omega L_{\mathrm{TAG}}}, \tag{2.3}$$

$$M = \sqrt{L_{\mathrm{TAG}} L_R}. \tag{2.4}$$

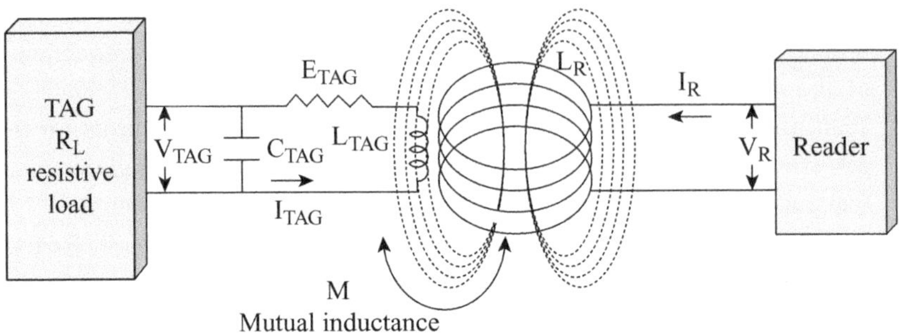

Figure 2.3 Resonant frequency in tags with inductive coupling.

At the resonance frequency the energy transfer between reader and tag is maximized. The formulae (2.2–2.3), showing the voltage and current induced in the tag, indicate that they are proportional to the resonant frequency and to the mutual inductance coefficient (M) between the tag and reader antenna coils.

The communication between reader and tag takes place by modulating the amplitude of the magnetic field generated by the reader antenna with the digital signal that must be transmitted. The receiver circuit of the tag recognizes the modulated field and decodes, from this, the information transmitted by the reader. Now, while on the reader side the energy for transmitting and modulating is not a problem, on the passive tag side it is not the same. The communication back from the tag to the reader is achieved by inductive coupling, as in a transformer in which the secondary winding (coil) varies its load ("load modulation") and the modulated signal is detected in the primary winding (reader).

In Figure 2.4 a block diagram of a passive tag with inductive coupling is illustrated.

The sequence of events in the reader–tag communication can be summarized as follows:

1. The reader control logic sends the data to the reader transmitter that generates the signal sent to the coils.
2. The current in the reader coil induces a magnetic field that concatenates with the tag coil.
3. A power supply is extracted by the tag circuit that rectifies and limits the received signal.
4. The detector/decoder provides the decoded data to the tag control logic (which can write it into the memory).
5. The detector/decoder indicates to the modulator the instants in which it is possible to enable the transmission back to the reader.
6. The control logic of the tag reads the data in its memory and reports the availability to the modulator which generates the control signal with appropriate timing, actually modulating the impedance of the tag antenna itself.
7. The reader (through the detector) senses the antenna tag impedance changes (since the coils of the antenna tag and reader behave as a transformer as mentioned before) and transmits the received data to its control logic.

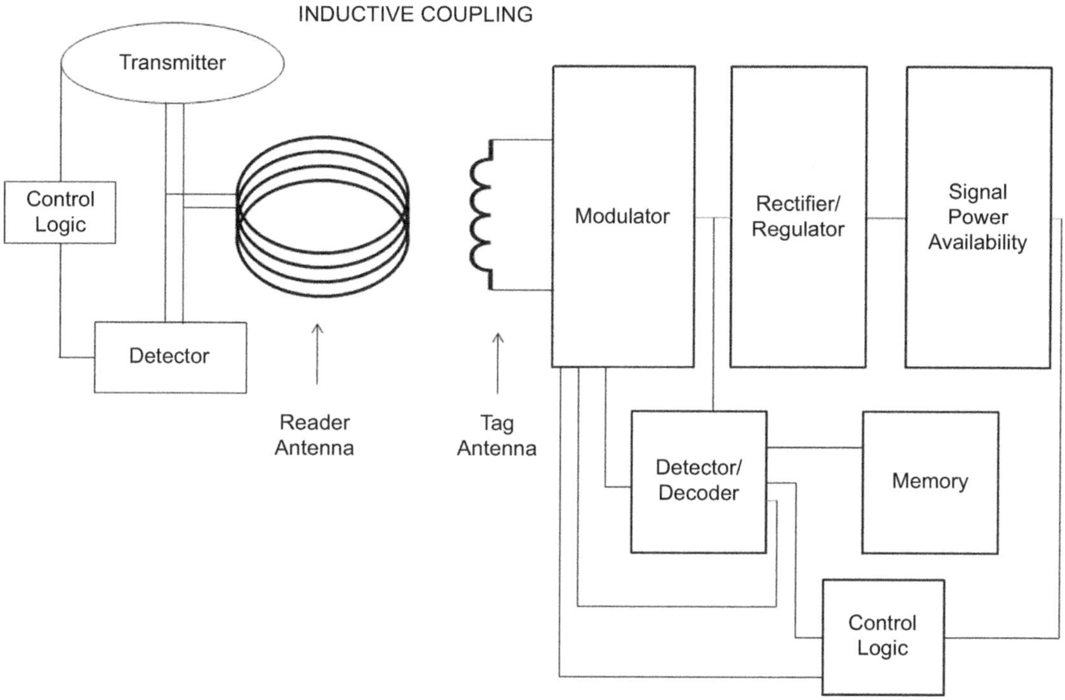

Figure 2.4 Passive tag with inductive coupling.

When an electromagnetic wave hits irregularities in the medium in which it propagates, it disperses randomly. This phenomenon is called scattering. In systems such as radar or RFID, a transmitter sends an electromagnetic wave and a receiver detects the scattering generated from a tagged object on which the wave impacts.

The passive tags that use the back-scatter effect, are therefore designed as a compromise between the capacity to absorb the energy to power up the tag, and the capacity to reflect the incident power, modulating the back-scatter signal, to respond to a query.

The tags operating in UHF or higher frequencies use techniques very similar to the ones used for tags operating at lower frequencies; they receive the power from the EM field emitted by the reader. The difference consists in the way in which energy is transferred and in the fact that antennas instead of coils are used. As mentioned above, the transfer of energy takes place in a "far-field" regime. In this case dipole antennas are usually used as in conventional radio systems. When the EM wave emitted by the reader antenna impinges on the tag, part of the energy is absorbed to provide power to the tag (through an ordinary AC to DC rectifier), but part is reflected back towards the reader as back-scatter.

Even in this case modulation of the back-scatter is obtained by variation of the antenna impedance during the reader transmission. The reader captures the resulting changes in reflected signal. The use of the back-scattered modulation technique in the far field introduces different problems with respect to those occurring in systems at lower frequencies. One is due to the fact that the field emitted by the reader is not only reflected

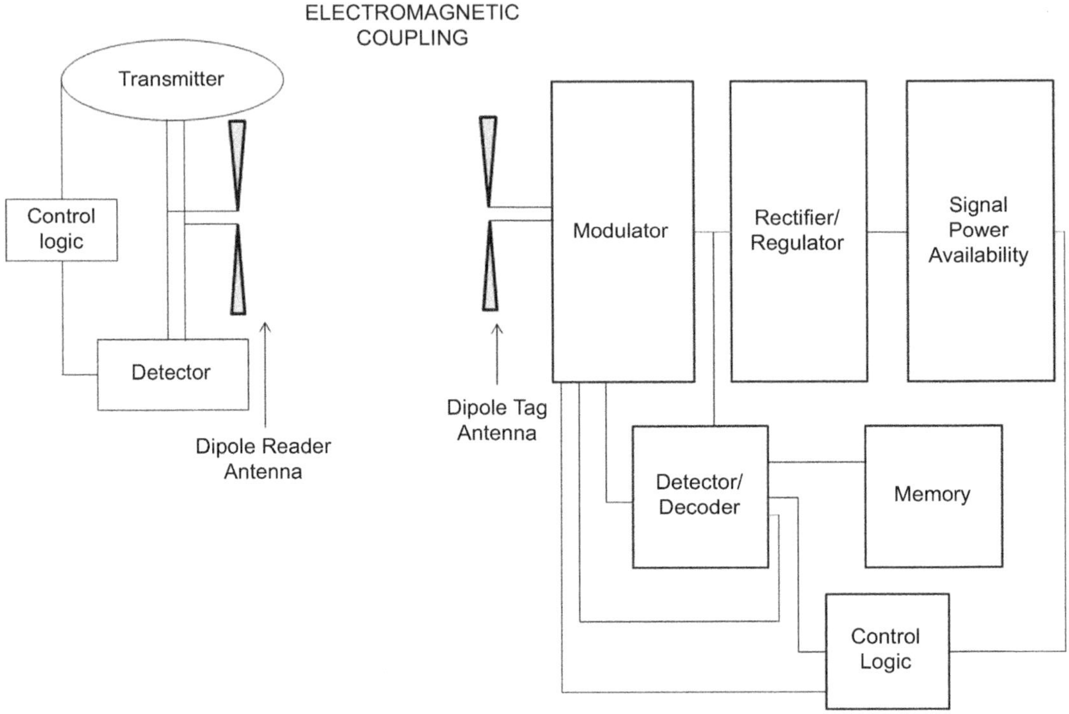

Figure 2.5 Passive tag with electromagnetic back-scattering.

by the tag antenna but also by all the surrounding objects with dimensions comparable to the wavelength employed. This reflected field overlaps the main field emitted by the reader and can cause damping or even cancelation.

The high frequency (UHF, SHF) passive tags operate on distances greater than those inductively coupled and have simpler antennas. Figure 2.5 shows a block diagram of a passive tag with electromagnetic coupling running (in back-scattering) in the UHF band.

In summary, the sequence of events in the query in the back-scatter tag by the reader is as follows:

1. The control logic of the reader sends the data to the transmitter that generates the RF signal to the dipole tag antenna.
2. The signal propagates in space (far-field) and is received by the dipole tag antenna.
3. A power supply is extracted by a rectifier/limiter circuit.
4. The detector/decoder provides the decoded data to the control logic (which eventually will write it into the memory).
5. The detector/decoder signals the oscillator that drives the modulator the instants in which it is possible to activate the transmission of data to the reader.
6. The control logic of the tag reads the data in its memory and communicates the availability to the oscillator that generates the driving signal of the modulator with specific timing (modulates the impedance of the tag antenna recognizing the back-scatter).

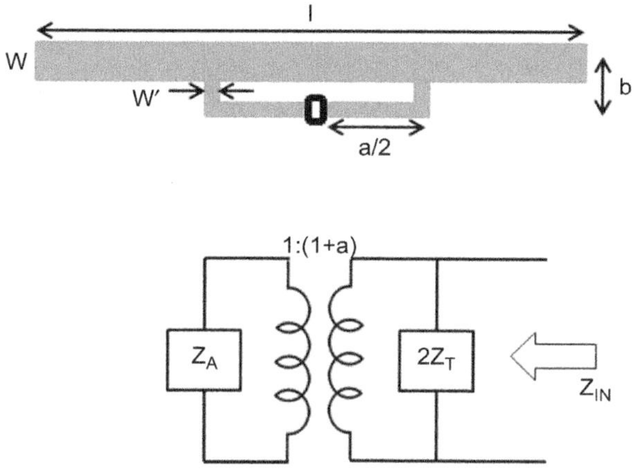

Figure 2.6 The T-match structure and the equivalent circuit.

7. The reader receives the back-scattered signal, decodes it through the receiver and transmits the received data to the control logic.

2.2 Fundamentals and advances in RFID antenna design

In an RFID system the antenna design is critical to ensure an optimized linkage performance. The tag and antenna design adopted for RFID systems is inspired by radar technology in which a reader transmits a signal to a tag and the tag sends back recorded data to the reader. Various aspects need to be considered in the design: the operating frequency, the antenna chip impedance, the overall size, the antenna gain and directivity, the reading range, and so on. Particular attention has also to be paid to the actual environment that affects the near-field region of reader and tag antennas.

In this section the RFID antenna impedance-matching techniques will be summarized by following the guidelines given in [11]. Simulations results, showing how the physical parameters influence the resistance and reactance value of the input impedance, will be described by the use of matching charts. For each design solution, the role of the main geometrical parameters over the complex impedance tuning is examined.

Because of the integrated on-chip energy-storage device, most of the available passive RFID IC chips exhibit an input capacitive reactance roughly ranging from 100 Ω to 400 Ω [7], while the real part is about an order-of-magnitude smaller. The tag antenna impedance should be inductive in order to achieve conjugate matching. To obtain low cost devices, it is not feasible to use external matching networks involving lumped components. Therefore, the matching mechanisms have to be embedded within the tag's antenna layout.

To introduce tunable parameters to a dipole of length l, a centered short-circuit stub and a second dipole of length a are added, forming the T-match structure. The IC chip is connected to the second dipole with a distance b from the larger dipole. T-match acts as an impedance transformer, as shown in Figure 2.6.

The impedance at the source point is given by [12] and given in [11]:

$$Z_{in} = \frac{2Z_t(1+a)^2 Z_A}{2Z_t + (1+a)^2 Z_A}, \tag{2.5}$$

$$Z_t = \mathrm{j}Z_0 \tan ka/2, \tag{2.6}$$

$$Z_0 = 276 \log_{10}(b/\sqrt{r_e r_{e'}}), \tag{2.7}$$

$$r_e = 0.25w, \tag{2.8}$$

$$r_{e'} = 8.25w', \tag{2.9}$$

$$\alpha = \ln(b/r_{e'})/\ln(b/r_e), \tag{2.10}$$

$$w/w' = 2. \tag{2.11}$$

Where, as reported in [7], (2.6) shows the input impedance of the short-circuit stub formed by the T-match conductors and part of the dipole; (2.7) reports the characteristic impedance of the two-conductor transmission line with spacing b; Z_A is the dipole impedance taken at its center in the absence of the T-match connection; (2.8) is the equivalent radius of the dipole, and (2.9) is the equivalent radius of the matching stub; (2.10) is the current division factor between the two conductors. The challenge is to give dimensions to a, b, and w' in order to match the antenna impedance to the complex chip impedance, Z_{chip}. Figure 2.7 shows the matching charts for the T-match structure. The ratio between the dipoles' cross-sections has been fixed to be as shown in (2.11).

It can be observed that even with small values of a and b, high values of input resistance are observed. This makes it difficult to match the impedance to IC chips. Therefore, a single T-match structure may not be capable of providing proper impedance matching.

The coupled structure consists of a feed loop with two terminals and a radiating body. The two terminals of the loop are connected to the IC, and the feed "communicates" with the antenna body through mutual coupling.

The inductively coupled structure can be modeled as a transformer. Figure 2.8 depicts the equivalent lumped element model, where R_{rb} and R_{loop} are the individual resistances of the radiating body and the feed loop. M is the mutual inductance and L_{loop} is the self-inductance of the feed loop.

Figure 2.9 shows the matching charts. The input reactance is nearly unaffected by the loop–dipole distance d. For a fixed loop size, the resistance reduces when the loop–dipole distance increases. Therefore, inductively coupled feeding structures present one of the theoretical optimum solutions to effectively match an antenna to arbitrary chip impedances.

A useful matching technique for tags fabricated with large planar dipoles is to employ a nested-slot structure, as shown in Figure 2.10, where by slot we mean an aperture or a window made by removing the area from the solid metal surface. The slot introduces an inductive component able to cancel the capacitive part of the chip impedance.

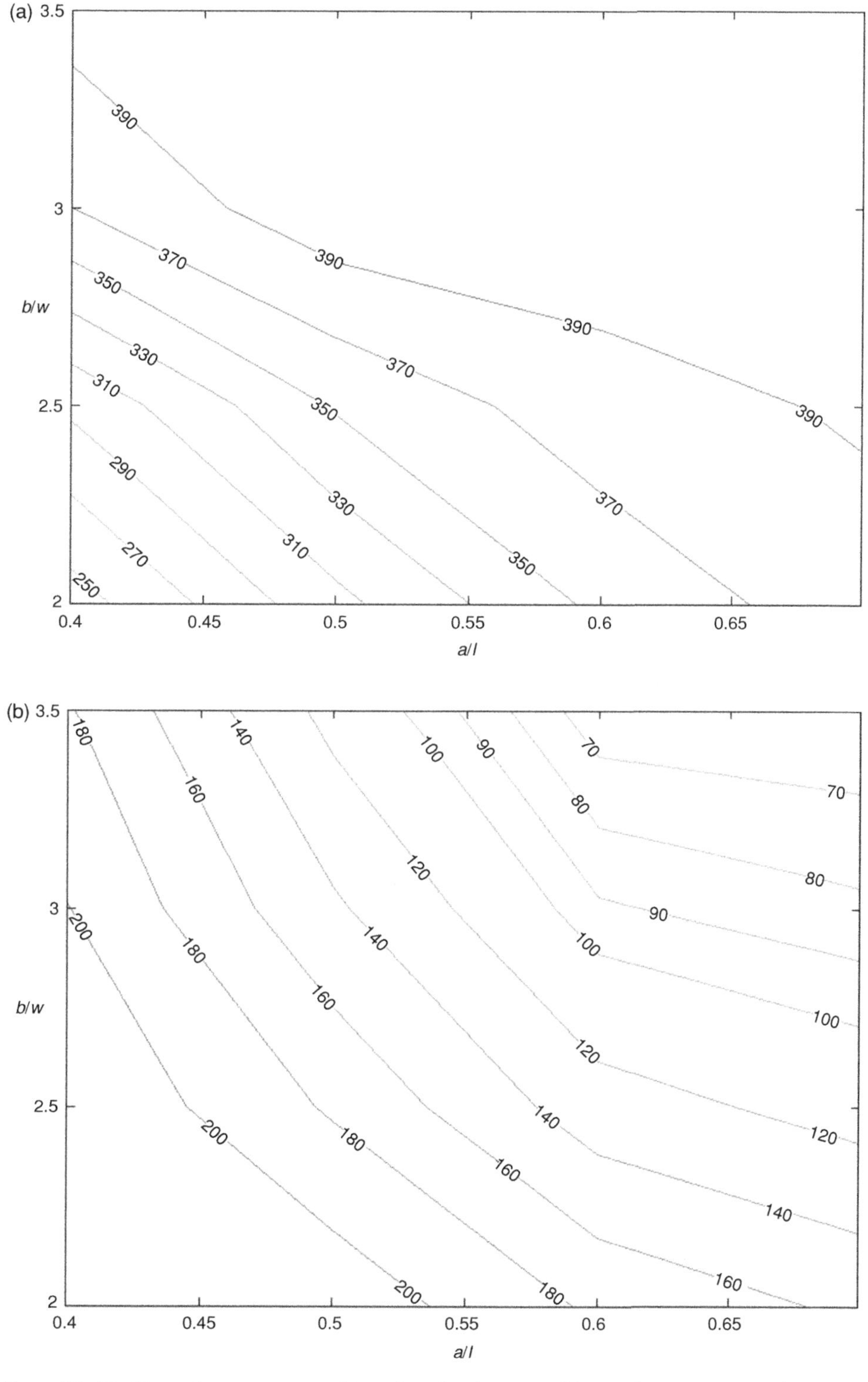

Figure 2.7 The T-match structure matching chart for the case of $l = \lambda/2$, $w = \lambda/100$, $w' = w/2$, (a) resistance (Ω) (b) reactance (Ω).

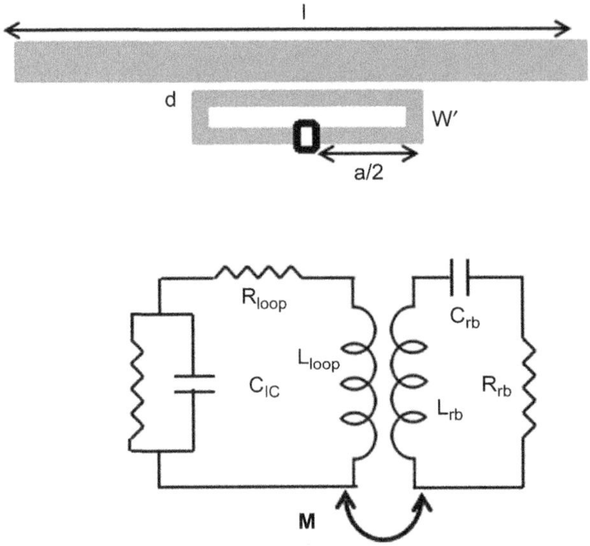

Figure 2.8 Layout of inductively coupled field and equivalent circuit.

Figure 2.11 shows the matching charts of the nested-slot structure. It can be seen that the resistance appears to be sensitive mainly to the slot's width *b*, while the reactance changes almost linearly along with the sizes of both *a* and *b*.

Together with impedance matching optimization, the reduction of the antenna size is assuming a relevant importance in design for applications that require low profile, light, portable RFID equipment. Two efficient ways to reduce size are usually adopted: the introduction of meander lines [12] and the application of the "inverted-F" structure [13].

With the increasing development of wireless technologies and application fields, the need to meet challenging requirements is continually becoming more critical.

As mentioned before, the polarization mismatch can lower the power received by the tag and reader, with a consequent degradation of the system quality. To avoid this phenomenon and to optimize the impedance matching, the reader antenna polarization is being kept circular. Moreover, the number of different frequency bands adopted for RFID communications in different countries is leading to the necessity of having tag antennas able to work at different frequencies by being multiband or at least dual band [14, 15].

2.3 Smart RFID tagged objects: from conventional RFID to networked RFID systems and green solutions

RFID is one of the key elements of the upcoming IoT (Internet of Things); it is often said that RFID is an enabling technology for IoT. This is because IoT is actually based on networked objects and yields simple and cost-effective systems able to connect them.

The conventional RFID structure is mutating toward innovative system architectures that will better fit the requirements of the next generation of pervasive networked and

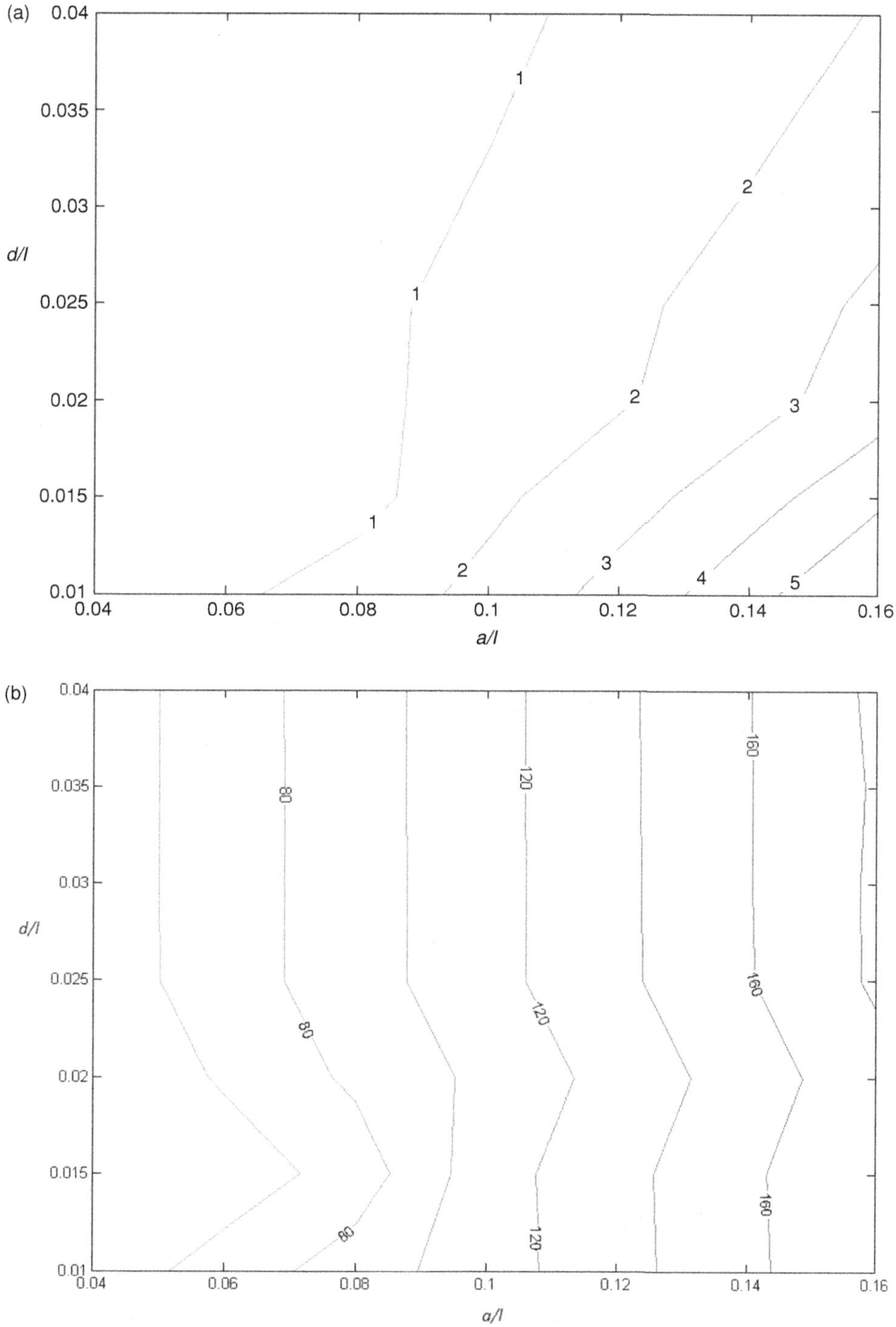

Figure 2.9 The inductive-coupled structure matching chart for the case of $a = b$, $l = \lambda/2$, $w = \lambda/100$, $w' = w/2$: (a) resistance (Ω) (b) reactance (Ω).

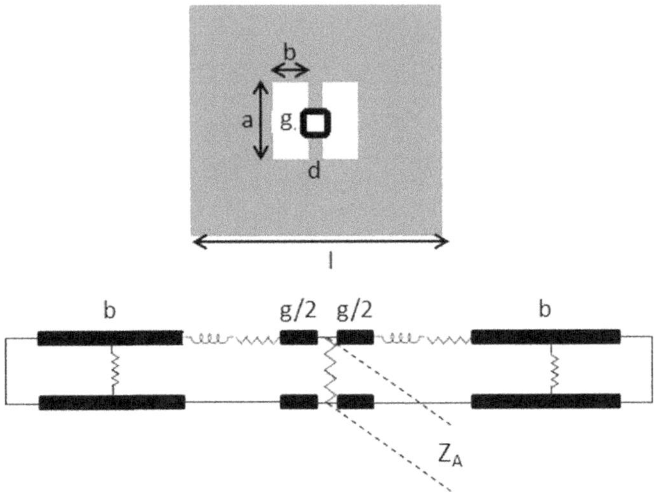

Figure 2.10 The nested-slot structure and the equivalent circuit.

interconnected systems. The global interconnections will need to support an expanding amount of information about identity, location, environmental conditions, and history of all the objects connected to the internet.

The first development that will help this evolution is the creation of systems able to sense, monitor, and adapt themselves to the surrounding environment; such systems can be called "smart RFID tags." All these features need to be combined into a single device that has to be as small as possible in order to be hosted by as many objects as possible.

Integration of smart devices into "things" implies the integration of chips and antennas into nonstandard substrates. It can be achieved by adopting System in Package (SiP) and System in Package on Paper (SiPoP) [16] technologies or similar; those technologies, in fact, based on the realization of controlled multilayer, heterogeneous, 3D structures, make feasible the low cost integration of antennas, lumped components as well as distributed circuits on a variety of materials (ceramic, liquid crystal polymers – LCP [17–19], paper and so on).

In order to achieve and guarantee an appropriate reading accuracy and reliable data transmission, the RFID tags on connected objects need to be as immune or adaptable as possible to the environmental conditions that can be hostile for electronic circuits. New solutions to implement insensitive/tolerant platforms are under investigation [20, 21]. First, the use of radiating elements placed above ground planes has been adopted; good impedance matching to the RFID chip and relatively high reading range are achieved, even when the tag is attached to lossy or conducting bodies. Second, the development of platform tolerant tags and RFID systems has been proposed for medical applications; platform tolerant tag designs are realized and measured to validate the performance of the tag in a human body.

A remaining big challenge for major development of IoT technologies is to enable operability between objects. Future tags must integrate different standards of

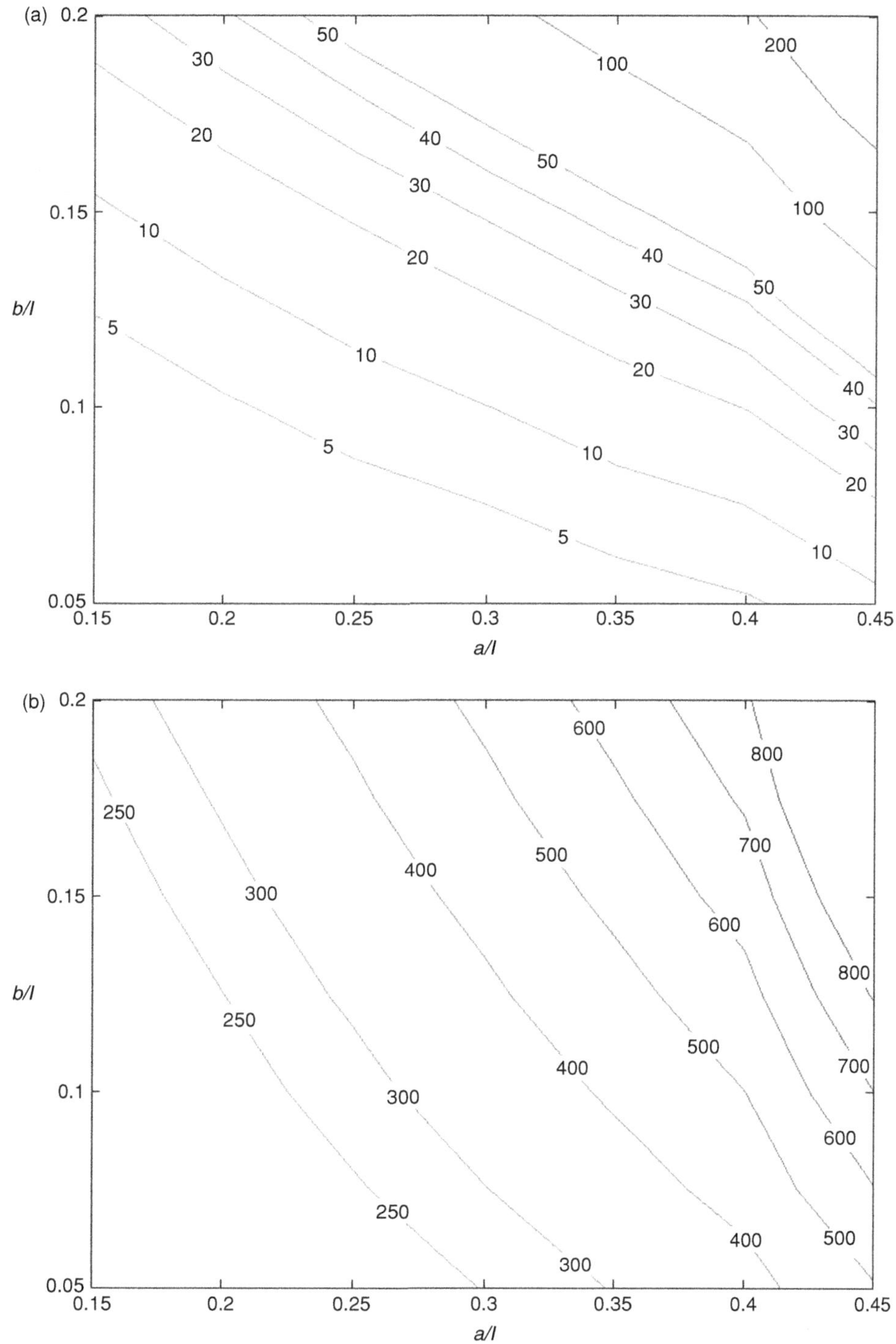

Figure 2.11 The nested-slot structure matching chart for the case of $l = \lambda/2$, $d = g = \lambda/120$, (a) resistance (Ω) (b) reactance (Ω).

communication to allow connection between different architectures, centralized or distributed, if a global standard cannot be established.

The main developments that will animate modern societies and technologies in the future are strongly related to cognitive intelligence and ubiquitous ad hoc networks in many different applications, such as logistics, anti-counterfeiting, supply chain monitoring, space, healthcare, pharmaceutical, military, etc. [22].

Regarding technological evolution, the increasing demand for low cost, flexible, reliable, low power consumption, and durable wireless modules and electronics is steering research towards the development of flexible devices, based on organic materials and substrates, with the goal of overcoming the limitations of ceramics, plastics, and silicon.

In this scenario, paper is an outstanding candidate as a low-cost substrate for radio frequency identification (RFID) and other RF applications. Paper, an organic-based substrate, is universally available and extremely cheap.

Paper also has many characteristics that make it suitable for a variety of applications; the most relevant is its low surface profile that makes paper suitable for fast printing processes such as direct write methodologies or inkjet printing of electronics instead of the traditional metal etching techniques. The combined use of inkjet printing, and other appropriate interconnecting technologies [23] can lead to the production of multilayer electronics in paper [24], theoretically enabling components such as antennas, ICs, memories, batteries, and sensors to be embedded in paper modules. A further achievement of inkjet printing technology is the printing of conductive polymers and functional materials on cellulose-based substrates (CBS). Organic materials can be printed and patterned using techniques such as flexo, gravure, offset, screen, inkjet, etc. [25–27].

Following these trends, mainly passive devices such as low-cost RFID transponders, sensors, memories, photo-voltaic cells, displays, or batteries have been developed for paper substrate integration, while the development of electronic components such as diodes and transistors has been demonstrated, so far, only in low UHF (RF) bands [28–31].

Consideration of future mass industrial production underlines again the importance of antenna design that is clearly to be the most critical element in terms of costs and dimensions. Antenna dimensions, in fact, determine the amount of substrate material to be manufactured and the quantity of conductive ink to be used in printing processes. Therefore, cost reduction of a tag device has to be achieved by antenna design optimization, meaning reduction of the antenna size for a certain fixed frequency [30] and reduction of ink used, i.e. a real metalized surface.

Moreover, in order to develop paper-based 3D multilayer circuits, the RF characterization of paper became a must for optimal designs. Some characterization work has been done in frequencies beneath the UHF band [31–34], and recently in or above the UHF band [12, 25]. These techniques can be applied to most of the multilayered materials foreseeable for SiP developments.

Paper-like materials are well suited for reel-to-reel processing of dimensions unthinkable for other materials; this feature, in combination with the development of the inkjet printing process and SiPoP integration at an industrial scale [23] enables a tremendous breakthrough in the mass fabrication of RFID tags for a networked society. This breakthrough will allow not only mass production of ultra-low-cost devices but also the realization of large arrays of devices thus enabling new solutions based on the wide area electronics (WAE) approach [35–38].

References

[1] Harry Stockman, "Communication by means of reflected power," *Proceedings of the IRE*, 1196–1204, (Oct.) 1948.

[2] D. B. Harris, "Radio transmission systems with modulatable passive responder," US 2927321, 1960.

[3] M. W. Cardullo and W. L. Parks, "Transponder apparatus and system," US 3713148, 1970.

[4] R. A. Hauslen, "The promise of automatic vehicle identification," *IEEE Trans. on Vehicular Technology*, **VT-26**, (*1*), 30–38, (Feb.) 1977.

[5] M. Hassan, M. Ali, and E. Aktas, "Radio frequency identification (RFID) technologies for locating warehouse resources: A conceptual framework," *Proceedings of 2012 European Conference on Smart Objects, Systems and Technologies (SmartSysTech)*, pp. 1–20, 12–13 June 2012.

[6] M. Merenda, C. Felini, and F. G. Della Corte, "Battery-less smart RFID tag with sensor capabilities," *IEEE International Conference on RFID-Technologies and Applications (RFID-TA)*, pp. 160–164, 5–7 Nov. 2012.

[7] ISO Standard. http://www.iso.org/iso/home/standards.htm.

[8] EPCglobal, "Radio-frequency identity protocols – HF version 2 RFID – draft version 0.0.9," EPCglobal Standard, Jul. 2006. [Online]. Available: http://www.epcglobalinc.org/standards/hfg2/.

[9] ISO/IEC SC31/WG 4, "ISO/IEC WD 18000–6 mode3; automatic identification – radio frequency identification for item management part 6: Mode 3 – physical layer, anti-collision system and protocols for ultra high frequency (UHF) systems," Feb. 2002.

[10] EPCglobal, "Radio-frequency identity protocols – class-1 generation-2 RFID v1.2.0," EPCglobal Standard, EPCglobal Inc., Oct. 2008, http://www.epcglobalinc.org.

[11] G. Marrocco, "The art of UHF RFID antenna design: impedance-matching and size-reduction techniques," *IEEE Antennas and Propagation Magazine*, **50**, (*1*), 66–79, 2008.

[12] C. Balanis, *Antenna Theory, Analysis and Design*, 3rd Edn., John Wiley & Sons, Inc. 2005.

[13] G. Orecchini, F. Alimenti, V. Palazzari, *et al.*, "Design and fabrication of ultra-low cost RFID antennas and tags exploiting paper substrates and ink-jet printing technology," *IET Microwaves, Antenna and Propagation*, **5**, (*8*), 993–1001, 2011.

[14] A. Babar, L. Ukkonen, and L. Sydanheimo, "Dual UHF RFID band miniaturized multipurpose planar antenna for compact wireless systems," *International Workshop on Antenna Technology (iWAT)*, pp. 1–4, 2010.

[15] C. Mariotti, V. Lakafosis, M. M. Tentzeris, and L. Roselli, "An IPv6-enabled wireless shoe-mounted platform for health-monitoring," *IEEE Topical Conference on Wireless Sensors and Sensor Networks (WiSNet)*, pp. 46–48, 2013.

[16] F. Lolli, M. Virili, G. Orecchini, *et al.*, "Electromagnetic characterization of paper-glue compound for system-in-package on paper (SiPoP) future developments," *Microwave and Wireless Components Letters, IEEE*, **22**, (*10*), 545–547, 2012.

[17] J.-H. Lee, S. Sarkar, S. Pinel, *et al.*, "3D-SOP millimeter-wave functions for high data rate wireless systems using LTCC and LCP technologies," *Proceedings 55th Electronic Components and Technology Conference*, pp. 764–768, 2005.

[18] A. Rida, A. Margomenos, J. S. Lee, *et al.*, "Integrated wideband 2-D and 3-D transitions for millimeter-wave RF front-ends," *Antennas and Wireless Propagation Letters, IEEE*, **9**, (*11*), 1080–1083, 2010.

[19] A. Rida, A. Margomenos, and M. M. Tentzeris, "Novel wideband 3D transitions on liquid crystal polymer for millimeter-wave applications up to 100 GHz," *Microwave Symposium Digest, MTT '09. IEEE MTT-S International*, pp. 953–956, 2009.

[20] M. A. Ziai and J. C. Batchelor, "Thin ultra high-frequency platform insensitive radio frequency identification tags," *Microwaves, Antennas & Propagation, IET*, **4**, (*3*), 390–398, 2010.

[21] H. Rajagopalan and Y. Rahmat-Samii, "Platform tolerant and capsule-pill RFID antenna designs: An overview of recent developments at UCLA," *IEEE International Workshop on Antenna Technology (iWAT)*, pp. 144–147, 2012.

[22] J. Buckley, Final report of the conference "From RFID to the Internet of Things," ftp://ftp.cordis.europa.eu/pub/ist/docs/ka4/au_conf670306_buckley_en.pdf, Brussels, Mar. 2006.

[23] F. Alimenti, M. Virili, G. Orecchini, *et al.*, "A new contactless assembly method for paper substrate antennas and UHF RFID chips," *IEEE Trans. on Microwave Theory and Technique*, **59**, (*3*), 2011.

[24] A. C. Siegel, S. T. Phillips, M. D. Dickey, *et al.*, "Foldable printed circuit boards on paper substrates," *Advanced Functional Materials*, **20**, (*1*), 28–35, 2010.

[25] L. Yang, A. Rida, R. Vyas, and M. M. Tentzeris, "RFID tag and RF structures on a paper substrate using inkjet-printing technology," *IEEE Trans. on Microware Theory and Technique*, **55**, (*12*), 2894–2901, 2007.

[26] E. Fortunato, N. Correia, P. Barquinha, *et al.*, "High-performance flexible hybrid field-effect transistors based on cellulose fiber paper," *IEEE Electron Device Lett.*, **29**, (*9*), 988–990, 2008.

[27] S. Couderc, O. Ducloux, B. J. Kim, and T. Someya, "A mechanical switch device made of a polyimide-coated microfibrillated cellulose sheet," *Journal of Micromechanical and Microengineering*, **19**, 1–11, 2009.

[28] S. Steudel, K. Myny, V. Arkhipov, *et al.*, "50 MHz rectifier based on an organic diode," *Nature Material*, **4**, 597–600, 2005.

[29] S. Steudel, S. D. Vusser, K. Myny, *et al.*, "Comparison of organic diode structures regarding high-frequency rectification behavior in radio-frequency identification tags," *J. Appl. Phys.*, **99**, (*11*), 114–519, 2006.

[30] V. Subramanian, J. M. J. Frechet, P. C. Chang, *et al.*, "Progress toward development of all-printed RFID tags: Materials, processes, and devices," *Proc. IEEE*, **93**, (*7*), 1330–1333, 2005.

[31] M. Virili, G. Casula, C. Mariotti, *et al.*, "7.5–15 MHz organic frequency doubler made with pentacene-based diode and paper substrate," Accepted for publication in the *International Microwave Symposium Digest* (MTT), 2014 IEEE MTT-S.

[32] S. Simula, S. Ikalainen, and K. Niskanen, "Measurement of the dielectric properties of paper," *Journal of Imaging Science and Tech.*, **43**, (*5*), (Sept.) 1999.

[33] H. Ichimura, A. Kakimoto, and B. Ichijo, "Dielectric property measurement of insulating paper by the gap variation method," *IEEE Trans. Parts, Materials and Packaging*, **PMP-4**, (*2*), (June) 1968.

[34] L. Apekis, C. Christodoulides, and P. Pissis, "Dielectric properties of paper as a function of moisture content," *Dielectric Materials, Measurements and Applications, Fifth International Conference*, pp. 97–100, 27–30 June 1988.

[35] http://www.prnewswire.com/news-releases/printed-electronics-for-healthcare-cosmetics-and-pharmaceuticals-2014–2024–226876611.html.

[36] http://www.businesswire.com/news/home/20131125005576/en/Research-Markets-Printed-Electronics-Healthcare-Cosmetics-Pharmaceuticals#.UsbGdNLuKdg.

[37] D. Dimitrakopoulos and P. R. L. Malenfant, "Organic thin film transistors for large area electronics, *Advanced Materials*, **14**, (*2*), 99–117, (Jan.) 2002.

[38] A. Nathan, A. Ahnood, T. C. Matthew, *et al.*, "Flexible electronics: the next ubiquitous platform," *Proc. IEEE (Centennial Issue)*, **100**, 1479–1510, 2012.

3 Energy scavenging and storage for RFID systems

Alessandra Costanzo, Diego Masotti, Aldo Romani, and Marco Tartagni

3.1 Introduction

Electronics is progressively penetrating more deeply into human lives: integration has made possible the wealth of many small mobile devices that society currently enjoys (e.g. smart phones, MP3 players, GPS navigation assistants, etc.). In this scenario, active RFID systems hold the promise of implementing smart environment and objects, and can ease processes in many applications fields, e.g. industrial processes, personal healthcare, environmental monitoring. Pervasive computing and wireless sensor networks are introducing their potential while power consumption has been greatly reduced thanks to energy-aware design techniques. The availability of low cost batteries has been one of the main drivers of these advances, even though it now represents one of the main limitations. In fact, power supplies still mainly rely on electrochemical cells with limited stored charge and are often impracticable to replace.

During the last years energy harvesting from ambient sources has proven to be a viable solution: the environment is an intrinsic source of low-density highly available energy [1] in either steady or intermittent and irregular forms such as, for example, vibrations [2], thermal gradients [3], indoor light [4], and electromagnetic radiation [5]. At the current state of the art, most energy harvesters can provide in practical cases an output power density of about 10–100 $\mu W/cm^3$ [6]. In this scenario, mechanical vibrations represent a viable solution for powering low power electronic systems (e.g. wireless sensor nodes, personal healthcare devices, etc.).

Piezoelectric transducers have proven to provide an efficient way to generate collectable electrical charge from mechanical stress and motion. Piezoelectric ceramics, such as PZT, offer the possibility of ease of fabrication, reduced production costs and high customizability. Other exploitable principles include electromagnetic induction, which requires the use of magnets and windings, or electrostatic energy generation, which requires initial amounts of energy to be provided to the system.

Heat and temperature gradients are also suitable sources for many applications. Temperature gradients can easily be found in electrical machinery, engines, and, to a lesser extent, between the human body and the environment. Transducers exploiting the Seebeck effect, such as for example, arrays of thermocouples, usually referred to as micro-thermogenerators [7], can convert into electrical form a part of the heat flowing between the two thermal sources at different temperatures. Up to now, excellent

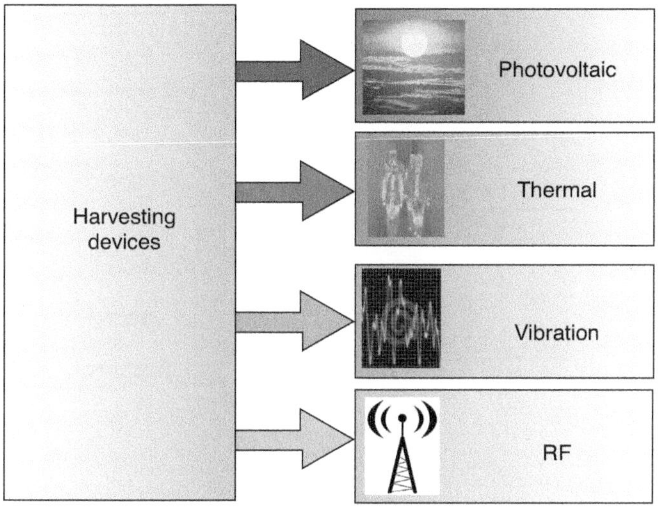

Figure 3.1 Multiple energy harvesting sources.

degrees of miniaturization have been achieved [8] together with unobtrusive implementations of energy transducers [9].

Electromagnetic radiation is also a suitable form of exploitable energy. Wireless and radio communications spread RF energy at low densities through almost all humanized environments. This technology is the basis of RFID systems, where far-field or near-field radiation can be exploited in order to extract energy. Specific RF devices, such as for example rectennas [10], i.e. rectifying antennas, can be used in order to optimize energy harvesting.

One of the main difficulties consists in the fact that many ambient vibrations are often variable and irregular (e.g. on transportation systems such as trains, cars, etc.) and usually have low amplitudes. In this case, energy harvesting techniques tailored to the state of the vibration source are essential to achieve an efficient power conversion. RF radiation is also highly variable and intermittent, e.g. in GSM communication where mobile terminals set their output power according to the directive imposed by the base station. In this context, reliable models of the above-mentioned energy sources may contribute to estimating the available power and the behavior of energy sources when subject to specific excitations. In this perspective, the physical principles behind different energy harvesting principles will be introduced and explained from an analytical point of view.

Energy harvesting is the conversion of ambient energy to electricity to power small electric and electronic devices. The ambient energy includes human power, vibration, light, heat, movement, etc., as schematically sketched in Figure 3.1.

Ambient energy harvesting is a need for all modern wireless communication-based humanized environments, e.g. 90% of envisaged uses of wireless sensor networks (WSN) are impractical without energy harvesting. Therefore the goal is to provide energy autonomous systems, i.e. systems designed to operate and/or communicate in

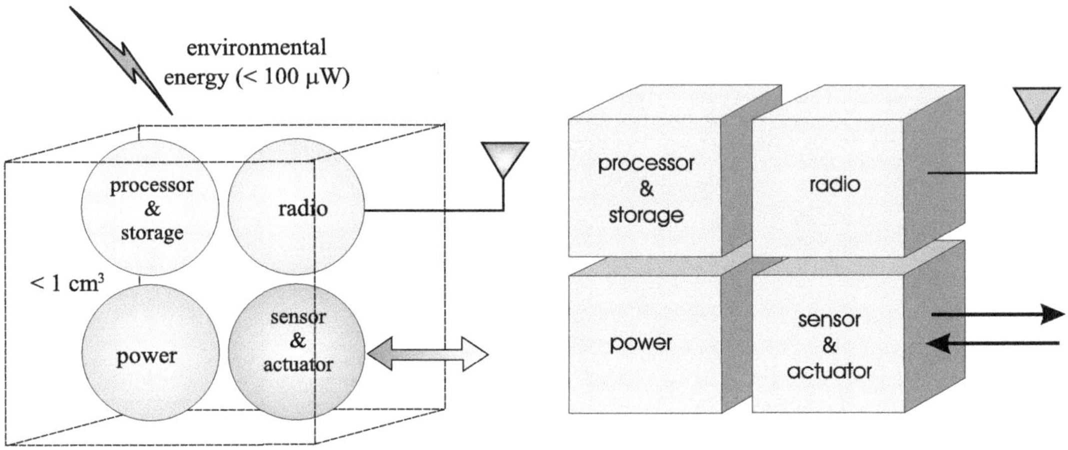

Figure 3.2 Common target specifics.

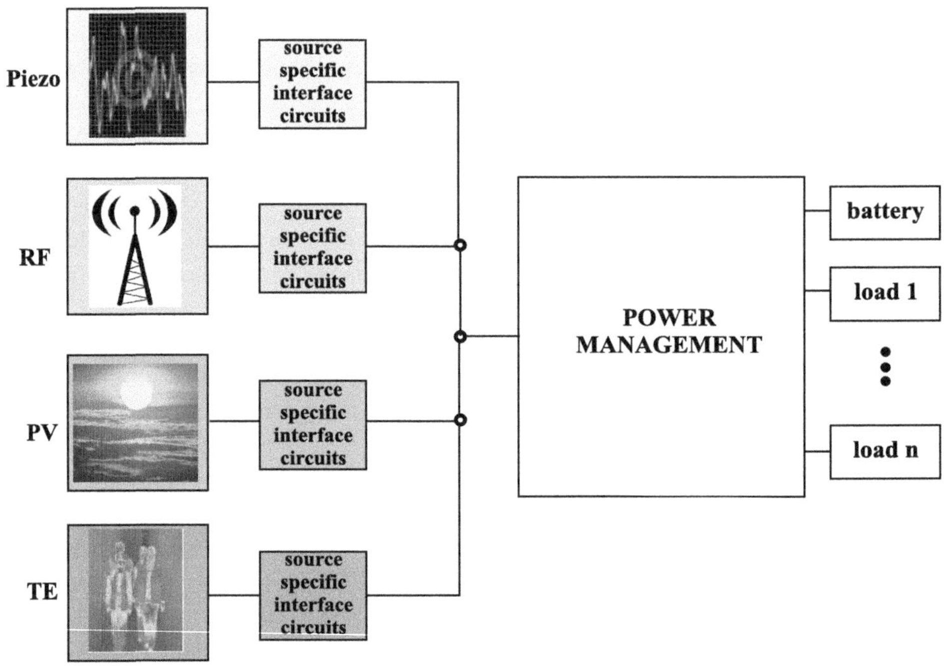

Figure 3.3 Multi-source energy harvester architecture.

known/unknown environments over their whole lifetime. These systems must have the ability to operate with less than hundreds of μW of power in some cm^3. Figure 3.2 summarizes a common target for harvesters: $< 100\ \mu W/cm^3$ on average, indefinitely.

The functional blocks of a multi-source energy autonomous system (shown in Figure 3.3) are:

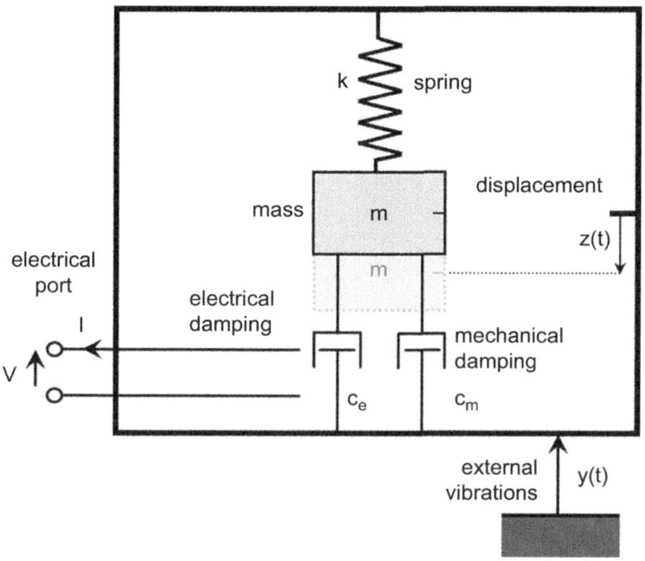

Figure 3.4 Lumped parameters representation of a vibrational energy harvesting device.

Energy generation section:

Energy harvesting: Devices or systems harvesting energy from correlated or uncorrelated sources of energy.

Energy sources and storage: Any kind of energy storage element that could be used to accumulate energy in excess from the harvester and provide it to the system in its location whenever the energy is insufficient.

Energy conversion and optimization section: Any energy conversion system that trades and optimizes the energy stored/harvested in the energy generation block to the energy consumption block.

Energy consumption section: Data acquisition, elaboration, storage, and transmission.

3.2 Modeling vibrational energy harvesters

Miniaturized vibrational energy harvesters are usually implemented with oscillating structures including a seismic mass. Commonly used structures include cantilevers, beams, and membranes. Since most of the mass of the whole structure is concentrated into the seismic mass, and since small oscillations are usually applied because of the relatively low amplitude of vibrations, a lumped parameter model is usually adopted [11–13]. In such cases, a lumped parameter model of a harmonic oscillator with a single degree of freedom holds to a close approximation, as shown in Figure 3.4. A seismic mass m is attached to an oscillating structure, whose elastic constant is denoted by k. Mechanical losses are modeled by a linear damper c_m. When an impressed external vibration $y(t)$ is applied the structure starts oscillations and a deflection $z(t)$ from the rest position is generally observed. In general, the above parameters depend on the type of mechanical structure considered and, in simpler cases, such as for example cantilevers, analytical expressions can be found [14, 15].

When an external harmonic vibration is applied, at steady state, energy is continuously bounced between the kinetic energy of m and the elastic energy in k, while input energy compensates the mechanical losses introduced by c_m. By using the laws of dynamics, the following differential equation describes the behavior of the system:

$$m\ddot{z} + c_m\dot{z} + kz = -m\ddot{y}. \tag{3.1}$$

The above equation refers to pure mechanical oscillators. When electromechanical devices are involved, as is the case in vibration energy harvesting, different physical effects are exploited in order to convert part of the energy from the system into electrical form. This subtraction of energy from the mechanical domain can be modeled as an additional damping factor, c_e, referred to as the electrical damping coefficient. This assumption holds whenever the damping force exerted by the energy harvesting principle used F_e is proportional to the oscillation velocity $\dot{z}$:

$$F_e = c_e\dot{z}. \tag{3.2}$$

In this case the energy harvesting system is modeled by the following differential equation:

$$m\ddot{z} + (c_m + c_e)\dot{z} + kz = -m\ddot{y}. \tag{3.3}$$

The instantaneous harvested power and the instantaneous mechanically lost power can be expressed respectively as:

$$p_e(t) = c_e\dot{z}(t)^2, \tag{3.4}$$

$$p_m(t) = c_m\dot{z}(t)^2. \tag{3.5}$$

As long as harmonic oscillations are involved, the corresponding average power values are obtained:

$$P_e = \frac{1}{2}c_e V_z^2, \tag{3.6}$$

$$P_m = \frac{1}{2}c_m V_z^2, \tag{3.7}$$

where V_z is the amplitude of the harmonic oscillation velocity $\dot{z}(t)$.

At the resonance frequency ω_0, defined as

$$\omega_0 = \sqrt{\frac{k}{m}} \tag{3.8}$$

in Equation (3.3), the reactive terms are canceled out and:

$$\left(\frac{c_m + c_e}{m}\right)\dot{z}(t) = -\ddot{y}(t).$$

Thus, we can obtain from (3.6) the following expressions for the harvested power at resonance [16]:

$$P_e = \frac{1}{2}\frac{c_e}{(c_e + c_m)^2}A^2m^2, \tag{3.9}$$

$$P_e = \frac{1}{2}\frac{c_e}{(c_e + c_m)^2}Y^2k^2, \tag{3.10}$$

where A is the amplitude of the external acceleration $\ddot{y}(t)$ and Y the amplitude of the external displacement $y(t)$, where it necessarily holds that $A = 2\omega_0 Y$.

Since the electrical damping is in many cases a design parameter, from (3.9) and (3.10) it can be found that the power extracted from the vibration source is maximized whenever $c_e = c_m$, setting an initial design constraint. If the external mechanical stimulus is able to set a constant oscillation amplitude Y, it can be seen that stiffness plays an important role in defining the power output. If the external vibration is instead an applied force, the higher the seismic mass m, the higher the power output.

However, the above considerations are quite general and are based on the assumption of damping forces proportional to the oscillation velocity. Actually, according to the energy harvesting principle, other relations may hold, upon which it will be necessary to define more specific results. This will be done in the following sections of this chapter.

3.2.1 Electromagnetic generators

Electromagnetic generators exploit the principle of electromagnetic induction. They normally include permanent magnets and coils subject to vibrations. The variations in the concatenated flux of the coil produce an electromotive force V which can be exploited in order to generate power. Normally, the housing is subject to vibrations, while an elastic element allows the seismic mass to oscillate. A schematic representation of an electromagnetic microgenerator is shown in Figure 3.5.

According to Faraday's law of induction, the open circuit voltage developed across the terminals of the microgenerator can be expressed as:

$$V_\infty(t) = -\frac{\mathrm{d}\Phi}{\mathrm{d}t} = \alpha\dot{z}(t), \tag{3.11}$$

where Φ is the magnetic field flux through the coil, whose variations are supposed, for small oscillations, to be proportional to the mass displacement through a coefficient α.

In contrast, when a resistive load R is connected as output load, the following equation must be considered:

$$V_\infty(t) = L\frac{\mathrm{d}i(t)}{\mathrm{d}t} + (R_L + R)I(t), \tag{3.12}$$

where L and R_L are respectively the inductance and the parasitic resistance of the coil and I is the electrical output current. In general, in environmental vibration energy harvesting, very low frequencies usually lower than 100 Hz are involved, and the miniature generators usually yield low values of inductance, so that $j\omega L \ll R$. In addition, the parasitic

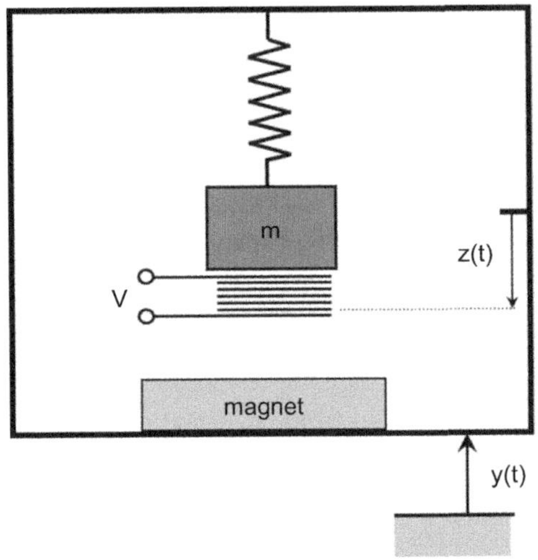

Figure 3.5 Representation of an electromagnetic microgenerator.

resistance R_L is usually much lower than that of the output load R. For this reason, I can be approximated as:

$$I = \frac{V_\infty}{R + R_L + j\omega L} \approx \frac{V_\infty}{R}. \tag{3.13}$$

The electrical damping force exerted when the current I flows through the external load is proportional to I through the coefficient α:

$$F_e = \alpha I. \tag{3.14}$$

According to (3.14), the motion differential equation becomes:

$$m\ddot{z} + c_m \dot{z} + \alpha I + kz = -m\ddot{y}. \tag{3.15}$$

By using (3.14), (3.13), and (3.11), we obtain from (3.15) that the electrical damping coefficient can be approximated as:

$$c_e \cong \frac{\alpha^2}{R}, \tag{3.16}$$

from which, for a given electromagnetic generator, the optimal load $R = \alpha^2/c_m$ can be found.

3.2.2 Piezoelectric generators

Piezoelectric materials have the property of accumulating electrical charge when subject to stress and deformations. This phenomenon is due to the formation of microscopic electric dipoles within the crystal structures because of mechanical deformations. During the fabrication of transducers, a special poling process is applied, usually consisting in the application of an intense electric field at a high temperature. This causes the

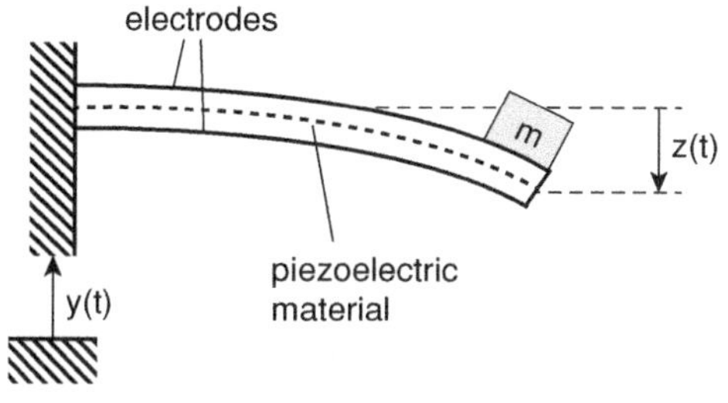

Figure 3.6 Structure of a piezoelectric bimorph cantilever beam.

alignment of all the microscopic electric dipoles along the direction of the poling electric field, so that macroscopic effects can be observed upon deformation of the mechanical structure. The resulting electric field produces a polarization charge of mechanical origin, so that a voltage is developed between electrodes located in the device. Several topologies of electrodes are usually employed, where they can be located on the surface of the device or internally, as for example in stacked transducers. Symmetrically, the application of external voltage to the transducer electrodes produces mechanical stress and deformations. These two aspects of the piezoelectric effect are usually referred to as direct and converse. The constitutive equations of piezoelectric materials are summarized by the following, involving mechanical and electrical quantities [17]:

$$\begin{cases} D = \varepsilon E + d\sigma, \\ \delta = dE + \dfrac{\sigma}{Y}, \end{cases} \tag{3.17}$$

where E is the electric field, D the electrical displacement, σ and δ the mechanical stress and strain, Y the Young's modulus, d the piezoelectric charge constant.

Normally several shapes of miniaturized transducers are employed in energy harvesting applications, such as for example cantilevers [17, 18] (Figure 3.6), membranes [19], shells [20], and complex shapes [21]. The elastic element is usually identified with the piezoelectric transducer, which stores elastic energy. A seismic mass is usually attached to the transducer, so that the generic oscillator model of Figure 3.4 usually applies. Since such a mass is usually much larger than that of the oscillating structure itself, a usual approximation is to identify the whole mass of the system with the seismic mass.

Going back to macroscopic quantities from (3.17), direct and converse piezoelectric effects can be summarized by the following relations [22]:

$$F = kz + \alpha V, \tag{3.18}$$

$$I = \alpha \dot{z} - C_P \dot{V}, \tag{3.19}$$

where C_P is the electrical capacitance measured between the electrodes, I the current flowing through an external load, k is the equivalent Hooke's elastic constant of the transducer, referred to mass displacement and is derived from α and from the geometrical properties of the transducer. In (3.18), which combines electrical and mechanical quantities, F is the external applied force, z the mass displacement, and V the voltage measured across the electrodes of the device. The following motion differential equation also holds:

$$m\ddot{z} + c_m\dot{z} + kz + \alpha V = -m\ddot{y}. \tag{3.20}$$

The above equations also lead to an equivalent electromechanical circuit representation, as shown in [22, 23], which is helpful to predict the joint behavior of transducer and load [24]. When an external resistor is connected as a load, it holds that $I = V/R$. In the frequency domain, from (3.19), the relation between voltage and oscillation velocity can be expressed as follows, where V and Z are the complex representative vectors of $V(t)$ and $z(t)$:

$$V = \frac{\mathrm{j}\omega\alpha R}{1 + \mathrm{j}\omega RC} Z. \tag{3.21}$$

Expression (3.21) leads to two limit expressions for the electrical damping force, respectively for $R \to \infty$ and for$R \to 0$, which are proportional to the elongation and to the oscillation velocity:

$$F_{e,\infty}(t) = \frac{\alpha^2}{C_P} z(t), \tag{3.22}$$

$$F_{e,0}(t) = \alpha R\dot{z}(t). \tag{3.23}$$

In the above two limit cases, two different forms of damping are experienced. Energy harvesting is fruitfully performed when (3.23) holds and power is transferred to the load. However, in general, for all intermediate loads, the damping will be dependent on both elongation and velocity. Equations (3.21), (3.22), and (3.23) also determine the optimum load for maximizing the harvested power.

3.2.3 State of the art and benchmarks in vibrational energy harvesters

Following the previous section on generic principles of vibrational harvesters, this section will list state of the art vibrational harvesters with specific focus on those using power management implemented using ultra-low-power integrated circuits. Then, we will review the most used benchmarks in vibrational harvesters, referring mainly to devices that are smaller than one third cubic centimeter and are suitable for miniaturized autonomous systems to be implemented in wireless sensor networks.

Vibration harvesters are the main candidates to end up in a first product for energy harvesting, namely as energy providers for tyre pressure monitor systems. However, a lot of research has still to be performed in order to get power levels up, cost down, and reliability issues solved. As stated in the previous chapter, there are three ways of converting vibration into energy:

1. Electrostatic: a capacitor is made consisting of two opposing metal structures. One of these structures is fixed, the other one moves in the presence of an external force. The change in voltage is proportional to the capacitance change.

2. Piezoelectric: where a mass is suspended by a beam, with a piezoelectric layer on top of the beam. When the mass starts to vibrate, the piezoelectric layer is deformed and a voltage is generated.
3. Electromagnetic or inductive: a mass of magnetic material moves through a magnetic field. The change in flux generates a voltage.

As far as electrostatic harvesters are concerned, microgenerators have been successfully implemented as in [25–27]. More recently, microelectronic interfaces for micromachined electrostatic harvesters have been implemented in CMOS technology as in [28–31]. Their performance with interface circuits was assessed in [32]. Several examples of micromachined piezoelectric harvester have been presented [11, 33, 34], together with recent advances in microelectronic device interface implementations as in [35–39]. From the point of view of micromachining, the electrostatic and piezoelectric harvesters are easier to fabricate. In general it is very difficult to make electromagnetic harvesters using micromachining, the reason being the need for integration of magnetic materials and/or the fabrication of coils with enough windings. However, interesting miniaturized implementations have been presented in [40–43]. A 2012 paper shows an interesting realization of an interface in CMOS technology [44].

To compare different solutions, the simplest figure of merit could be derived considering that, as illustrated in the previous section (see Equation (3.9)), the maximum harvested power is proportional to the square of the acceleration and to the mass, which is proportional in turn to the volume. Thus, the normalized power density (NPD) is defined as [35]:

$$\text{NPD} = \frac{\text{Power output}}{\text{Volume} \cdot \text{Acceleration}^2}. \tag{3.24}$$

This figure is easy and practical to calculate, but it does not take into account second order effects such as the maximum displacement of the mass within the volume and the damping factor. According to the above constraints, the maximum theoretical power could be calculated as [36]:

$$P_{\text{MAX}} = \frac{1}{2} Y_0 Z_l \omega^3 m, \tag{3.25}$$

where Y_0 is again the amplitude of oscillations, and Z_l is the maximum displacement of the mass within the harvester volume. Thus, the *energy efficiency* of the harvester could be calculated as [36]:

$$E_H = \frac{\text{Power output}}{\text{Maximum theoretical output}} = \frac{\text{Power output}}{\frac{1}{2} Y_0 Z_l \omega^3 m}. \tag{3.26}$$

The figure of merit defined above is mainly a measure of how closely a specific design approaches its ideal performance. However, it does not distinguish between implementations of different proof mass density or geometry. For this reason, a variant of this metric has been introduced, referred to as volume figure of merit (FoMV) [36]. However, its calculation is complicated and more insights are discussed in the original paper. A large comparison table between vibrational harvesters using both energy efficiency

Table 3.1 Comparison of performance of existing energy harvesting systems.

Ref.	Harvester size	P_{out}	NPD [$\mu W/mm^3/g^2$]	Power consumption	Volume
[35]	27.3 mm^3	68μW@1g	1.49	< 1μW	0.3 cm^3
		1.8μW@1g	6.45		
[37]	28.5 mm^3	85μW@1g	0.97	10μW	1 cm^3
		17μW@0.6g	1.45		
[38]	130 mm^3	@3.3g	–	2μW	–
[39]	228 mm^3	40μW	–	0.9μW	–

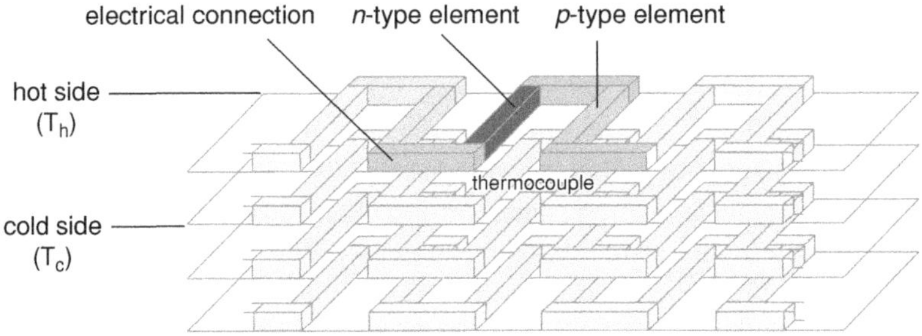

Figure 3.7 Structure of a thermocouple and of a thermogenerator.

and volume figure of merit is also given in [36] and the main conclusions are: i) harvester effectiveness values are mostly in the 1–10% range, with best values over 30%, ii) among different types, the best performances are given by (in this order): piezoelectric, electrostatic, and magnetic harvesters.

As far as microsystems are concerned, due to the miniaturization of the devices, it is better to compare them using the normalized power density figure of merit. A recent paper summarizes the performance of the most recent implementations using integrated electronics power management systems [35, 37–39] (see Table 3.1).

3.3 Thermoelectric generators

Micro-thermoelectric generators exploit the Seebeck effect in order to convert heat into electrical power. They are usually based on arrays of thermocouples, which are the main constitutive element. An exhaustive discussion of such technology is presented in [45]. A thermocouple is usually made by connecting together an electron conducting material (n-type) and a hole conducting material (p-type). All thermocouples are thermally connected in parallel and electrically connected in series, in order to obtain higher output voltages. Series elements can also be connected in parallel in order to increase the output current. The conventional structure of a thermoelectric generator is shown in Figure 3.7. Heat flows from the hot junction to the cold junction, because of the impressed thermal gradient.

As a consequence, because of the different concentrations of charge carriers at the two junctions, diffusion phenomena cause an increase in the voltage barrier between the two junctions. A voltage difference ΔV can be measured between the two terminals of a thermocouple, proportional to the thermal gradient:

$$\Delta V = \alpha(T_H - T_C), \tag{3.27}$$

where α is the Seebeck coefficient, T_H and T_C are respectively the higher and the lower temperatures.

However, besides the Seebeck effect, other effects influence the behavior of the thermogenerator, such as the Peltier, Joule, and Thomson effects, as summarized also in [45–47]. The Peltier effect is observed when an external load is connected and a current I flows through the thermogenerator. The effect consists in the production of a heat flow between the two junctions:

$$\dot{Q}_P = \alpha I T_H. \tag{3.28}$$

When an electrical current I flows, an additional amount of heat is generated because of the Joule effect:

$$\dot{Q}_J = R_{TEG} I^2, \tag{3.29}$$

where R_{TEG} is the intrinsic resistance of the thermocouple, which can be expressed as:

$$R_{TEG} = \rho_P \frac{h}{S_P} + \rho_N \frac{h}{S_N}, \tag{3.30}$$

where ρ_P and ρ_N are the electric resistivities of the two legs, h the thickness of the thermocouple, S_P and S_N the surface areas of the two legs. With respect to (3.29), it is reasonable that the generated heat is absorbed by both sides of the thermocouple. For this reason, the following expression for the net heat flow at the hot side holds:

$$\dot{Q}_H = \dot{Q}_P - \frac{\dot{Q}_J}{2} + K_{TEG}(T_H - T_C), \tag{3.31}$$

where K_{TEG} is the Fourier coefficient for thermal conduction through the two legs of the thermocouple, expressed as:

$$K_{TEG} = \lambda_P \frac{S_P}{h} + \lambda_N \frac{S_P}{h}, \tag{3.32}$$

where λ_P and λ_N are the thermal conductivities of the two legs.

An equivalent electrothermal circuit of thermogenerators can be developed from the above equations. As reported in [46–48] a common model is shown in Figure 3.8.

Efficiency of thermogenerators is also an important issue. In standard devices it normally ranges around 10%. As discussed in [7], an upper limit is set by Carnot efficiency, and the following expression holds:

$$\eta = \frac{T_H - T_C}{T_H} \cdot \frac{\sqrt{1 + ZT} - 1}{\sqrt{1 + ZT + T_C/T_H}}, \tag{3.33}$$

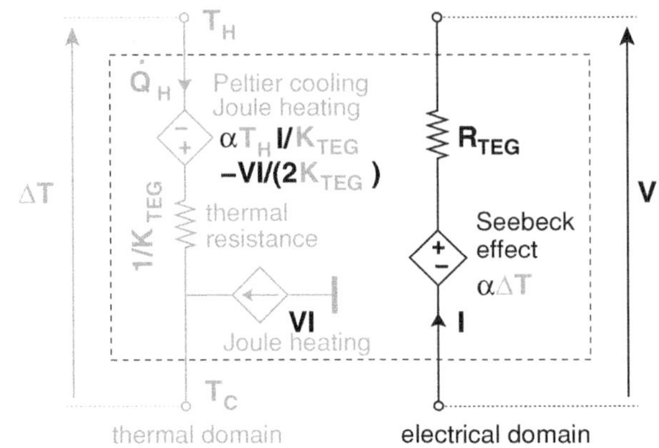

Figure 3.8 Equivalent electrothermal circuit of a thermocouple element, as presented in [46–47].

where the term ZT is usually referred to as a figure of merit of the thermogenerator, which can be approximated as:

$$ZT = \frac{\alpha^2 T}{\rho_{TEG} K_{TEG}}. \tag{3.34}$$

According to (3.34), in order to increase the figure of merit, and consequently the efficiency, high temperatures or high Seebeck coefficients are required, as well as a low electrical resistivity and a low thermal conductivity.

3.4 Scavenging architectures for vibrational and thermoelectric energy harvesters

The main building blocks of a vibrational energy harvester include elements requiring mechanical design and components targeting the electrical specifications of the transducer. Figure 3.9 shows the typical energy conversion chain for vibrational energy harvesters, and all the related design topics.

First of all, it is essential to analyze the available input vibrations and, more specifically, the typical amplitudes and frequency components. Based on this, an electromechanical transduction principle has to be chosen and a component has to be designed. Among the available technologies, the most suitable are piezoelectric and electromagnetic transducers [49]. Piezoelectric transducers have the advantage of a relative ease of fabrication [17] and the possibility of making components with customized sizes and geometries to meet specific application needs [18, 20, 21]. Piezoelectric transducers are also available as flexible composite materials [50], which is a desired feature for many applications. Relatively high voltages are normally generated by piezoelectric transducers, especially with stacked multilayered transducers. In contrast, electromagnetic transducers have a more complex structure since a permanent magnet and a miniaturized coil are required on the oscillating mechanical structure [51]. With respect to piezoelectric transducers,

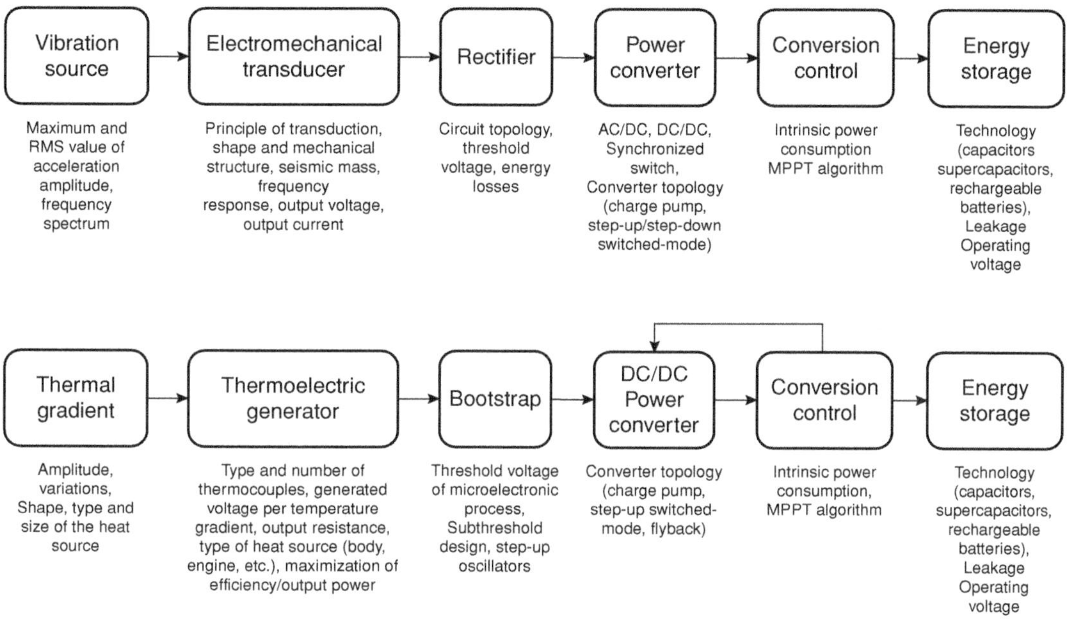

Figure 3.9 Energy conversion chain for vibrational and thermoelectric energy harvesters.

lower voltages are generated. However, higher currents are usually available in comparison. During the design or the choice of the transducer, additional parameters have to be defined, such as for example the seismic mass, the size, and a specific oscillating structure (e.g. cantilever, simple beam, membrane, etc.). From these parameters a typical resonance frequency of the transducer is defined, which should be chosen in order to match the energy content of input vibrations. Other important matters are, in addition, the output voltages and currents, which are necessary elements to be assessed before designing power converter circuits.

Since oscillations are usually involved in electromechanical energy harvesters, a rectifier block is normally included in order to spare designers the difficulties of handling negative voltages. The rectifier is the first block of electric power conversion. Several choices are possible, depending on the generated voltages and on the subsequent power conversion block architecture. The most frequent choices include full bridge rectifiers or voltage doubler rectifiers. However, such solutions have limited efficiencies in the case of low voltages. Solutions based on negative voltage converters [52] or active rectifiers [53] are known to increase conversion efficiency. However, some drawbacks such as respectively bidirectional conduction and additional power consumption are present.

The power converter is usually the core of the conversion chain. Its task is to ensure operation in the proximity of the maximum power point (MPP) of the transducer under actual operating conditions. Several architectures have been successfully employed, among which are capacitive charge pumps [54], switched-mode power converters [55], and synchronized power converters [56–58]. The choice depends both on transducer features and on the electrical requirements on the load. When high voltages are involved, high efficiencies can be obtained with switched-mode step-down topologies [57]. In the case of low and irregular input voltages, high efficiency is achieved with synchronized switch step-up topologies [59].

Conversion control can be performed with digital control [57] or by means of analog circuits [60]. It is essential to reduce as much as possible the intrinsic power consumption, which can be done by choosing the most efficient control algorithm or with specific energy-aware design techniques.

Energy storage can be implemented with rechargeable batteries or with simple capacitors or supercapacitors. The use of a battery provides a stable reference voltage and a known power supply for the conversion control circuits, whereas capacitors have a highly variable voltage as they are charged by the converter or discharged by the load. In addition, differently from batteries, capacitors may also be fully discharged, so that designers must cope with bootstrapping power conversion circuits without stable reference or supply voltages.

In thermoelectric energy harvesting, it is essential to analyze the properties of the heat source. As an example, a wearable body heat source has different properties and requirements than an engine waste heat source. In the first case, the transducer should maximize the efficiency and limit heat flow, in order not to be uncomfortable to the wearer [61]. On the contrary, waste heat sources may be exploited with the purpose of maximizing the output power rather than conversion efficiency [62]. For this reason it is important to properly design the thermogenerator and its thermal resistance, which determines the heat flow from the energy source to the heatsink. The number of series connected elements determines the open circuit voltage, which should be high enough to be handled by the subsequent power conversion circuits. However, the higher the number of series elements, the higher the output resistance of the thermogenerator.

A bootstrap circuit may be required when no batteries or stable voltage references are available, since thermogenerators typically output voltages as low as tens of mV, which is much lower than the typical threshold voltage of all available diodes or transistors. Solutions for starting up down to 40 mV have been recently developed and make use of a transformer to increase voltage [63].

Power conversion blocks for thermoelectric generators usually are based on step-up converters, such as for example boost or flyback converters. Maximum power point tracking can be performed either by modulating the duty-cycle of activation or with voltage control [64].

3.5 Scavenging architecture for RFID rectenna and voltage multiplier

In principle the main building blocks of a wireless link for RF energy harvesting are the same as those for radio communication links, but design specifications are different: for the present case the system conversion efficiency and thus the efficiency of each constitutive subsystem is the principal goal. The main building blocks to be designed are shown in Figure 3.10.

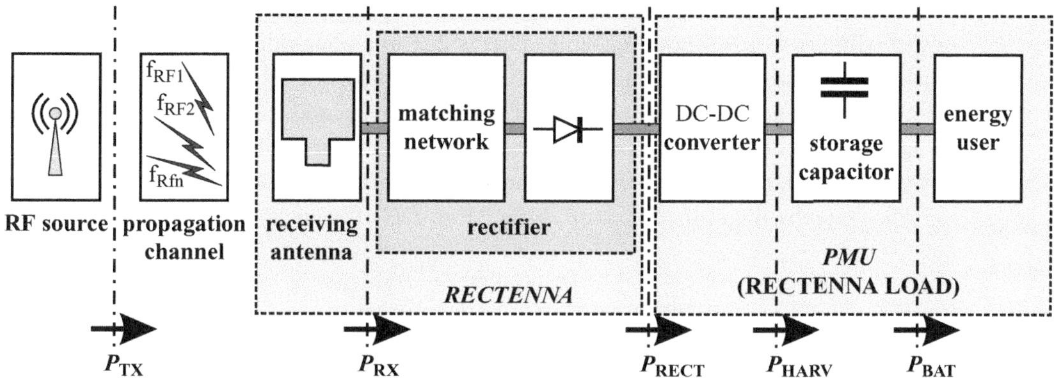

Figure 3.10 Block diagram of an RF energy harvesting system exploiting far-field radiated EM energy.

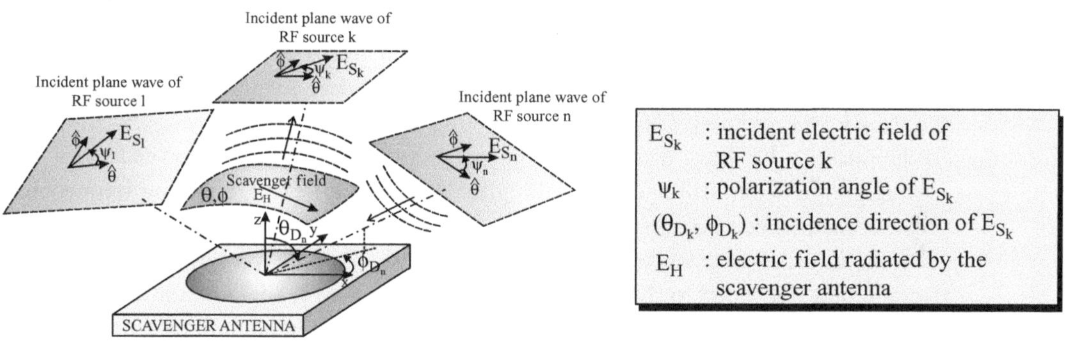

Figure 3.11 General scenario of a rectenna in the presence of several radiating sources.

The source power (P_{TX}) is converted into radiated electromagnetic RF energy by the transmitting antenna. The receiver side is located in the far-field region of the transmitter and is designed to receive the largest part of the RF energy (P_{RX}). The RF energy to be converted into DC, for power supply purposes, should be totally available to the input of a rectifier stage with no losses and reflections. To reach this goal a matching network is needed between the antenna and the rectifier ports: the subsystem, consisting of the antenna and the rectifier, is commonly referred to as a rectenna (rectifying antenna). RF sources may be radiated on purpose or they are made available by the ambient, depending on the application target, while the converted DC power is either used right away to turn on the devices and start communication operations, or is stored for future usage. The system load represents the input impedance of a power management and storage unit to be used as the system rechargeable battery.

The RF power harvesting system design is a quite challenging problem since the source locations and characteristics (frequency, polarization, and direction of arrival) are randomly variable. The general scenario of the rectenna operating in such conditions is represented in Figure 3.11. One of the major challenges in RF energy scavenging is the need to deal with unknown sources characterized by an extremely variable link budget, especially for mobile phone systems. That is due to an unpredictable distance between base station and mobile terminal, and to a carrier frequency arbitrarily allocated within a specified range, according to multiple-access policies and traffic block probabilities. The final target is to energy-scavenge from all the RF sources at the same time, for any possible polarization, angle of arrival, and power intensity [64–66].

The rectenna target specification is the RF-to-DC efficiency:

$$\eta_{\text{RF-DC}} = \frac{P_{\text{RECT}}}{P_{\text{TX}}}. \tag{3.35}$$

With reference to Figure 3.10, P_{TX} is the RF power available at the transmitting antenna input port and P_{RECT} is the DC power available to the rectifier load. It is noteworthy that the efficiency η_{DC} may be more clearly represented as the product of two concurrent factors belonging to different computational domains:

$$\eta_{\text{RF-DC}} = \eta_{\text{LINK}}\,\eta_{\text{RECT}} = \frac{P_{\text{RX}}}{P_{\text{TX}}}\,\frac{P_{\text{RECT}}}{P_{\text{RX}}}. \tag{3.36}$$

The first factor, η_{LINK} is evaluated at RF and is mainly related to the radio-link efficiency; the second one, η_{RECT} is the RF–DC system conversion that is highly dependent on the RF received power level and thus is computed and optimized by nonlinear design techniques. In RF links used for harvesting purposes the received power P_{RX} is also strongly time-dependent and may belong to sources at different operating frequencies. As a consequence, the optimum rectenna DC load may vary accordingly. In such situations, in order to preserve maximum efficiency, a DC–DC power management and storage unit is needed to dynamically track the optimum loading condition. The conversion block has its own power consumption and needs to be optimized for the highest conversion efficiency. In this case the system is designed for the optimum harvesting system efficiency, that is computed as:

$$\eta_{\text{HARV}} = \eta_{\text{RF-DC}}\,\eta_{\text{CONV}} = \eta_{\text{LINK}}\,\eta_{\text{RECT}}\frac{P_{\text{HARV}}}{P_{\text{RECT}}}. \tag{3.37}$$

The term η_{CONV} involves the base band part of the system and is dependent on the transient performance of the power management circuit. The goal is to dynamically keep the rectenna at its optimum loading condition with minimum power consumption. For this purpose the term P_{HARV} is computed after considering the power consumed by the converter. The DC–DC conversion efficiency is described in Section 3.6.

The term η_{LINK} can be evaluated by means of the well known Friis formula (under the assumption of TX and RX antennas lying within the Fraunhofer region of each other); this formula should also incorporate the frequency-dispersive channel attenuation due to fading and multipath phenomena and polarization attenuation due to the mismatch between the TX and RX antennas since they could both degrade the link efficiency as well. To accurately account for these effects both deterministic and statistical channel models may be used [67, 68].

The rectenna efficiency term η_{RECT} can be evaluated in the steady-state nonlinear regime and represents the figure of merit of the rectenna design, including its load, to

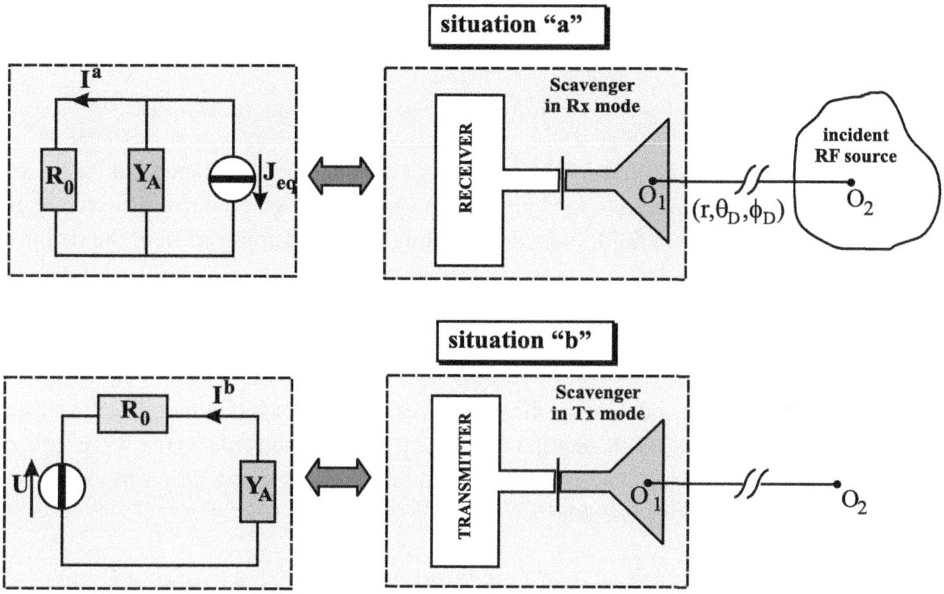

Figure 3.12 Situations "a" and "b" considered for the application of the reciprocity theorem.

provide the maximum DC power. The circuit excitation is the actual power received by the antenna and must be accurately accounted for in the design process. A possible rigorous procedure is described in the following subsection.

3.5.1 General purpose approach to the design of the rectenna

A general method based on the harmonic balance technique is introduced in this section for the rectenna nonlinear design in steady-state operating conditions. For this purpose a rigorous circuit equivalent representation of the RF sources must be created based on a combination of electromagnetic theory and numerical techniques. Let us consider a number of RF sources available in the ambient, described in terms of their respective radiated fields incident on the rectenna. If we refer to typical wireless communication standards this corresponds to power densities. Frequency, direction of arrival, and polarization of each incident field may be arbitrary. Assuming that the RF sources are located in the far-field region of the rectenna, a kth incident field at the harvester location may be described as a uniform plane wave and thus by a constant complex vector E_{inc_k}. For each source the reciprocity theorem can be used to exactly compute the RF excitation of the rectifier circuit [69]. For this purpose two electromagnetic configurations of the harvester antenna, labeled "a" and "b" in Figure 3.12, are considered, corresponding to the receiving and transmitting modes of operation, respectively. The corresponding equivalent circuits are also shown in the figure.

Let us denote by $\mathbf{E}_A$ $(r, \theta, \phi, \omega)$ the far field radiated by the harvesting antenna operating in the transmitting mode when powered by a voltage source of known amplitude U, internal resistance R_0, and angular frequency ω, as in Figure 3.12(b). The variables r,

θ, ϕ are spherical coordinates in a reference frame having its origin in the antenna phase center O_1. We may write:

$$\mathbf{E}_A(r, \theta, \phi, \omega) = U\frac{\mathrm{e}^{-\mathrm{j}\beta r}}{r}\mathbf{e}_A(\theta, \phi, \omega), \tag{3.38}$$

where the normalised field $\mathbf{e}_A$ does not depend on the distance r or on the source voltage U. It is now possible to rigorously compute the Norton current source equivalent to the kth RF incident field. Indeed, by a straightforward application of the reciprocity theorem we get

$$\left[\frac{1}{Y_A(\omega_k)} + R_0\right] I_k^a I_k^b = \mathrm{j}U\frac{2\lambda_k}{\eta}\mathbf{E}_{\mathrm{inc}_k} \bullet \mathbf{e}_A(\theta_{Dk}, \phi_{Dk}, \omega_k), \tag{3.39}$$

where θ_{Dk}, ϕ_{Dk} define the direction of arrival of the field radiated by the kth environmental RF source, ω_k is its angular frequency, and η is the free-space wave impedance. $Y_A(\omega)$ is the frequency-dependent antenna admittance computed by EM analysis. From (3.39) and Figure 3.12 we obtain:

$$J_k(\omega_k) = \mathrm{j}[1 + R_0\, Y_A(\omega_k)]\frac{2\,\lambda_k}{\eta}\,\mathrm{E}_{\mathrm{inc}_k} \bullet \mathbf{e}_A(\theta_{Dk}, \phi_{Dk}, \omega_k). \tag{3.40}$$

In this way several crucial phenomena, such as power mismatch between antenna and receiver front-end, polarization conflicts between incident electric fields and harvester antenna, and the simultaneous presence of different RF sources, are automatically included in all the evaluations. It is also possible to provide a more realistic description of the radio-channel link in order to include an accurate evaluation of η_{LINK} (see Equation (3.37)) in the equivalent current source representation of the received power (3.40).

Taking advantage of the linearity of the excitation mechanism, the contributions of all ambient sources may be represented by the Norton equivalent current sources shown in Figure 3.13, which provides a complete circuit description of the rectifier excitation. It is noteworthy that the generator impedance $Y_A(\omega)$ rigorously represents the antenna complex impedance which significantly varies with the frequency of operation. This description task is fairly easily accomplished either by direct measurements of the antenna impedance or by electromagnetic simulation based on the exact description of both the antenna layout and of the antenna materials.

The maximum power that could be received is expressed as the sum of the individual current sources under conjugate match, which is given by:

$$P_{\mathrm{RF}} = \sum_k \frac{|J_k(\omega_k)|^2}{8\mathrm{Re}[Y_A(\omega_k)]}, \tag{3.41}$$

and represents the maximum RF available power to be used as a reference for the definition of the system efficiency. Actually this is the physical upper bound that could be available to the rectifier in the optimum operating condition of the radio link. The circuit equivalent of Figure 3.13 is used as the current generator of the rectifier. This approach makes it easy to handle single- or multi-source (multi-frequency) incident fields and, most of all, to straightforwardly compute the rectenna nonlinear regime for

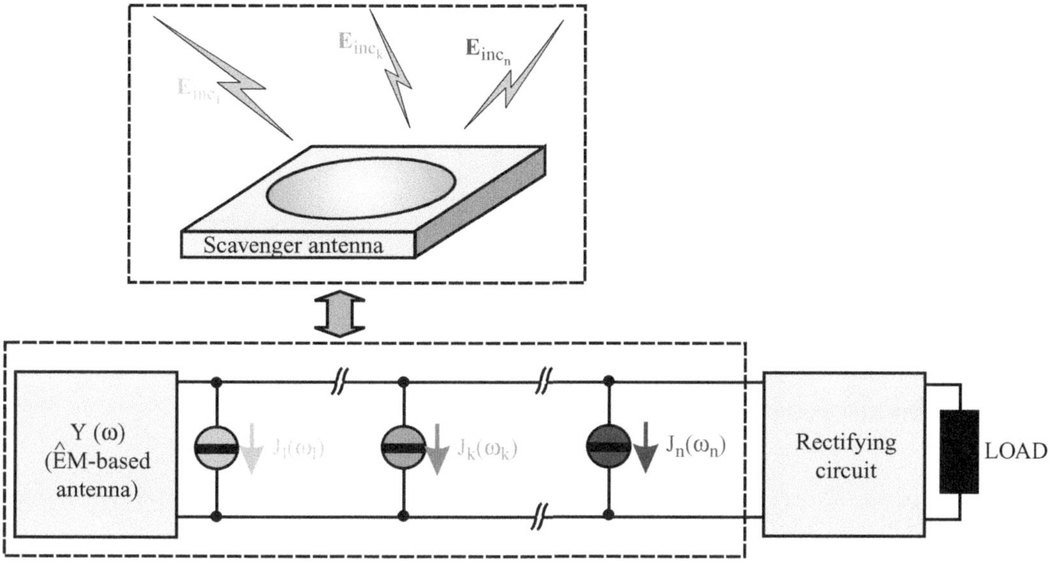

Figure 3.13 Norton equivalent circuit of the rectenna system under multiple excitations.

incidence of single or multiple ambient RF sources by means of a periodic or quasi-periodic HB analysis, respectively [70].

3.5.2 Selection of rectenna components

3.5.2.1 Antenna topology

Due to the simultaneous presence of different RF sources in all human-affected environments, the receiving antennas of RF harvester subsystems have to be able to collect energy from each of the corresponding RF operating frequency bands. Moreover the typical low-power level of the incoming RF signals forces us to devote particular attention to the antenna efficiency. The usually unknown direction of arrival of the RF source fields means that on the one hand there is not an absolutely broadside radiation pattern, while on the other hand it strongly suggests having a circularly polarized antenna or at least the capability to receive vertically and horizontally polarized signals.

Having these design goals in mind, different choices are possible in the selection of the harvester radiating element(s).

Wideband antenna

A wideband antenna for energy harvesting usually covers the wireless standards frequency bands (typically from 900 MHz up to 2.45 GHz). This kind of antenna can be quite easily designed by using stepped or smoothly curved topology (as is the case of the planar monopoles in Figure 3.14), providing a nonresonant behavior all over the operating bandwidth.

The performance limit of this kind of broadband antenna is the dispersive behavior from both the near- and the far-field points of view. But if the variable shape of the

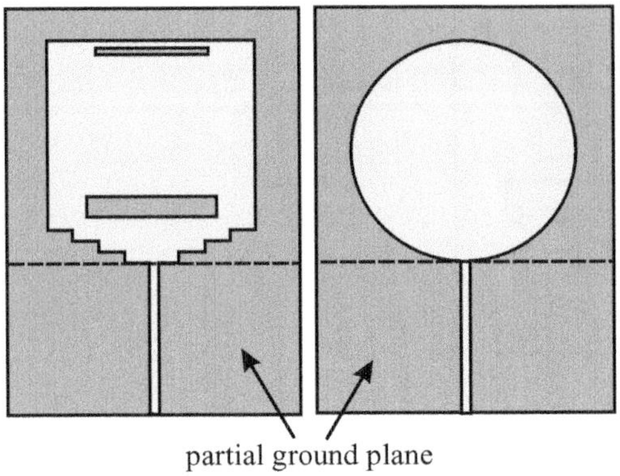

Figure 3.14 Examples of wideband planar monopoles.

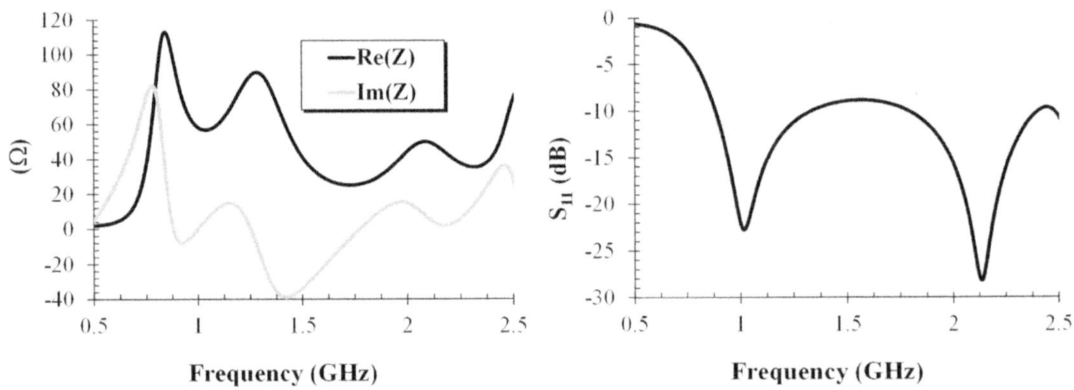

Figure 3.15 Stepped-dipole impedance and reflection coefficient.

radiation pattern is not a big concern in harvesting applications, a highly dispersive input antenna impedance can be matched to the rectifier section only with difficulty (see Figure 3.15 for reflection coefficient and antenna impedance frequency behavior of the stepped-dipole of Figure 3.14). Also circular polarization and efficiency could not be sufficiently guaranteed in all the bands of interest, due to the nonresonant behavior. The partial ground plane (for excitation purposes, only) and the consequent almost omnidirectional radiation pattern can represent a problem in wearable applications.

Multi-resonant antenna

This kind of antenna is a single radiator able to resonate at the frequencies of interest. Several solutions have been proposed in the literature for multi-standard mobile phones, typically by cutting edges and/or removing a portion of the metallization (dual-band examples are shown in Figure 3.16). The unavoidable EM-couplings between different portions of the metallization could affect the efficiency, while circular polarization properties are almost unattainable due to the difficulty in having a 90°-rotation invariant planar topology.

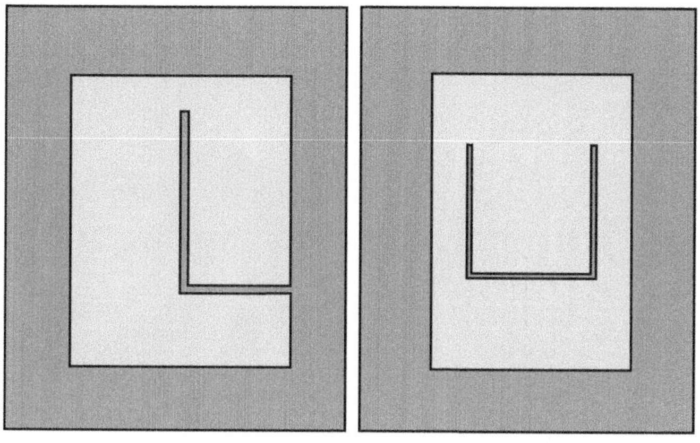

Figure 3.16 L- and U-shaped dual-band patch antennas.

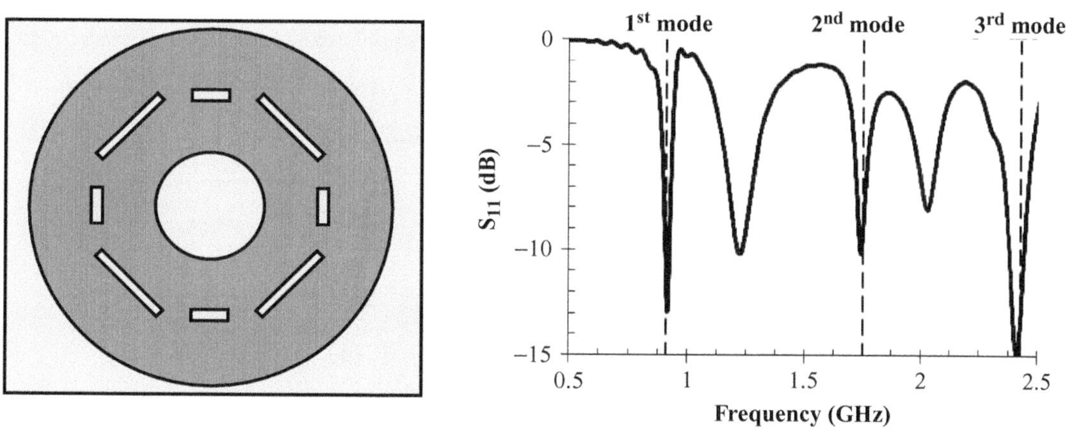

Figure 3.17 Tri-resonant slotted annular ring antenna and its reflection coefficient.

A more suitable solution can be provided by the exploitation of several resonant modes of the same radiating element: superior modes could not provide a broadside main lobe, but this does not affect the harvesting properties. For this kind of antenna the exact tuning of the resonant frequencies represents a demanding EM problem: the inclusion of slots in the metallization can help from this point of view, as is the case for the tri-band annular ring antenna shown in Figure 3.17, together with the multiband reflection coefficient. The different input impedance levels offered by these antennas at the different frequencies could be a problem in the design of the matching section, especially when the number of resonant modes is greater than two. The resonant behavior of this solution offers high efficiencies; circular polarization is also possible, provided that the corresponding resonant mode surface currents allow orthogonal modes of excitation.

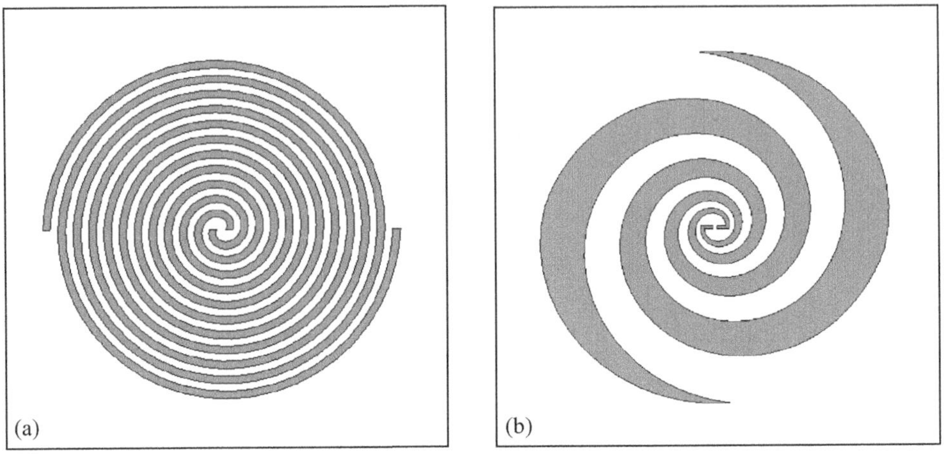

Figure 3.18 Self-complementary spiral antennas: (a) Archimedean; (b) logarithmic.

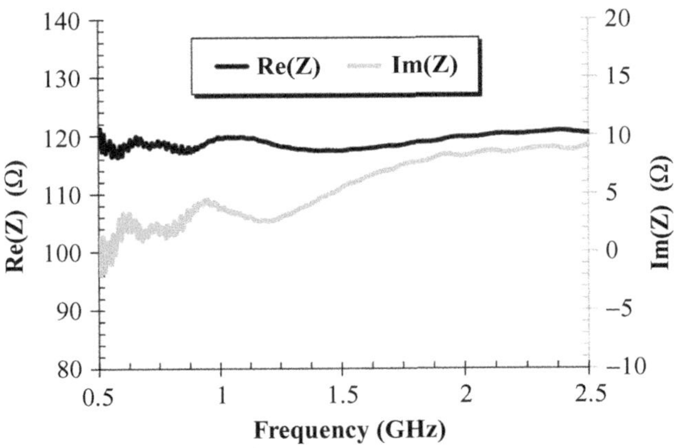

Figure 3.19 Archimedean spiral input impedance.

The ground plane presence in all these solutions automatically allows a direct wearable application.

Self-complementary antenna

Archimedean or logarithmic spiral antennas (Figure 3.18) are examples of this antenna family. They naturally exhibit an ultrawide-band behavior, depending on their dimension. The property of the self-complementarity (equal repartition of metallized and non-metallized areas) provides a very useful constant value of the antenna impedance (shown in Figure 3.19 for the Archimedean solution), and the circular polarization is automatically guaranteed. The absence of a ground plane does not make these solutions directly usable for on-body applications: the inclusion of a back shield must take into account a suitable distance from the radiating element (depending on the frequency

Table 3.2 Low-threshold voltage diodes parameter comparison.

	Skyworks SMS7630	Skyworks SMS7621	Skyworks SMS3922	Skyworks SMS1546	Macom MA4E2054	Agilent HSMS2852	Agilent HSMS8101
I_s [μ]	5	0.04	0.03	0.3	0.03	3	0.046
C_{j0} [pF]	0.14	0.1	0.7	0.38	0.13	0.18	0.18
V_j [V]	0.34	0.51	0.595	0.51	0.4	0.35	0.5
N	1.05	1.05	1.08	1.04	1.05	1.06	1.09
R_s [Ω]	20	12	9	4	11	25	6
I_{bv} [μ]	100	10	10	10	10	300	100
V_{bv} [V]	2	3	20	3	5	3.8	7.3
$\eta_{RF\text{-}DC}$	13%	7.68%	3.7%	11.5%	7.4%	10.54%	8.25%

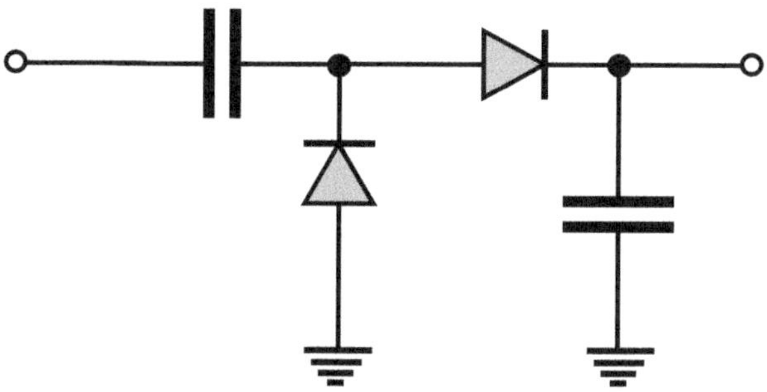

Figure 3.20 Circuit topology of a single-stage full-wave peak-to-peak rectifier.

range), in order not to adversely affect the radiating properties, and also cause a significant worsening of input impedance flatness.

3.5.2.2 Rectifier topology

Another important degree of freedom for the rectenna design is the number of rectifier stages. When designing rectenna for RFID transceivers this number is established by the voltage level required to turn on the active devices for the back-scattering operation, regardless of the achievable conversion efficiency. When the rectenna is adopted for harvesting purposes the main goal is conversion efficiency, which is a nonlinear function of the available power levels [68]. A single-stage full-wave peak-to-peak rectifier, whose topology is sketched in Figure 3.20, is usually recommended in such situations. In fact where low power levels (P_{RF}) are concerned, diode losses may significantly degrade the conversion efficiency, which is the primary performance index: therefore the multi-stage rectifier topology must be avoided.

Schottky diodes are usually adopted for their low forward voltage drop compared to other nonlinear devices like diode-connected MOSFETs. Particular care is needed in the selection of the proper diode. The primary objective is to make use of devices whose turn-on is guaranteed at ultra-low incident power levels. In addition, in order to obtain good conversion efficiency the lowest possible zero-bias junction capacitance (C_{j0}) and the highest saturation current (I_s) are required. In Table 3.2 a list of suitable diodes and

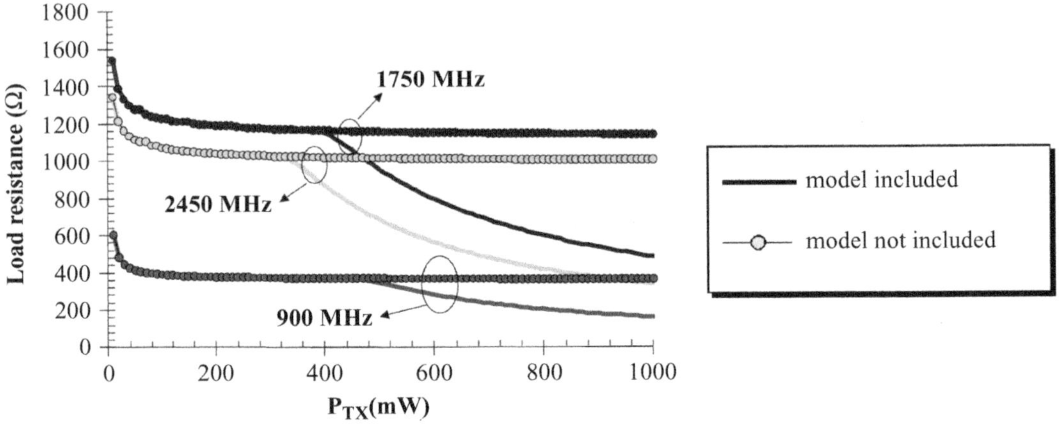

Figure 3.21 Optimum annular ring rectenna load as a function of P_{TX}, with and without diode breakdown model.

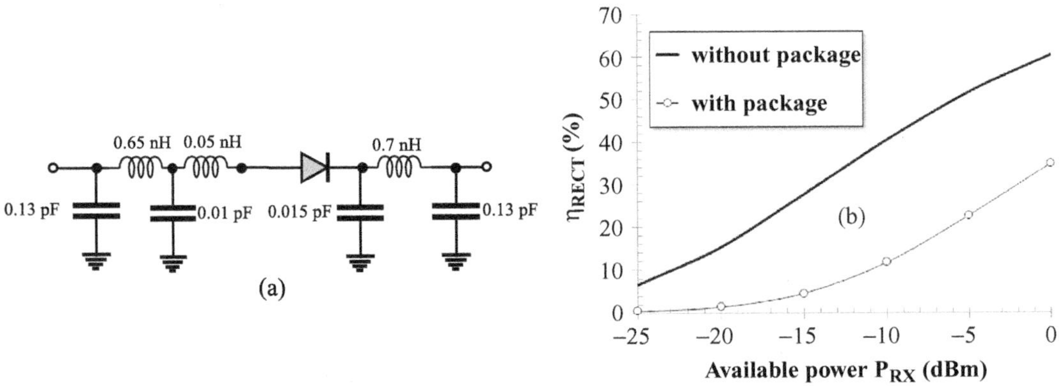

Figure 3.22 (a) Skyworks SMS7630 package model; (b) Conversion efficiency of the tri-band annular ring rectenna with and without diode package at 2.45 GHz.

their electric properties is reported: in the authors' opinion the best candidate among these choices is the Skyworks SMS7630 for its very low threshold voltage.

As in most RF circuit design, in a rectenna project an accurate model of the nonlinear diodes is of fundamental importance. In the rectenna case, due to the very low power budgets involved, the nonlinear model must be extremely accurate. As an example the effects of diode breakdown are shown in Figure 3.21, where the optimum dispersive loading condition of the tri-band annular rectenna is plotted as a function of transmitted power for all the operating frequencies.

For the same reasons, the diode package needs to be accounted for also in rectenna nonlinear design [71]. In fact package parasitics can significantly influence the entire rectenna performance, especially at higher frequencies. Figure 3.22 (a) shows a model of

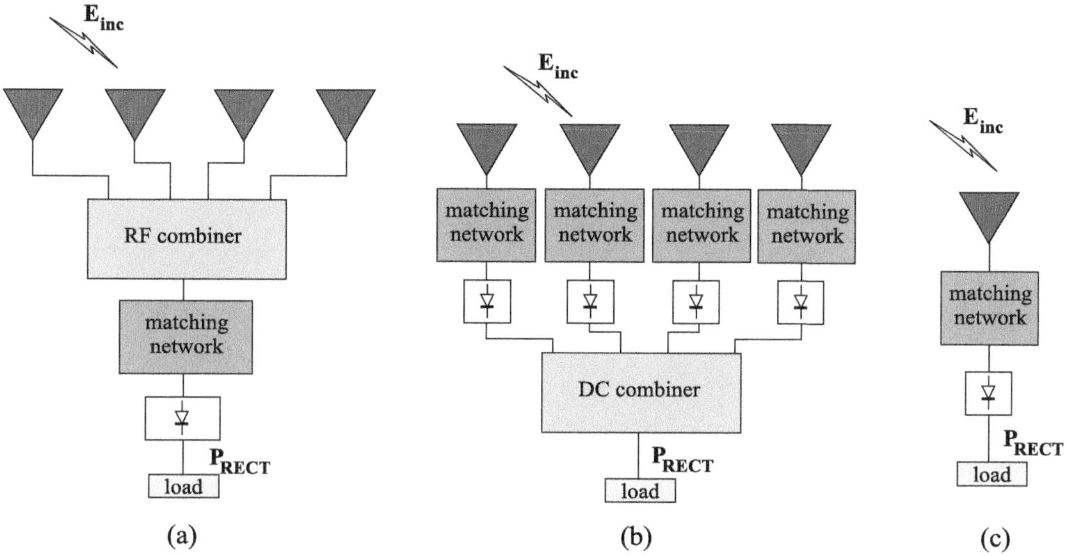

Figure 3.23 RF energy harvesters architecture.

the Skyworks SOT-23 diode package used in the tri-band rectenna design; in Figure 3.22 (b) the corresponding effects of package parasitics on the rectenna performance at 2.45 GHz can be observed.

3.5.2.3 Antenna(s)/rectifier(s) architecture

A crucial issue in the design of RF energy harvesters is the choice of the system architecture. In fact several topologies combining the same elements are available, and they can be summarized in the schemes shown in Figure 3.23. Solution (a) exploits the arrangement of several antennas in an array configuration: the RF combiner represents the so-called corporate feed network of the array. In this case the high gain of the receiving array allows the single rectifier to work in a high-efficiency region, but this is true just for a limited space portion of directions of arrival. In practice this solution is suitable for far-field wireless power transfer (WPT) applications, i.e. when the RF source location is known. The architecture (b) makes use of the same amount of radiating elements as the previous one, but signals combination is moved to the DC section, thus losing the array factor contribution: therefore each antenna works loaded by the other "parasitic" antennas, with a low-gain radiation pattern, strongly affected by the electromagnetic couplings with the other elements. In this case each rectifier is forced to operate in a low-efficiency region, but this is true for a broader range of incoming field directions. Moreover the matching network between each antenna and its rectifier must be replicated, thus increasing the cost and the area of the whole harvesting system. Case (c) is the simplest, since the circuitry and the antenna elements are reduced to a minimum: the absence of parasitic coupling effects allows the single antenna to operate at its best, in terms of efficiency and radiation pattern shape. As in case (b), the rectifier works in a low-efficiency region for a wide range of directions of arrival, depending,

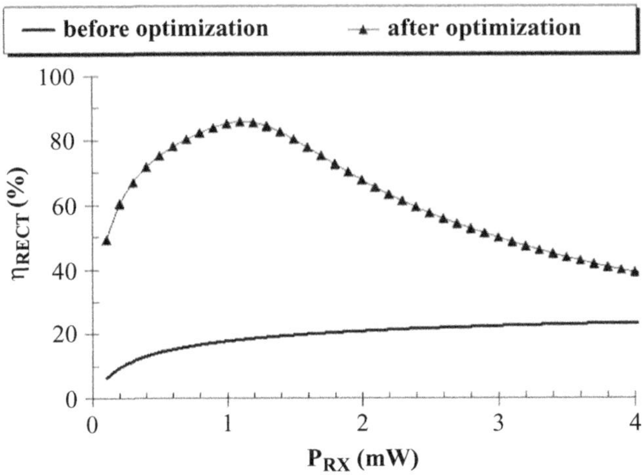

Figure 3.24 Starting point and optimized conversion efficiency for a single-frequency rectenna [72].

of course, on the shape of the antenna radiation surface. In conclusion, the choice of harvester architecture is mainly between topologies (b) and (c), since the RF source location is typically unknown. Solution (c) is more suitable for RFID-tag or ultra-low-power battery-less wireless sensors applications, for its reduced dimensions and cost. On the other hand solution (b) is more convenient in those scenarios where the available energy is sufficiently high, thus justifying the higher losses to be sustained due to the use of a multiple-rectifier configuration.

3.5.3 Rectenna design results

Following the procedure described in the previous subsection, the rectified power may then be accurately computed by a full nonlinear harmonic balance (HB) analysis of the receiver, including the antenna described by EM simulation. In this way the overall figure of merit η_{RECT} can be successfully used as the design goal of a comprehensive system optimization.

Some representative results obtained by the described procedure, focusing on incident low-power levels, typical of scavenging scenarios, are discussed below. In Figure 3.24 the starting point and final point of a rectenna optimization are compared: η_{RECT} is plotted versus the received power P_{RF}, for a single-frequency rectenna operating for the UMTS band [72]: an electromagnetic conversion efficiency better than 50% is obtained as the available RF power P_{RX} exceeds 100 μW.

To verify the rectenna behavior a reference scenario, with defined transmitting sources, (frequency, polarization, distances) should be built up. As an example the following two figures show the performance of the circularly polarized, triple-resonator rectenna of Figure 3.17 [73–75], in the presence of resonant dipoles placed at a distance of 30 cm and excited by different levels of RF power (P_{TX}) at the three frequencies. In

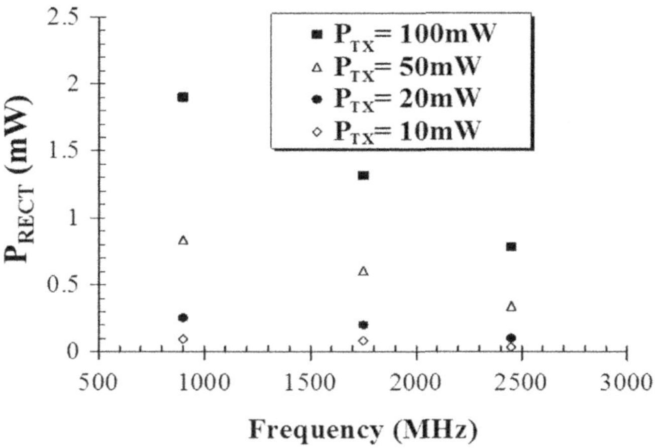

Figure 3.25 Rectenna RF–DC conversion efficiency for a range of incoming RF fields radiated by a resonant dipole.

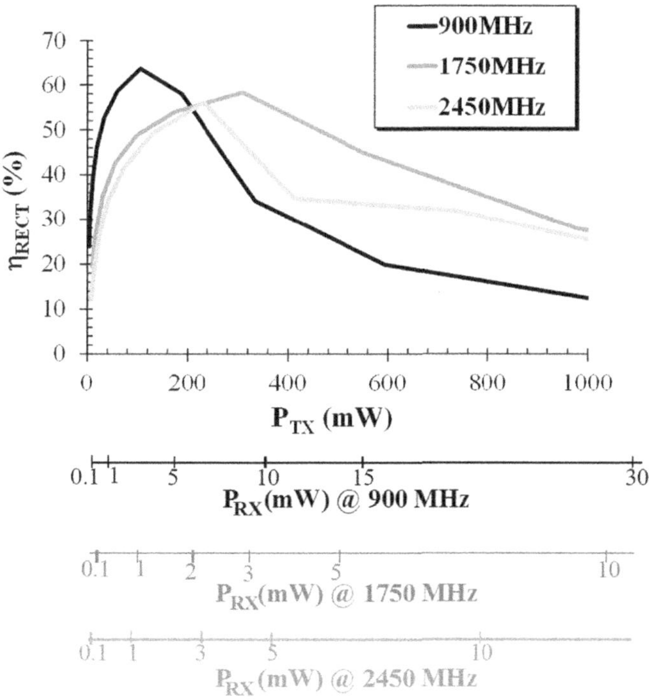

Figure 3.26 Rectified power for a range of RF source levels.

Figure 3.25 the rectified powers delivered to the rectenna optimum load are plotted for RF sources at three frequency bands, corresponding to the GSM 900, 1800, and the WiFi standards. In Figure 3.26 the rectenna efficiency is considered and plotted again versus the transmitted power. For this example the efficiency peak is searched by optimizing the system for scenarios where the transmitted power is typically of the order of 100 mW (and thus the received power is of the order of tens of microwatts), which is the case of indoor environments populated by several RF sources.

Figure 3.26 shows the axes corresponding to the available RF power P_{RX} at the three operating frequencies. It is of great importance to note the strong impact of frequency on the P_{RX} scale, given the same P_{TX} level. This is due to both the channel effect (ideal Friis formula in this case) and the dispersive behavior of the antenna: the higher gain-pattern of the antenna at 2450 MHz rather than at 1750 MHz can be deduced from inspection of the figure.

3.6 Design of power conversion circuits

Besides the design of efficient energy transducers of the types presented in the previous sections, it is also essential to design power conversion and management circuits which, in the case of strict power constraints, have to be carefully tailored to the specific applications of interest. Power conversion, storage, and management are key issues for successfully exploiting environmental energy sources in order to power external applicative loads.

Depending on the type of energy source, electrical AC/DC or DC/DC conversion may be required. As an example, vibrational energy sources produce low frequency AC voltages or currents. In contrast, photo-voltaic modules, thermoelectric generators, or rectified RF sources produce DC outputs and require different types of power conversion optimization.

In this application field, the main design concern is to cope with very low available input power levels, down to 10 $\mu W/cm^3$ in many practical cases [16]. This amount of power should allow a supply to the power converter and, at the same time, storage of the remaining power for later use by the application circuits [64]. This means that the best trade-off between intrinsic power consumption and efficiency must be found. For this purpose, the choice of components plays a key role. Nano-power components such as operational amplifiers, comparators, linear voltage regulators have been available on the market for some years. In addition, microcontrollers and sensors with average power requirements down to a few μW make it possible to perform data acquisition and processing at very low energies. As a reference, commercial microcontrollers with a consumption down to 100 μA/MIPS have been announced recently [76]. Besides this, other important parameters are related to leakage and bias currents of individual components. All these aspects have to be taken into account when designing micro-power harvesting circuits, while at the same time necessary trade-offs have to be defined in choosing the most suitable power conversion scheme and maximum power tracking algorithm. These aspects will be made more clear in this section, where, as case studies, the design of two power conversion circuits for harvesting power respectively from piezoelectric and RF sources will be presented.

In the case of vibrational energy harvesting, a low frequency AC/DC conversion is involved. In this field specific optimizations of power conversion have been reported in the literature. The simplest solution for achieving this purpose is the use of diode rectifiers. However, diode voltage drops do not allow us to exploit the full voltage swing and the output voltage cannot exceed the input voltage. Switched-mode power conversion schemes based on step-down converters [55] have also been successfully deployed. In the specific case of piezoelectric transducers, activating switching converters in a way synchronous with vibrations [77] has also been shown to significantly increase the output power. Synchonized switching techniques such as synchronous electrical charge extraction (SECE) [78], synchronized switch on inductor (SSHI) [79], or single-supply pre-biasing (SSPB) [80] have been deeply discussed and analyzed in the literature.

The power converter circuit described in this section is based on SECE. With this technique, electrical charge is removed from transducers through a resonant circuit when voltage maxima or minima are detected. Since piezoelectric transducers have a mainly capacitive impedance, a resonant inductor is used as a temporary energy buffer. When all charge has been removed, the inductor is switched to a storage capacitor, in order to form a second resonant circuit used for transferring energy to the output. This technique is well described in [56, 78].

In order to implement a self-starting circuit implementation of this type of power conversion, several aspects have to be considered. First, since battery-less implementations are usually preferred in power harvesting applications, the circuit should be able to extract power passively in the initial phases. Then, specific circuits for tracking vibration maxima/minima and for precisely controlling the timings of control signals have to be designed. Finally, since each conversion is activated only twice per period, all sources of leakage currents have to be minimized. Figure 3.27 shows a possible complete circuit implementation.

In this circuit, since handling the negative voltages generated by the transducers with no stable voltage power supply would increase the complexity of the design, a MOSFET bridge composed of M1–4 and configured as a negative voltage converter (NVC) is used for generating its absolute value V_{RECT} with no voltage drops. The passive startup is ensured by the Schottky diode D1, which allows passive energy storage onto C_{O} which should be chosen with the minimum available self-discharge rate, as is the case, for instance, with polypropylene plastic film capacitors. When the minimum required voltage is stored on C_{O}, a low drop out (LDO) regulator is enabled and a stable supply voltage V_{DD} for the control circuits is generated. Control circuits include a peak detector and an ultra-low-power microcontroller (MCU). The peak detector can be implemented by comparing a scaled version of the rectified voltage and a slightly delayed version. A nano-power comparator, such as for example the LPV7215 from Texas Instruments, ensures the lowest possible leakage current, down to 580 nA, and switching delays of 10 μs, still much lower than the typical vibration periods. The MCU is normally left in a 0.1 μA sleep mode, and is woken up by interrupt requests issued by the peak detector. As the MCU wakes up, its duty is to first issue the EN and EN_P signal to activate charge extraction from the transducer. As the current in L becomes maximum, STORE is asserted and current flows onto C_{O}. At the end of conversion, the residual charge on the transducer is reset by asserting RES and then the sleep mode is entered again. Other important parameters are the leakage currents through the power MOSFETs, which,

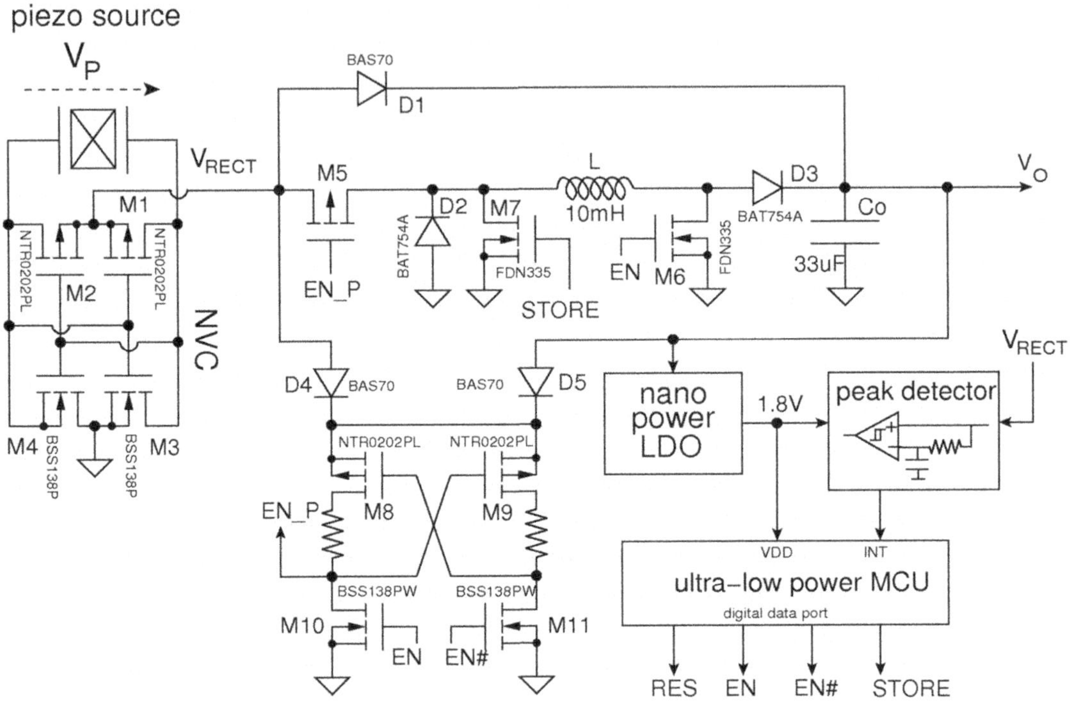

Figure 3.27 Self-starting circuit implementation of a synchronous charge extractor for piezoelectric transducers.

in this kind of application where the circuit is inactive for most of the time, should be minimized. The average power consumption of such a control unit (i.e. voltage regulator, MCU, peak detector) was determined by measuring the supply currents. With a 50 Hz input vibration, as the stored voltage V_O varies between 2 V and 5.5 V, the corresponding power consumption ranges from 3 μW to 8.25 μW. Such low power values allow a positive power budget even in the case of weak and irregular vibrations.

In the literature, several types of power converter circuits have been presented to date, and a detailed review is presented in [81].

Besides kinetic energy harvesters, which require low frequency AC/DC conversion, many energy transducers require DC/DC conversion, such as for example, thermoelectric generators, photo-voltaic modules and RF power sources. In this case, given the very limited available power budgets, it is essential to extract the maximum available power from the energy transducers. In this context, maximum power point tracking (MPPT) techniques have to coexist with the search for an optimal trade-off between harvested power and the energy required for handling the conversion. Also in this case, an energy-aware design allows us to minimize the energy overhead introduced by the electronic power converter.

As a case study, Figure 3.28 shows a DC/DC power converter circuit based on an adaptive open circuit fractional voltage method [10, 76], suitable for use with rectified RF sources in the 900 MHz, 1800 MHz, and 2400 MHz frequency bands.

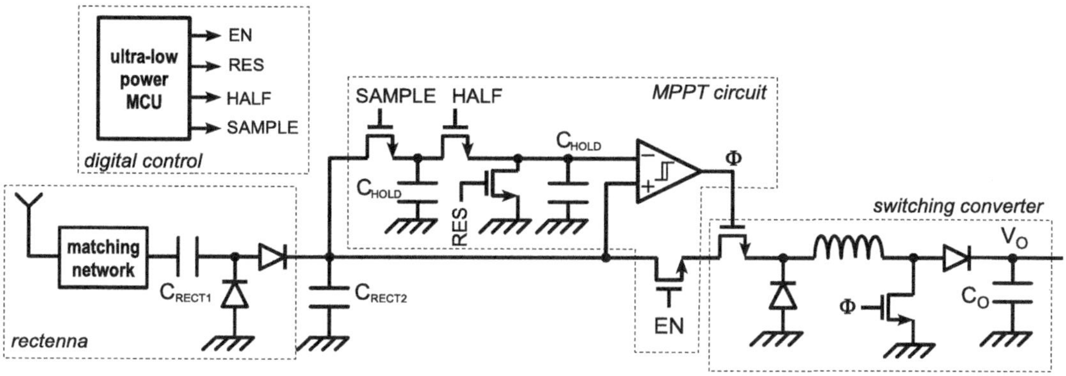

Figure 3.28 An ultra-low-power DC/DC converter.

The circuit is also suitable for use with thermoelectric generators, photo-voltaic modules, or generic DC sources. In such environments the received power is subject to frequent variations. For instance, the power level of each transmission slot in the GSM protocol is controlled in real time by the base station and may quickly change. As an example, in the GSM standard each phone operates in a 1/8 time multiplexed channel where each slot has a duration of 577 μs within frames of 4.6 ms. The converter presented is designed for rapidly responding to the variations in received power occurring in every slot.

In this kind of application, consistent with the current state of the art set out in the literature, a rechargeable battery is used as an energy buffer and for providing an initial voltage. Such a device is normally required in this kind of RF harvesting application because of the very low output voltage of rectennas in practical situations, which is usually comparable to the threshold voltage of commercial MOSFETs, making it extremely difficult to build any kind of switching converter or charge pump. The converter has a boost topology and consists of two pairs of synchronous switches ($SW_1 - SW_2$, $SW_3 - SW_4$), an inductor L, a diode D_3, and a storage capacitor C. A simple control circuit is used for driving the switches. The operation of the converter may be described as follows. Initially ENABLE $= 0$ V, so that the rectenna is disconnected from the boost converter and the hold capacitor charge is reset. The voltage V_{RECT} rises to its open circuit value. Both the rise time and the final voltage value depend on the actual rectified power, on the frequency and the number of RF sources, and on the output impedance at the operating frequency. Once V_{RECT} reaches a steady-state value, it is sampled on C_{HOLD}, and is halved by subsequent pulses on the SAMPLE and HALF signals ($V_{\mathrm{W}} = \max\{V_{\mathrm{RECT}}(t)\}/2$). This sets the desired operating point, which is usually located approximately at half the open circuit voltage. Different operating points can be selected by simply changing the ratio between the two hold capacitors. As soon as ENABLE is activated, the sampled voltage V_{W} and the rectenna output voltage V_{RECT} are input to a hysteresis comparator whose output determines the control signals Φ and $\bar{\Phi}$. The converter operates in two phases, in which power is respectively extracted from the rectenna and transferred to the output. In the latter phase, the rectenna is disconnected from the converter and V_{RECT} rises because of the received RF power.

The voltage V_W should be periodically updated after several commutations, when a different power level is expected, for example within a few GSM slots. The maximum update period is mainly limited by the rise time of the open circuit output voltage V_{RECT} of the rectenna, which depends on its equivalent internal resistance and on the value of the filter capacitor C_{RECT2}. The minimum update period is determined by the leakage currents discharging C_{HOLD}, which are usually negligible. Suitable design parameters are $L = 10$ mH, $C_{RECT1} = 1$ nF, $C_{RECT2} = 47$ nF, $C = 0.1$ μF.

As far as the intrinsic consumption of the power converter is concerned, special care needs to be taken in sizing parameters such as L, C, C_{RECT2}, and the hysteresis width ΔV. For proper operation, the time interval T_{iL2} (during which current flows through L into C) should be shorter than the rise time T_R of the rectified voltage V_{RECT}, which depends on the received RF signal, on C_{RECT2}, and on the hysteresis width ΔV. The constraint $T_{iL1} < T_R$ guarantees that $i_L = 0$ by the end of each phase so as to prevent unwanted storage of energy in L, where T_{iL1} is the time interval during which the current flows from the rectenna into L. However, since usually $T_{iL2} \ll T_R$, T_{iL1} should be high enough to guarantee feasible rates of switching control. Another parameter to be carefully set is the hysteresis width ΔV, since higher values tend to move the system away from its desired operating point. On the other hand, a lower ΔV increases the number of commutations and thus the dynamic power required for control. Thus an accurate trade-off has to be defined. Dynamic power consumption is also directly proportional to the total capacitances of the nodes toggling during Φ and $\bar{\Phi}$ phases. These values depend on the transistor sizes, i.e. smaller transistors and lower control voltages reduce dynamic switching power. However, this reduces the switch conductance as well. The achievement of a good design trade-off is thus necessary for tuning the overall efficiency. In general, since the series resistances associated with the source, the inductor, and the diodes are usually higher than those of MOSFETs, the main design parameter to minimize is their leakage current.

References

[1] J. Paradiso and T. Starner, "Energy scavenging for mobile and wireless electronics," *IEEE Pervasive Computing*, **4**, (*1*), 18–27, 2005.

[2] S. Roundy, P. K. Wright, and J. Rabaey, "A study of low-level vibrations as a power source for wireless sensor nodes," *Computer Communications*, **26**, (*11*), 1131–1144, 2003.

[3] Z. Wang, V. Leonov, P. Fiorini, and C. Van Hoof, "Realization of a wearable miniaturized thermoelectric generator for human body applications," *Sensors and Actuators A*, **156**, (*1*), 95–102, (Nov.) 2009.

[4] A. Nasiri, S. Zabalawi, and G. Mandic, "Indoor power harvesting using photovoltaic cells for low-power applications," *IEEE Transactions on Industrial Electronics*, **56**, (*11*), 4502–4509, 2009.

[5] T. Paing, J. Shin, R. Zane, and Z. Popovic, "Resistor emulation approach to low-power RF energy harvesting," *IEEE Transactions on Power Electronics*, **23**, (*3*), 1494–1501, (May) 2008.

[6] A. Chandrakasan, D. Daly, J. Kwong, and Y. Ramadass, "Next generation micro-power systems," *2008 IEEE Symposium on VLSI Circuits*, pp. 2–5, June 2008.

[7] G. Snyder, "Small thermoelectric generators," *The Electrochemical Society Interface*, (*Fall*), 54–56, 2008.

[8] H. Bottner and J. Nurnus, "New high density micro structured thermogenerators for standalone sensor systems," *ICT 26th International Conference on Thermoelectrics*, pp. 306–309, 2007.

[9] GreenTEG Web Site, http://www.greenteg.com.

[10] A. Costanzo, M. Fabiani, A. Romani, D. Masotti, and V. Rizzoli, "Co-design of ultra-low power RF/microwave receivers and converters for RFID and energy harvesting applications," in *Microwave Symposium Digest (MTT),* IEEE MTT-S International, pp. 856–859, 2010.

[11] M. Marzencki, Y. Ammar, and S. Basrour. "Integrated power harvesting system including a MEMS generator and a power management circuit," *Int. Conf. on Solid-State Sensors, Actuators and Microsystems, Transducers 2007*, pp. 887–890, 2007.

[12] S. P. Beeby, N. Torah, M. J. Tudor, *et al.*, "A micro electromagnetic generator for vibration energy harvesting," *Journal of Micromechanics and Microengineering*, **17**, (*7*), 1257–1265, 2007.

[13] N. G. Stephen, "On energy harvesting from ambient vibration," *Journal of Sound and Vibration*, **293**, (*1–2*), 409–425, (May) 2006.

[14] A. Badel, M. Lagache, D. Guyomar, E. Lefeuvre, and C. Richard, "Finite element and simple lumped modeling for flexural nonlinear semi-passive damping," *Journal of Intelligent Material Systems and Structures*, **18**, (*7*), 727–742, (Feb.) 2007.

[15] A. Erturk and D. Inman, "An experimentally validated bimorph cantilever model for piezoelectric energy harvesting from base excitations," *Smart Materials and Structures*, **18**, (*2*), 025009, 2009.

[16] S. Roundy, P. Wright, and J. Rabaey, "A study of low level vibrations as a power source for wireless sensor nodes," *Computer Communications*, **26**, (*11*), 1131–1144, (Jul.) 2003.

[17] S. Roundy and P. Wright, "A piezoelectric vibration based generator for wireless electronics," *Smart Materials and Structures*, **13**, (*5*), 1131–1142, (Oct.) 2004.

[18] S. Mehraeen, S. Jagannathan, and K. Corzine, "Energy harvesting from vibration with alternate scavenging circuitry and tapered cantilever beam," *IEEE Transactions on Industrial Electronics*, **57**, (*3*), 820–830, 2010.

[19] J. Palosaari, M. Leinonen, J. Hannu, J. Juuti, and H. Jantunen, "Energy harvesting with a cymbal type piezoelectric transducer from low frequency compression," *Journal of Electroceramics*, **28**, (*4*), 214–219, (Mar.) 2012.

[20] B. Yang and K.-S. Yun, "Piezoelectric shell structures as wearable energy harvesters for effective power generation at low-frequency movement," *Sensors and Actuators A: Physical*, **118**, 427–433, (Mar.) 2012.

[21] G. Taylor and J. Burns, "The energy harvesting eel: a small subsurface ocean/river power generator," *IEEE Journal of Oceanic Engineering*, **26**, (*4*), 539–547, 2001.

[22] E. Lefeuvre, A. Badel, C. Richard, and D. Guyomar, "Energy harvesting using piezoelectric materials: Case of random vibrations," *Journal of Electroceramics*, **19**, (*4*), 349–355, (Feb.) 2007.

[23] R. P. Paganelli, A. Romani, A. Golfarelli, *et al.*, "Modeling and characterization of piezoelectric transducers by means of scattering parameters. Part I: Theory," *Sensors and Actuators A: Physical*, **160**, (*1–2*), 9–18, (Mar.) 2010.

[24] A. Romani, R. P. Paganelli, E. Sangiorgi, and M. Tartagni, "Joint modeling of piezoelectric transducers and power conversion circuits for energy harvesting applications," *IEEE Sensors Journal*, **13**, (*3*), 916–925, 2013.

[25] T. Sterken, P. Fiorini, K. Baert, R. Puers, and G. Borghs, "An electret-based electrostatic micro-generator," *Proceedings of the 12th Int. Conf. On Solid-State Sensors, Actuators and Microsystems Transducers*, pp. 1291–1294, 2003.

[26] P. Miao, P. D. Mitcheson, A. S. Holmes, *et al.*, "MEMS inertial power generators for biomedical applications," *Microsyst Technology*, **12**, 1079–1083, 2006.

[27] G. Despesse, J. J. Chaillout, T. Jager, F. Cardot, and A. Hoogerwerf, "Innovative structure for mechanical energy harvesting," *Int. Conf. on Solid-State Sensors, Actuators and Microsystems, Transducers 2007*, pp. 895–898, 2007.

[28] B. Andò and S. Baglio, "Investigation on mechanically bistable MEMS devices for energy harvesting from vibrations," *IEEE Journal on Microelectromechanical Systems*, **21**, (*4*), 779–790, 2012.

[29] A. Kempitiya, "Analysis and optimization of asynchronously controlled electrostatic energy harvesters," *IEEE Transactions on Industrial Electronics*, **59**, (*1*), 456–463, 2012.

[30] E. Torres and G. Rincón-Mora, "A 0.7 μm BiCMOS electrostatic energy-harvesting system IC," *IEEE Journal of Solid-State Circuits*, **45**, (*2*), 483–496, 2010.

[31] E. Torres and G. Rincón-Mora, "Self-tuning electrostatic energy-harvester IC," *IEEE Transactions on Circuits and Systems II: Express Papers*, **57**, (*10*), 808–812, 2010.

[32] P. Mitcheson and T. Green, "Maximum effectiveness of electrostatic energy harvesters when coupled to interface circuits," *IEEE Transactions on Circuits and Systems I: Regular Papers*, **59**, (*12*), 3098–3111, 2012.

[33] P. Glynne-Jones, S. P. Beeby, E. P. James, and N. M. White. "The modeling of a piezoelectric vibration powered generator for microsystems," *Int. Conf. On Solid-State Sensors, Actuators and Microsystems, Transducers 2001*, pp. 46–49, 2001.

[34] R. Elfrink, T. M. Kamel, M. Goedbloed, *et al.*, "Vibration energy harvesting with aluminum nitride-based piezoelectric devices," *Proc. of the PowerMEMS Int. Workshop*, Sendai, pp. 249–252, 10–11 Nov. 2008.

[35] E. E. Aktakka, R. L. Peterson, and K. Najafi, "A self-supplied inertial piezoelectric energy harvester with power-management IC," in *Proceedings of ISSCC*, **18**, (*10*), 120–121, 2011.

[36] B. P. D. Mitcheson, M. Ieee, E. M. Yeatman, *et al.*, "Energy harvesting from human and machine motion for wireless electronic devices," *Proceedings of the IEEE*, **96**, (*9*), 1457–1486, 2008.

[37] R. Elfrink, V. Pop, D. Hohlfeld, *et al.*, "First autonomous wireless sensor node powered by a vacuum-packaged piezoelectric MEMS energy harvester," *IEEE International Electron Devices Meeting (IEDM),* pp. 1–4, Dec. 2009.

[38] Y. K. Ramadass and A. P. Chandrakasan, "An efficient piezoelectric energy harvesting interface circuit using a bias-flip rectifier and shared inductor," *IEEE Journal of Solid-State Circuits*, **45**, (*1*), 189–204, 2010.

[39] D. Kwon and G. A. Rincon-mora, "A single-inductor AC-DC piezoelectric energy-harvester/battery-charger IC converting ±(0.35 to 1.2V) to (2.7 to 4.5V)," in *ISSCC 2010*, **22**, (*1*), pp. 494–495, 2010.

[40] S. P. Beeby, R. N. Torah, M. J. Tudor, *et al.*, "A micro electromagnetic generator for vibration energy harvesting," *J. Micromech. Microeng.*, **17**, 1257–1265, 2007.

[41] Neil N. H. Ching, H. Y. Wong, Wen J. Li, Philip H. W. Leong, and Zhiyu Wen, "A laser-micromachined multi-modal resonating power transducer for wireless sensing systems" *Sensors and Actuators A: Physical*, **97–98**, 685–690, 2002.

[42] D. Spreemann, Y. Manoli, B. Folkmer, and D. Mintenbeck, "Non-resonant vibration conversion," *Journal of Micromechanics and Microengineering*, **16**, S169–173, 2006.

[43] T. Sterken, P. Fiorini, and R. Puers. "Motion-based generators for industrial applications," *Proceedings of the International Conference on Design, Test, Integration and Packaging of MEMS/MOEMS*, pp. 328–331, 2000.

[44] D. Maurath, P. F. Becker, D. Spreemann, and Y. Manoli, "Efficient energy harvesting with electromagnetic energy transducers using active low-voltage rectification and maximum power point tracking," *Journal of Solid State Circuits*, **47**, 1369–1380, 2012.

[45] D. M. Rowe, *CRC Handbook of Thermoelectrics*, CRC Press, 1995. ISBN: 978–0–8493–0146–9, e-ISBN: 978–1–4200–4971–8.

[46] S. Lineykin and S. Ben-Yaakov. "Spice compatible equivalent circuit of the energy conversion process in thermoelectric modules," *23rd Israel Convention*, Tel Aviv, pp. 346–349, 2004.

[47] S. Lineykin and S. Ben-yaakov, "Analysis of thermoelectric coolers by a spice-compatible equivalent-circuit model," *IEEE Power Electronics Letters*, **3**, (*2*), 63–66, 2005.

[48] S. Lineykin and S. Ben-Yaakov, "Modeling and analysis of thermoelectric modules," *IEEE Transaction on Industry Application*, **43**, (*2*), 505–512, (Mar./Apr.) 2007.

[49] A. Khaligh, P. Zeng, and C. Zheng, "Kinetic energy harvesting using piezoelectric and electromagnetic technologies – State of the art," *Industrial Electronics, IEEE*, **57**, (*3*), 850–860, 2010.

[50] H. Sodano, J. Lloyd, and D. J. Inman, "An experimental comparison between several active composite actuators for power generation," *Smart Materials and Structures*, **15**, (*5*), 1211–1216, (Oct.) 2006.

[51] C. Williams, C. Shearwood, M. Harradine, *et al.*, "Development of an electromagnetic micro-generator," *Circuits, Devices and Systems*, IEEE Proceedings, 2001.

[52] T. Hehn, F. Hagedorm, D. Mourath, *et al.*, "A fully autonomous integrated interface circuit for piezoelectric harvesters," *IEEE Journal of Solid-State Circuits*, **47**, (*9*), 2185–2198, 2012.

[53] C. Peters, J. Handwerker, D. Maurath, and Y. Manoli, "A sub-500 mV highly efficient active rectifier for energy harvesting applications," *IEEE Transactions on Circuits and Systems-I*, **58**, (*7*), 1542–1550, 2011.

[54] D. Maurath and Y. Manoli, "A self-adaptive switched-capacitor voltage converter with dynamic input load control for energy harvesting," in *Proceedings of ESSCIRC'09*, pp. 284–287, 2009.

[55] G. K. Ottman, H. F. Hofmann, and G. Lesieutre, "Optimized piezoelectric energy harvesting circuit using step-down converter in discontinuous conduction mode," *IEEE Transactions on Power Electronics*, **18**, (*2*), 696–703, (Mar.) 2003.

[56] A. Romani, C. Tomburini, A. Golfarelli, *et al.*, "Dynamic switching conversion for piezoelectric energy harvesting systems," *IEEE Sensors*, 689–692, (Oct.) 2008.

[57] A. Romani, R. P. Paganelli, and M. Tartagni, "A scalable micro-power converter for multi-source piezoelectric energy harvesting applications," *Procedia Engineering*, **5**, 782–785, 2010.

[58] A. Romani, M. Filippi, and M. Tartagni, "Micropower design of a fully autonomous energy harvesting circuit for arrays of piezoelectric transducers," *IEEE Transactions on Power Electronics*, **29**, (*2*), 729–739, 2014.

[59] A. Romani, R. Paganelli, E. Sangiorgi, and M. Tartagni, "Joint modeling of piezoelectric transducers and power conversion circuits for energy harvesting applications," *Sensors Journal, IEEE*, **13**, (*3*), 916–925, 2013.

[60] N. Krihely and S. Ben-Yaakov, "Self-contained resonant rectifier for piezoelectric sources under variable mechanical excitation," *IEEE Transactions on Power Electronics*, **26**, (*2*), 612–621, 2011.

[61] V. Leonov and R. J. M. Vullers, "Thermoelectric and hybrid generators in wearable devices and clothes," *Wearable and Implantable Body Sensor Networks, 2009, Sixth International Workshop on BSN*, pp. 195–200, 2009.

[62] J. Langley, M. Taylor, G. Wagner, and S. Morris, "Thermoelectric energy harvesting from small aircraft engines," *SAE Technical Paper* 2009–01–3093, 2009, doi:10.4271/2009–01–3093.

[63] J. Im, S. Wang, and K. Lee, "A 40mV transformer-reuse self-startup boost converter with MPPT control for thermoelectric energy harvesting," *IEEE International Solid-State Circuits Conference Digest of Technical Papers*, no. 2010, pp. 104–105, 2012.

[64] A. Costanzo, A. Romani, D. Masotti, N. Arbizzani, and V. Rizzoli, "RF/baseband co-design of switching receivers for multiband microwave energy harvesting," *Sensors and Actuators A: Physical*, **179**, (*1*), 158–168, 2012.

[65] J. P. Curty, N. Joehl, C. Dehollain, and M. J. Declercq, "Remotely powered addressable UHF RFID integrated system," *IEEE Journal of Solid-State Circuits*, **40**, (*11*), 2193–2202, 2005.

[66] J. A. Hagerty, F. B. Helmbrecht, W. H. McCalpin, R. Zane, and Z. Popovic, "Recycling ambient microwave energy with broad-band rectenna arrays," *IEEE Trans. Microwave Theory Tech.*, **46**, (*3*), 1014–1024, 2004.

[67] V. Rizzoli, A. Costanzo, D. Masotti, *et al.*, "Integration of non-linear, radiation, and propagation CAD techniques for MIMO link design," *International Journal of Microwave and Wireless Technologies*, **4**, (*5102*), 223–232, (Apr.) 2012.

[68] J. Essel, D. Brenk, J. Heidrich, and R. Weigel, "A highly efficient UHF RFID frontend approach," in *IEEE MTT-S International Microwave Workshop on Wireless Sensing, Local Positioning and RFID (IMWS 2009)* Croatia, pp. 1–4, 2009.

[69] A. Costanzo, A. Romani, D. Masotti, N. Arbizzani, and V. Rizzoli, "RF/baseband co-design of switching receivers for multiband microwave energy harvesting," *Sensors and Actuators A: Physical*, **179**, 158–168, Jun. 2012.

[70] V. Rizzoli, D. Masotti, F. Mastri, and E. Montanari, "System-oriented harmonic-balance algorithms for circuit-level simulation," *IEEE Trans. on Computer-Aided Design of Integrated Circuits and Systems*, **30**, (*2*), 256–269, 2011.

[71] C. Emili, F. Alimenti, P. Mezzanotte, L. Roselli, and R. Sorrentino, "Rigorous modeling of packaged Schottky diodes by the nonlinear lumped network LN-FDTD approach," *IEEE Trans. Microwave Theory Tech.*, **48**, 2277–2282, 2000.

[72] D. Masotti, A. Costanzo, and S. Adami, "Design and realization of a wearable multi-frequency RF energy harvesting system," *Proc. 5th European Conference on Antennas and Propagation*, Rome, pp. 517–520, Apr. 2011.

[73] M. Dini, M. Filippi, A. Costanzo, *et al.*, "A fully-autonomous integrated RF energy harvesting system for wearable applications," *Proc. 2013 IEEE European Microwave Conference*, accepted for publication.

[74] V. Rizzoli, A. Costanzo, D. Masotti, and F. Donzelli, "Integration of numerical and field-theoretical techniques in the design of single- and multi-band rectennas for micro-power generation," *EuMA International Journal of Microwave and Wireless Technologies*, (*3–4*), 293–303, July 2010.

[75] A. Costanzo, D. Masotti, F. Donzelli, and S. Adami: "Device to convert radiofrequency electromagnetic energy," patent WO/2012/042348, PCT/IB2011/002253, Apr. 2012.

[76] J. Borgeson, "Ultra-low-power pioneers√: TI slashes total MCU power by 50 percent with new 'Wolverine' MCU platform," Texas Instruments White Paper, http://www.ti.com/ww/en/mcu/wolverine/wolverine-whitepaper.pdf.

[77] D. Guyomar, A. Badel, E. Lefeuvre, and C. Richard, "Toward energy harvesting using active materials and conversion improvement by nonlinear processing," *IEEE Transactions on Ultrasonics, Ferroelectrics, and Frequency Control*, **52**, (*4*), 584–595, 2005.

[78] E. Lefeuvre, A. Badel, C. Richard, L. Petit, and D. Guyomar, "A comparison between several vibration-powered piezoelectric generators for standalone systems," *Sensors and Actuators A: Physical*, **126**, (*2*), 405–416, 2006.

[79] Y. C. Shu, I. C. Lien, and W. J. Wu, "An improved analysis of the SSHI interface in piezoelectric energy harvesting," *Smart Materials and Structures*, **16**, (*6*), 2253–2264, Dec. 2007.

[80] J. Dicken, P. Mitcheson, A. Elliott, and E. Yeatman, "Single-supply pre-biasing circuit for low-amplitude energy harvesting applications," *Proc. PowerMEMS*, 2011.

[81] J. Dicken and P. Mitcheson, "Power-extraction circuits for piezoelectric energy harvesters in miniature and low-power applications," *IEEE Transactions on Power Electronics*, **27**, (*11*), 4514–4529, 2012.

4 Technologies for RFID sensors and sensor tags

Thomas Ussmueller, Amelie Hagelauer, and Robert Weigel

4.1 RFID sensor concept and constraints

4.1.1 RFID sensor architecture

During the previous decades RFID systems have infiltrated the market for various everyday uses. The systems are used for access control, mass transit ticketing, and baggage tagging to name a few. So far most of the systems only include a unique ID for identification or have a small amount of memory included on the tag. The next step forward for RFID systems is the availability of information about the environment of the tag. On the one hand this includes the position of the tag, on the other hand sensor values like temperature, humidity, or pressure are of interest.

RFID systems are made up of one or more RFID interrogators/readers and numerous passive RFID tags, see Figure 4.1. For communication with the tag, the reader generates a high-frequency modulated electromagnetic field. The tag receives the information modulated in the field and it can answer to the request from the reader with a back-scatter operation, i.e. the reflection coefficient is changed with respect to the requested information. In addition to the transmission of information between the tag and the reader, the tag can receive the necessary energy for operation from the electromagnetic field of the reader.

4.1.2 Active and passive RFID tags

Due to a wide range of application scenarios in this field, there exists a multitude of different sensor nodes. The most widespread developments have active sensor nodes, i.e. sensor nodes with their own energy supply, e.g. with a battery, and are for high performance and/or high range use cases [1].

As opposed to this, passive RFID tags use the ambient energy for their power supply.

Typically the basis of the received wireless power on the rectification is the energy-harvesting approach. Thus, for low cost and maintenance-free operation, passive RFID tags are well suited. Other approaches work with thermal or mechanical energy harvesting.

Due to regulatory and physical constraints, the used frequency band of the passive RFID tag influences the available data rate and sets an upper limit for its operation range. HF-RFID tags operate in the near field of the reader antenna and so the operational

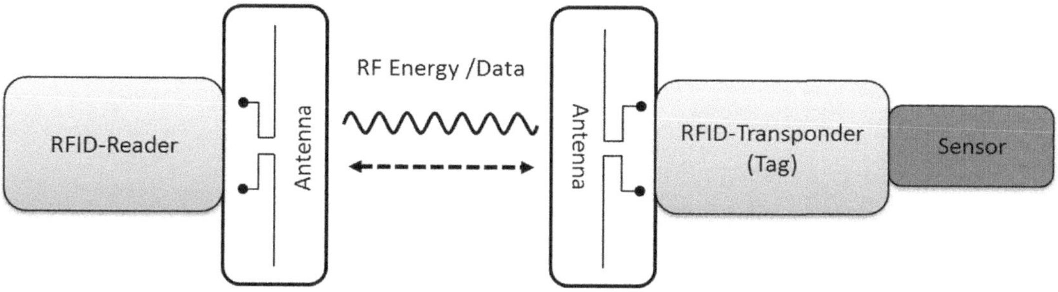

Figure 4.1 Passive RFID wireless concept with sensor functionality.

distance is limited between several centimeters up to one meter. By comparison with HF-RFID systems, UHF-RFID tags transmit their data and energy by means of the electrical far field and thus offer a higher read range up to 10 m [2].

4.1.3 RFID sensor technology for wireless sensor networks

Working at both HF and UHF frequencies, multi-standard RFID transponders allow a flexible and ubiquitous use in maintenance-free wireless networks. A high acceptance rate in the market of this multi-standard ability is ensured, because the compatibility of existing HF and UHF readers is guaranteed.

In addition to the RFID tag's basic functionality, the transmission of a unique identifier, additional functionality can be provided by future RFID systems. An integrated passive sensor interface and a locatable RFID tag are two of the most interesting features.

Several enhancements to existing solutions are allowed by the additional integration of a sensor interface. One application example would be the integration of a temperature sensor, which could monitor a cold chain. Other possible use cases include a monitoring device for the remote and maintenance-free acquisition of tyre pressure values.

An integrated localization capability of RFID systems provides multiple opportunities in different new application areas, like tracking of persons and objects. In logistics the automation of the management of storerooms is possible, because the position of a container with an affixed UHF transponder can be detected. An example concerning automotive application is the activation of several functions of a car, before a person is moving towards it. With active RFID tags having a separate battery for their power supply, all this functionality is available. In contrast, the aim of this work is the presentation of a completely passive solution covering the aforementioned functionality.

Consequently, when operating in the UHF mode only a few μW are available as a supply for all functional blocks, which includes the circuits for sensing and localization, and thus a highly efficient rectifier and low power circuit design technique have to be used for all components of the RFID tag [3].

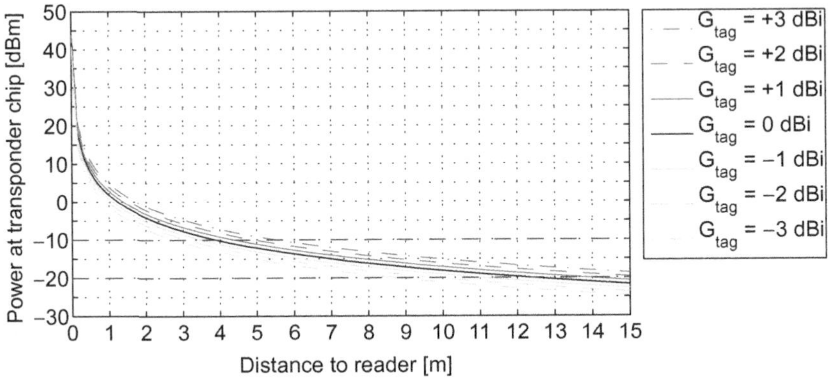

Figure 4.2 Simulated available RF power at a transponder chip in UHF mode at 868 MHz. A P_{ERP} of 33 dBm is assumed. Current UHF RFID transponders become operational in the range between −20 and −10 dBm (marked lines).

4.1.3.1 Power constraints

Only available at an arbitrary transponder antenna, the low level of power is surely the major limitation for passive UHF operation. P_{tag} can be defined as

$$P_{\mathrm{tag}} = P_{\mathrm{reader}} \frac{A_{\mathrm{er}} A_{\mathrm{et}}}{\lambda^2 d^2}, \tag{4.1}$$

where P_{reader} is the output power of the reader power amplifier, λ is the wavelength, and d is the distance between reader and tag. Respectively, A_{er} is the effective area of the reader antenna and A_{et} the effective area of the tag antenna. The relation between the effective area A_{e} of an antenna and its gain G can be expressed as

$$A_{\mathrm{e}} = \frac{\lambda^2 G}{4\pi}. \tag{4.2}$$

Replacing the effective antenna area in (4.1) with this definition yields

$$P_{\mathrm{tag}} = P_{\mathrm{reader}} G_{\mathrm{reader}} G_{\mathrm{tag}} \left(\frac{\lambda}{4\pi d} \right)^2. \tag{4.3}$$

Some assumptions have to be made, in order to determine the actual available power for base band processing and sensor interface. P_{reader} and G_{reader} together are assumed to be 33 dBm, which is the maximum equivalent radiated power (P_{ERP}) allowed to be transmitted (in Europe) in the UHF range around the operating frequency $f = 868$ MHz [4]. The wavelength λ thus is $c/f = 0.3456$ m. The value of G_{tag} is estimated in a range of −3 dBi to +3 dBi. Here, the variable antenna polarization and the variable matching between antenna and chip caused by nonlinearities of the analog front-end (AFE) are of course taken into account. Due to these assumptions, an estimation of the available RF power at a UHF transponder chip is possible. The results illustrated in Figure 4.2 were inserted into (4.3) and calculated over a distance of 15 m. The reading distance for passive UHF transponders is commonly targeted at 10 m. To enable operability of the tag at a distance of 10 m, an RF power of −18 dBm is needed at a given tag antenna

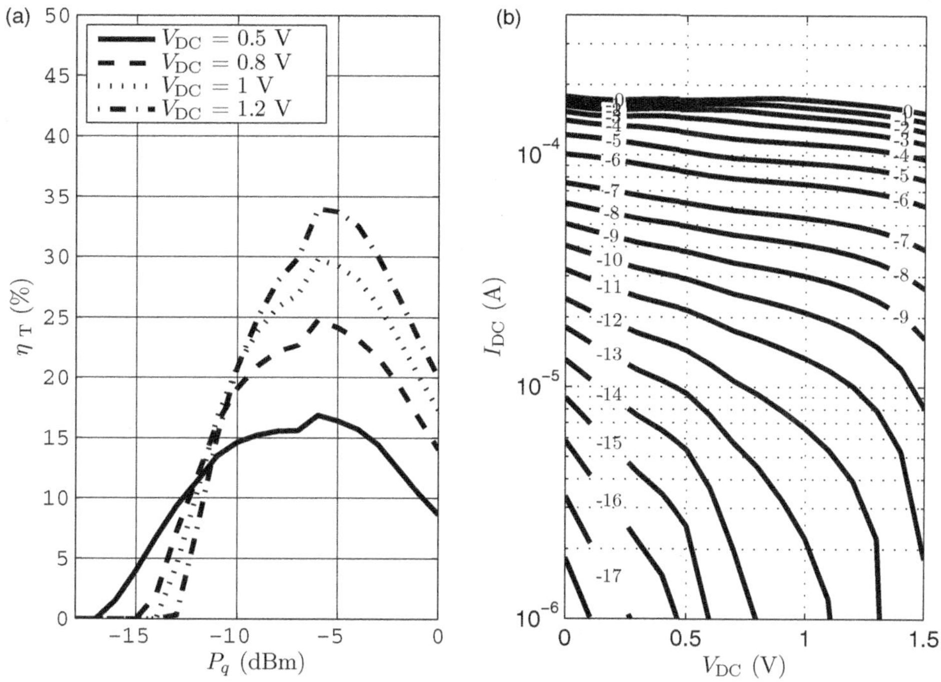

Figure 4.3 (a) Measured efficiency; and (b) I–V characteristic; of the implemented multi-standard AFE at UHF according to [7].

gain of 0 dBi. In fact this limit is reached by current implementations of UHF-only passive transponders [5]. The efficiency η_{AFE} of the AFE has to be considered, in order to determine the effective available power $P_{a,DC}$ for all other components besides the AFE, η_{AFE} is expressed as [6]

$$\eta_{AFE} = \frac{P_{a,DC}}{P_{RF} - P_{RF}^{refl}}, \tag{4.4}$$

where $P_{a,DC}$ is the available DC power excluding the AFE power consumption, P_{RF} is the incident RF power at the interface between the antenna and AFE, and P_{RF}^{refl} is the reflected RF power that is not available due to mismatch between antenna and AFE η_{AFE} has to be added to (4.3) in order to obtain $P_{a,DC}$, which results in

$$P_{a,DC} = P_{tag}\eta_{AFE} = P_{reader}G_{reader}G_{tag}\left(\frac{\lambda}{4\pi d}\right)^2 \eta_{AFE}. \tag{4.5}$$

The UHF efficiency η_{AFE} for the implemented multi-standard AFE is between 10% and 20% for an incident RF power of −13 dBm to −10 dBm, cf. the previous publications of Essel *et al.* [6, 7], and Figure 4.3(a). Caused by this additional RF power to the measured load lines of the AFE in UHF mode in Figure 4.3(b), a determination of power limits for the base band and sensor operation is possible. A supply current of 10 μA is feasible, if an incident RF power of −11 dBm and a generated supply voltage of V_{DC} = 1 V–1.2 V are given. At a distance of 5 m and an antenna gain of 0 dBi, the RF power is approximately

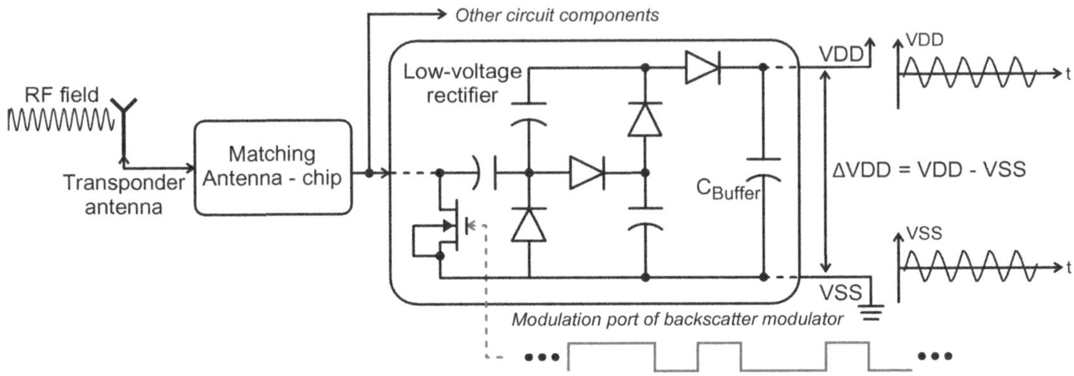

Figure 4.4 Rectified supply voltage of an RFID transponder circuit.

−12.5 dBm, according to Figure 4.2. This results in an available power of 4 μW at a supply of 1 V, as illustrated in Figure 4.3(b). A value of −11.5 dBm and thus an *I–V* pair of (10 μA/1 V) is reached in a distance of 4.5 m. In conclusion, all circuits of the multi-standard transponder should consume less than 10 μW, in order to achieve an acceptable reading range.

An interface implemented as ADC is needed for multipurpose wireless sensor nodes that contain multiple sensors. Solutions with per-sensor AD conversions are reported [8], but they need much more energy, compared to a single ADC with an input multiplexer. By using current 65 nm CMOS processes, ultra-low-power consumptions can be reached [9], but the costs thereby incurred are very high. Well-established 0.35 to 0.13 μm processes cause lower costs for RFID tags. On one hand, the absolute power consumption of the ADC has to be minimal, when implementing a passive sensor RFID tag in 130 nm bulk CMOS. But on the other hand, protocol timeout conditions have to be met, in order to prohibit minimizing the conversion rates to save power. The optimum of a classical successive approximation register (SAR) ADC architecture fulfills the power, timing, and quality requirements for passive UHF RFID sensing.

4.1.3.2 Current technological challenges

Circuit design for passive transponders involves additional boundary conditions, which often extend to the basic circuit design considerations, meaning that passive transponders are only powered by the RF field of the reader. The whole scenario is illustrated in Figure 4.4. After passing the transponder antenna and the matching network, the RF signal is converted to DC power by rectifier circuits in the transponder which supply power to all implemented components of the chip. The maximum transmitted RF power from the reader is limited by regulations to avoid frequency interference. Passive transponders can usually provide only a few microwatts of power and the respective supply voltage VDD is between 1.0 V to 1.5 V [10] without the use of more advanced techniques like charge pumps. However, these circuits have their own clock signal that must enable and disable the related CMOS switches of this component and may further increase power consumption [11]. The load capacitance C_{Buffer} at the output of the rectifier circuit is used for storing the rectified power. It must guarantee that the chip can collect sufficient energy to stay operational during back-scattering. During the logical high state, the back-scatter modulator shorts the input of the rectifier. Power cannot be transmitted in this case. This implies that only basic circuit architectures that are optimized for low voltage applications and low current draw can be considered. If security mechanisms should be implemented, they are also limited by the available power and the low clock frequency of the tag (few megahertz). The circuit design is also dependent on the transistors available in the semiconductor process used, which should have a low threshold voltage to be suitable for these RFID requirements. Finally, the technology used itself has an influence on the design: while bulk CMOS and Bi-CMOS processes offer advantages for low-cost processes, silicon on insulator (SOI), or silicon on sapphire (SOS) are optimized for low-power applications. This discussion is extended in [10, 12–15].

A popular method for reducing the power consumption of the transponder chip is clock gating, meaning that digital components that are not used in the current operating state are disabled. Clock gating avoids charging and discharging the load capacitances of the related digital gates. This method can be enhanced by using power switches that can separate complete subcircuits from the supply rail. Therefore, bias and leakage currents of those components can be saved. The duty cycle of the modulation signal can also be varied. A shortening of the duration in which the back-scatter modulator is activated also reduces the time interval in which the rectifier circuit cannot work. However, this increases the requirements for the reader circuit, which must recover the digital states sent by the transponder. This method is not possible for UHF transponders that must comply with the EPC protocol. Here, a duty cycle of 50% ±5% for the clock signal is required [16].

For RFID systems, every possible energy source in the environment of the transponder can support power generation in the transponder. Possible sources are sunlight (solar cells on the transponder) or body heat, for example. Integrating solar cells on a wireless sensor node has been demonstrated [17] but, although transferable to RFID technology in principle, there are no known applications at the moment. In Figure 4.4, the generated DC supply voltage VDD usually has a ripple caused by the rectifier. The magnitude of this ripple is also dependent on the implemented load capacitance C_{Buffer}. While the digital part of the transponder is relatively insensitive to the ripple, analog components such as reference circuits or the tag oscillator can be influenced. The ground node VSS of the transponder circuit is floating internally and its voltage level is created by the rectification. It has no fixed value, as well as a ripple caused by the rectifier circuit. Depending on the relative phase of the ripples at VDD and VSS, the differential supply voltage ΔVDD can achieve a better characteristic, which is actually used by the succeeding circuit components. If the two ripples are in common mode, the ripple on ΔVDD is suppressed. With increasing relative phase, a ripple on ΔVDD also exists. Furthermore, the available power supply and the resulting supply voltage VDD of the transponder are not constant. They vary with the distance between reader and transponder because of free-air losses, which also limit the operating range, and their orientation (polarization of the antennas) to each other. Therefore, reader antennas often have circular polarization to sustain a communication link independent of the orientation, but this further reduces the available power by 3 dB [18]. Power matching between the transponder antenna and the nonlinear input impedance of the transponder, as shown in Figure 4.4, is also a challenging task, which can usually only be optimized for certain operating conditions. Moreover, the gain of the related antenna is often dependent on the characteristics of the material to which the transponder is attached. Last but not least, the unit price of such a transponder must also be accounted for. As with every low-cost design, the whole chip area of the transponder circuit, which is proportional to the fabrication costs, should be as small as possible to be worth mass production. To be suitable for typical barcode applications, the unit price should be around a few US cents.

4.2 SAW-based RFID sensors

In this section SAW (surface acoustic wave)-based RFID sensors will be explained. These sensors are mainly based on a combination of SAW RFID tags [19] with traditional SAW sensor technologies for physical or chemical parameters [20, 21]. This combination results in passive transponders that are connected to their read-out unit solely by a wireless radio link. They do not require any power supply.

4.2.1 Basic principles of SAW RFID sensors

The SAW transponder technology is based on traditional surface acoustic wave technology. Therefore the antenna of the passive SAW transponder picks up a small portion of the RF signal from the transceiver. An interdigital transducer (IDT), which is connected to the antenna, converts the received signal into a surface acoustic wave via the converse piezoelectric effect. So basically there is a similarity of the technology used to SAW filters. The SAW device stores the micro-acoustic wave. This storing can be done using either a delay line [20] or a resonator configuration [22]. Moreover, according to the code number (or to the sensor variable) of interest, the micro-acoustic wave is coded. The coded RF micro-acoustic signal is re-converted into electrical signals by the IDT and re-transmitted to the transceiver by the transponder antenna. This response not only contains information about the number and geometries of the coding structures, but also the propagation and reflection properties of the SAW. The extraction of the desired information, for example a specific ID number or/and the sensor effect for a certain measurement, is enabled by using small SAW chips, which achieve long delay times in the range of some microseconds, based on the low velocity of SAW. Therefore electromagnetic multipath propagation phenomena cause environmental echoes at VHF/UHF frequencies, which have already safely faded away when the sensor response arrives at the transceiver. Thus the problem of the separation of the sensor response and the environmental echoes in the time-domain is easily solved [20].

Passive SAW transponders do not require any power supply, and their antennas are usually of different type, such as a dipole, patch, slot, or loop type. They cannot be addressed individually, because they are passive components without any active logic on chip. Frequency division multiple access (FDMA), time division multiple access

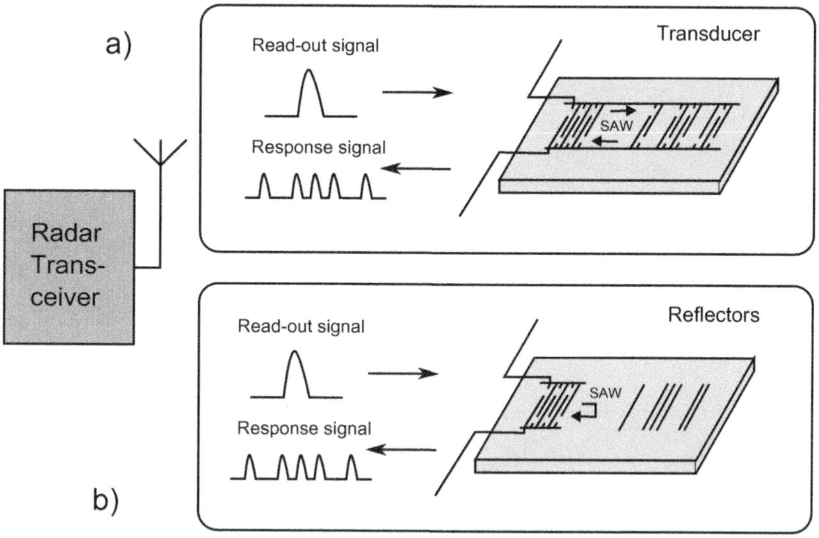

Figure 4.5 SAW RFID system concept: (a) using transducers; or (b) using reflectors.

(TDMA), code division multiple access (CDMA) [23], space division multiple access (SDMA), or combinations are required to access more than one transponder.

4.2.2 Design of SAW RFID sensors

A SAW RFID tag usually incorporates a coupling IDT and a coding area. To design the code area, there are two options, see Figure 4.5: either several coding transducers are wired together with a common busbar, Figure 4.5(a), or alternatively reflectors can be used, Figure 4.5(b). There is a similarity in the layout of a delay line with several coding transducers to that of a tapped delay line, where a long busbar wires the tapped transducer to the coupling transducer, building up a one-port device. In principle the coded transducer could also be wired to the coupling transducer and the antenna via a circulator.

Most SAW RFID tags are designed using a reflective delay line [20], for example like the one sketched in Figure 4.5(b). The SAW in a reflective delay line propagates towards reflectors, which are distributed in a characteristic barcode-like pattern, and is partially reflected at each reflector. Compared to the non-reflective arrangement of Figure 4.5(a) the chip size can be halved by the use of reflectors. Furthermore, the number of coding elements and the actual code do not alter the electrical impedance of the transducers. Materials like lithium niobate $LiNbO_3$ and substrates with a high piezoelectric constant are used. The coding is done by ASK, PSK, or PPM.

It is also possible to use an external sensor as an electrical load for an IDT [24], see Figure 4.6. The reflection characteristics of the IDT are changed by the load impedance. The value of the impedance is changed by the measuring signal of the external sensor.

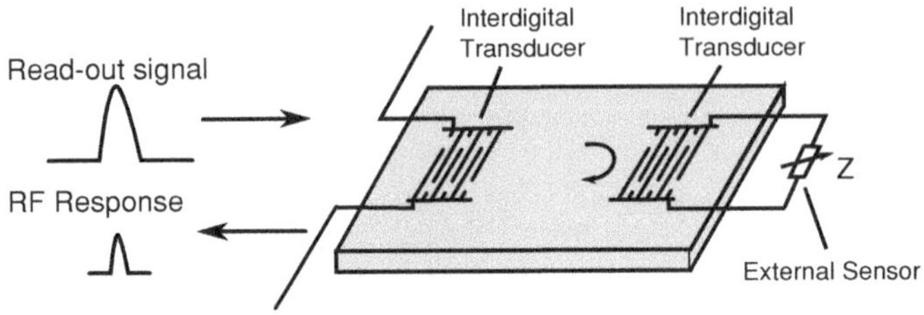

Figure 4.6 SAW device with an external sensor [20].

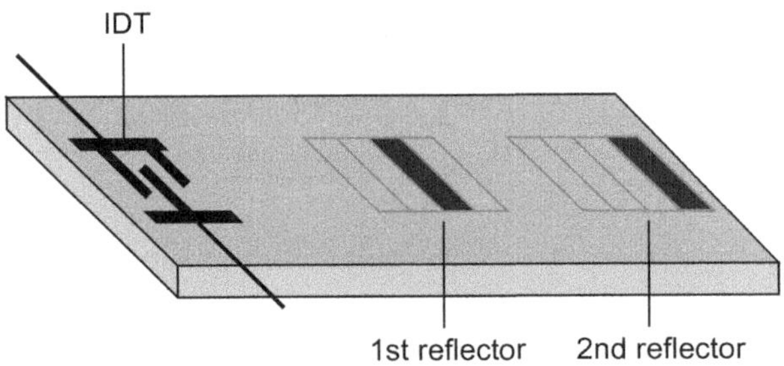

Figure 4.7 Pulse position coding schema.

The coding of a SAW ID tag depends on the applied modulation technique and is obtained by a specific coding arrangement. In the simplest case, the coding is realized by a binary amplitude shift keying ASK (on/off): each predetermined possible symbol position is set either by a transducer or reflector (on-bit) or not (off-bit). 2^{32}different tags can be coded with 32 symbol positions.

Other modulation techniques like phase shift keying (PSK) outperform the ASK on/off keying. Using a PSK modulation the phases of the time response are evaluated and not their amplitudes. A binary phase shift keying (BPSK) modulation obtains the same bit error rate with a 6 dB lower signal-to-noise ratio compared to an ASK modulation, thus enhancing the maximum read-out distance. Fewer symbols (reflectors) are used by higher-order PSK, like quadrature PSK (QPSK), and therefore it needs less chip size and undergoes lower insertion attenuation, but for the detection a higher signal-to-noise ratio is needed.

In a phase modulation system the requirements for a SAW transponder become more stringent, because all influences which concern the change of delay time of a reflector, e.g. the SAW velocity or the accuracy of the fabrication process, must be controlled within a small fraction of one single SAW wavelength.

A pulse position modulation (see Figure 4.7) might be a good choice for the modulation technique used in SAW transponders, because it achieves the code-density of a higher phase modulation and avoids their acute sensitivity to small inaccuracies during the fabrication process. In a pulse position modulation scheme the symbol is set at one out of several time slots. The accuracy in the time resolution of two targets is less than the time separation of one target in most radar systems. Therefore, compared to the symbol

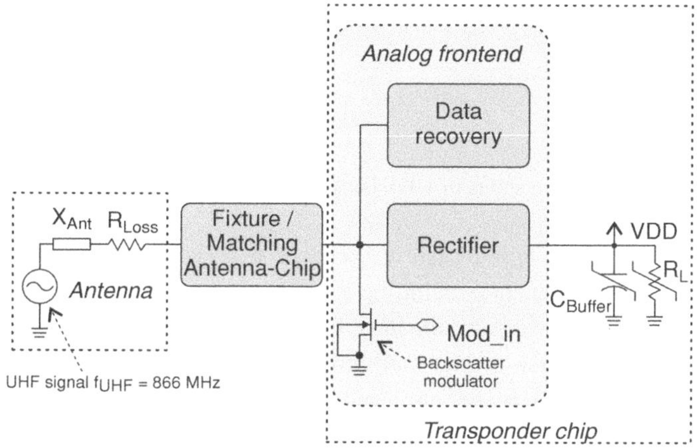

Figure 4.8 Transponder antenna and analog frontend.

separation in an ASK modulation, the time slots can be arranged with higher density. If each reflector is set in one out of 2^n time slots, each symbol codes for n bits.

4.2.3 Fields of application

There are several fields of application. Typical system applications are temperature sensors on lithium niobate $LiNbO_3$, with a large TCF and a high electro-acoustic coupling factor, working up to 1000 °C [23, 25, 26].

Other fields of application are sensors for measuring mechanical parameters like pressure (tyre pressure sensor) [27, 28], acceleration (accelerometer) [29], strain (torque sensor) [21, 30], or force.

Impedance changing sensors as shown in Figure 4.6 can, for example, be used to make water content sensors [31].

4.3 CMOS-based RFID sensors

4.3.1 System architecture of a CMOS RFID system

The evaluation of the chip design for the transponder starts without considering a certain antenna architecture. According to [32], an arbitrary antenna structure can be represented by a single equivalent circuit. A block diagram of the transponder antenna and the analog frontend are shown in Figure 4.8. R_{LOSS} describes the resistive losses of the antenna (here assumed 5 Ω for simulation). X_{ANT} equals the radiation impedance. The conversion of the wire-bound energy to radiated is represented by a fictitious impedance. The pad capacitance of the chip, the inductance of the bond wire, and resistive losses of the stripline between the antenna and chip are examples of the description of the parasitic effects of the transponder setup by the fixture. Usually made from passive lumped

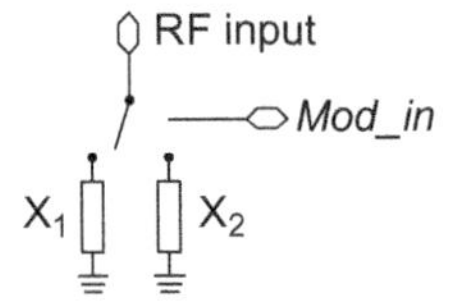

Figure 4.9 Common modulation states of UHF back-scatter modulator.

components, the matching circuit has to achieve a power matching of the antenna impedance to the chip impedance including the fixture.

The designs of all presented circuits have been created with respect to the EPC protocol. Referring to it, the back-scatter modulator for data communication from the UHF transponder back to the UHF reader can use amplitude shift keying (ASK) and/or phase shift keying (PSK) modulation [16]. The achievement of the respective modulation techniques is possible by using different combinations of lumped components [10, 12, 13]. In general ASK modulation is less efficient than PSK modulation, as the available power is almost constant at both modulation states [33–35]. In Figure 4.9 the switching of the back-scatter modulator between its two common modulation states is shown. Resistors or capacitors mostly provide the impedances. In this approach, Figure 4.8 demonstrates a shunt transistor, used as the back-scatter modulator followed by the rectifier circuit. Enabling a mixture of ASK and PSK modulation for the UHF transponder, it mainly performs a resistive modulation between the on and off resistance of the transistor [13]. In order to store the rectified power of a passive transponder, the load capacitance in Figure 4.8 at the output of the rectifier circuit is used. It is absolutely necessary that the chip can collect sufficient energy to stay operational during back-scattering, which is usually performed by the oxide capacitance of MOS transistors. Its nonlinear behavior is no longer of interest, since it is only used for storing energy. The DC power consumption of all implemented transponder circuits is represented by the nonlinear load.

The antenna, the only external component of this multi-standard tag (Figure 4.10), can be used at UHF and HF. This test chip has been integrated in a 0.13 μm digital CMOS technology. The two combined highly efficient power rectifiers convert the incoming electromagnetic RF wave into DC for supplying all active circuits on the transponder. If the DC supply voltage reaches a defined level, the power-on-reset generates a reset signal to place the digital part into a known state. The information, which is ASK modulated on the carrier by the interrogator, will be recovered by the demodulator.

By varying the impedance of the RF node in order to modulate the reflected power (UHF) or by changing the load of the inductively coupled transformer network, the back-scatter modulator at UHF and the load modulator at HF enable the tag-to-reader communication. At high RF levels the integrated voltage limiter protects the tag circuits. In addition voltage, current, and frequency reference circuits have been designed.

Among the classical tag functionality an ultra-low-power sensor interface has been implemented as well. It includes a four-to-one multiplexer, an integrated temperature sensor, and an SAR analog-to-digital converter. Either the internal temperature sensor, the received signal strength indicators of the UHF or the HF frontend, or an external analog sensor can be connected to the ADC by the multiplexer.

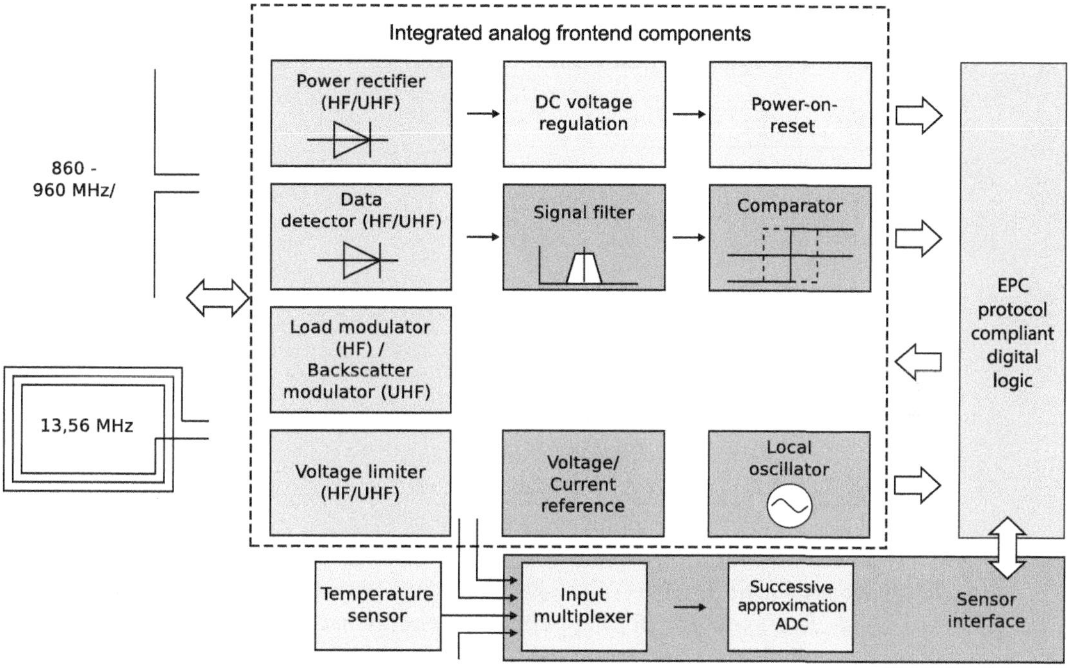

Figure 4.10 System architecture of a passive multi-standard RFID tag with sensor interface and switched passive reflector.

As the other on-chip components like the protocol logic or the oscillator and bandgap reference circuit need about 7 μW in total, the available power for the sensor interface ADC is less than 2 μW. The sensor interface itself consists of the ADC and several auxiliary circuits such as an input multiplexer with unity-gain buffer and a temperature sensor [36]. Table 4.1 lists the power which is needed by the different tag circuits as well as the remaining power for sensing purposes [37]. Depending on multipath effects, the available RF power in a specific distance depends on environmental conditions.

4.3.2 Multi-standard analog frontend

This section focuses on presentation of the most important analog building blocks of the RF frontend. It is necessary to know the behavior of the devices and the technology used for an ultra-low-power frontend design.

4.3.2.1 Ultra-low-power rectifier

The physical limits of wireless power transmission mainly determine passive communication systems. Due to regulatory and physical constraints, the frequency band used determines the data rate and limits the operation distance. In order to achieve the transmission of data and energy in UHF RFID systems, the average of the electrical far field must be used and the operational range should be about 8–9 m [2]. As supply for all functional blocks only a few μW are available and hence a highly efficient rectifier

Table 4.1 Power consumption of all component of the sensor-enhanced multi-standard RFID transponder.

Component	Available DC power at 4.5 m
Supplier:	
UHF&HF analog frontend	$\approx 10\ \mu$W
Consumers:	
Component	
EPC compl. digital logic	4.5 μW
Oscillator	1.2 μW
Bandgap reference	1.2 μW
Sensor interface	3.1 μW
Sensor interface breakdown:	
ADC	2 μW
Auxiliary circuits	1.1 μW

has to be used [38]. Designs of integrated circuits with ultra-low-power consumption and novel circuit concepts for high-efficiency energy harvesting are required, in order to achieve a higher reading range and to improve the operational reliability of passive RFID tags.

Two different types of rectifiers for use at UHF are presented in the following section. In passive UHF RFID transponders, the single-ended voltage multiplier is often used at ultra-low-power RF levels [33, 39]. Even at the fundamental frequency, the threshold voltage of the rectifying diodes used causes the problem of a high input impedance [40]. Devices with a low forward voltage drop are required for a highly efficient rectifier design. In order to achieve these essential devices, Schottky diodes are used in many frontend designs. These diodes have a low Schottky junction capacitance and, beside the low forward voltage drop, a low series resistance [33]. But often standard MOS devices have to be used, because in a standard CMOS process Schottky diodes are often not available. The efficiency of such rectifiers is lower, as simple MOS diodes have a higher forward voltage drop. Research activities are still ongoing in the area of highly efficient rectifier design with MOS devices for ultra-low-cost RFID transponders [41, 42]. The overall efficiency (η_{T}) compares these rectifiers. This indicator is expressed by

$$\eta_{\mathrm{T}} = \frac{P_{\mathrm{DC}}}{P_{\mathrm{q}} - P_{\mathrm{RF}}^{\mathrm{refl}}}, \tag{4.6}$$

where P_{DC} is the available output power, P_{q} is the incident RF power and $P_{\mathrm{RF}}^{\mathrm{refl}}$ the reflected RF power.

In Figure 4.11 the topology of a single-ended voltage multiplier is illustrated. The nonlinear behavior of the diode-connected PMOS transistors is the basis for the functionality of this rectifying structure. Amongst others the connected load, the available technology, and the nonlinear devices used are factors which determine the optimal number of rectifying stages. The conversion efficiency of a multi-stage rectifier depends

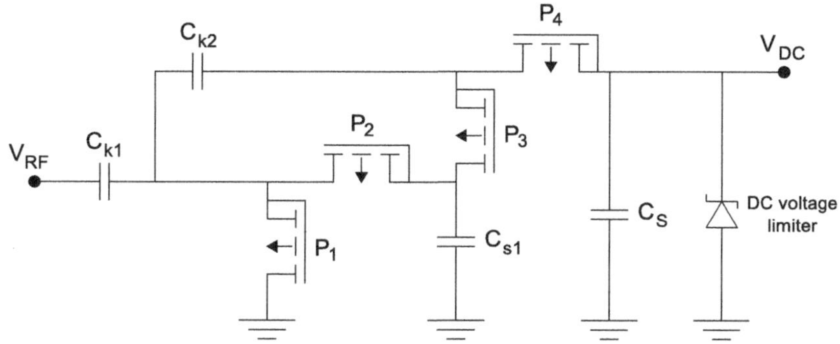

Figure 4.11 Single-ended two-stage voltage multiplier with diode-connected PMOS transistors.

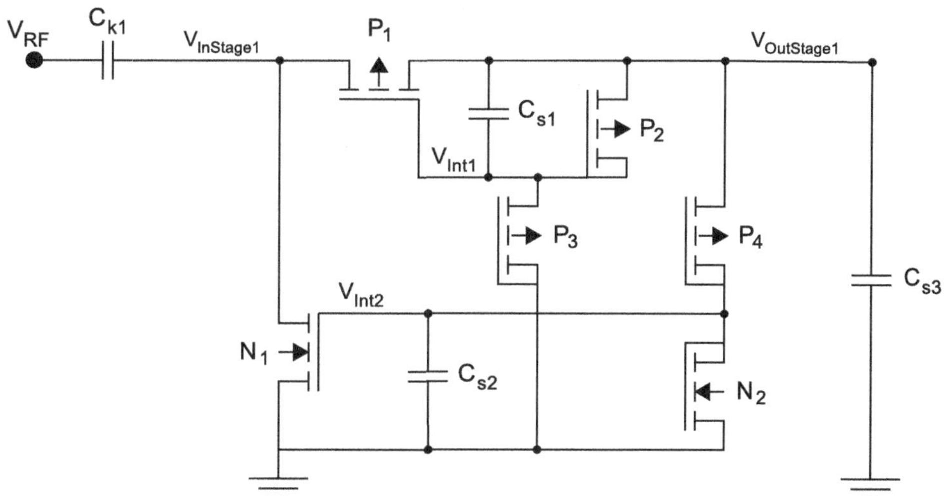

Figure 4.12 First stage of the implemented rectifier with threshold cancelation circuit for UHF.

on the matching between the source and the load impedance. An example for a typical task during the design is to find the optimum number of rectification stages [12]. The additional implementation of a DC voltage limiting structure is required to protect necessary auxiliary devices or digital logic against overvoltage. In [13] this rectifier topology is discussed in more detail.

Higher conversion efficiency without using Schottky diodes can be achieved by modification of the simple two-stage voltage multiplier. The first stage (input stage) of the implemented rectifier with a threshold cancelation circuit for the rectifying transistor pair P_1 and N_1 is shown in Figure 4.12.

Two PMOS transistors P_2 and P_3 and the capacitance C_{s1} carry out the cancelation of the threshold voltage of the PMOS transistor P_1, while P_4, N_2, and C_{s2} carry out the cancelation of the threshold voltage of the NMOS transistor N_1. The two transistors P_3 and P_4 are used in their subthreshold operating region as active high impedance

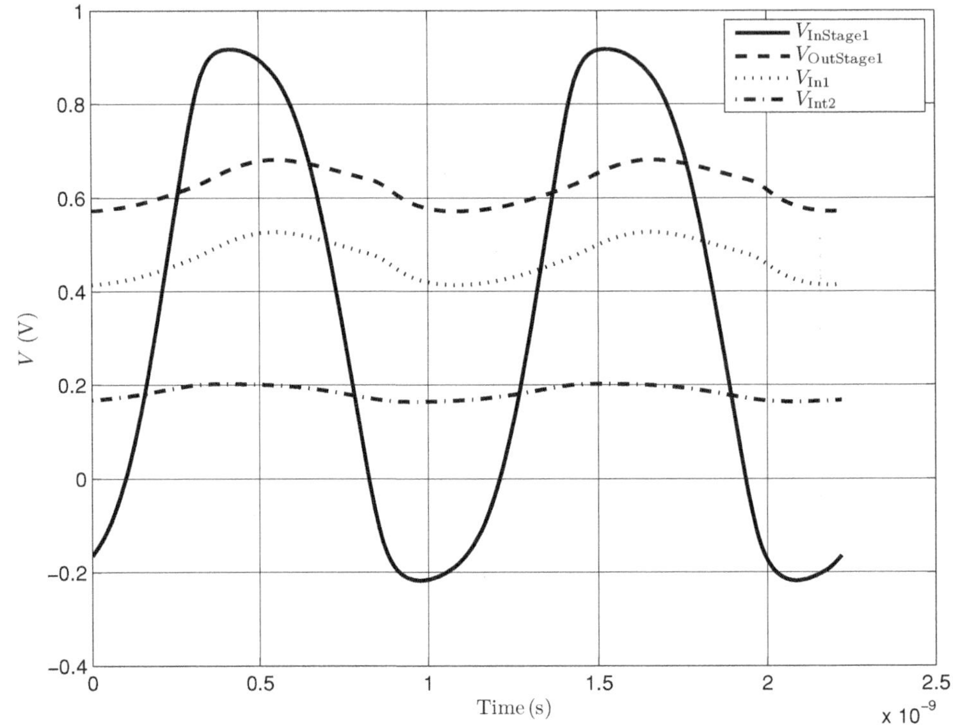

Figure 4.13 Internal signals of the first rectifier stage.

resistors. The bulk connection of the transistors can affect the value of the resistance. Integrated resistors with a comparable value as transistors are very large. To save chip area and thereby costs, it is helpful to use transistors as high impedance resistors.

In Figure 4.13 the simulated signal levels for the input voltage $V_{InStage1}$, the output voltage $V_{OutStage1}$, and the internal signals V_{Int1} and V_{Int2} can be seen. These simulated signals are plotted for two periods of a 900 MHz input signal at an input power level of about −15 dBm.

In comparison to $V_{OutStage1}$, the signal V_{Int1} is reduced about the threshold value of the transistor P_1. The threshold voltage of N_1 increases the signal V_{Int2}. There is no additional reference necessary for this threshold cancelation technique and also voltages or currents to adjust these internal voltages are not required. Therefore this self-cancelation technique is already working at very low input power levels even if the DC voltage at the output of the rectifier is too low to supply auxiliary circuits.

The complete two-stage rectifier is depicted in Figure 4.14 and implemented in a 0.13 μm CMOS technology. Analogous to the first stage of the rectifier, P_6, P_7, and C_{s5} compensate the threshold voltage of the transistor P_5. Furthermore there is a compensation of the threshold voltage of the transistor N_3 by P_8, N_4, and C_{s4}.

In Figure 4.15(a) measured efficiency and in Figure 4.15(b) the measured IV-characteristic of the rectifier circuit at UHF are shown. For a DC output power of

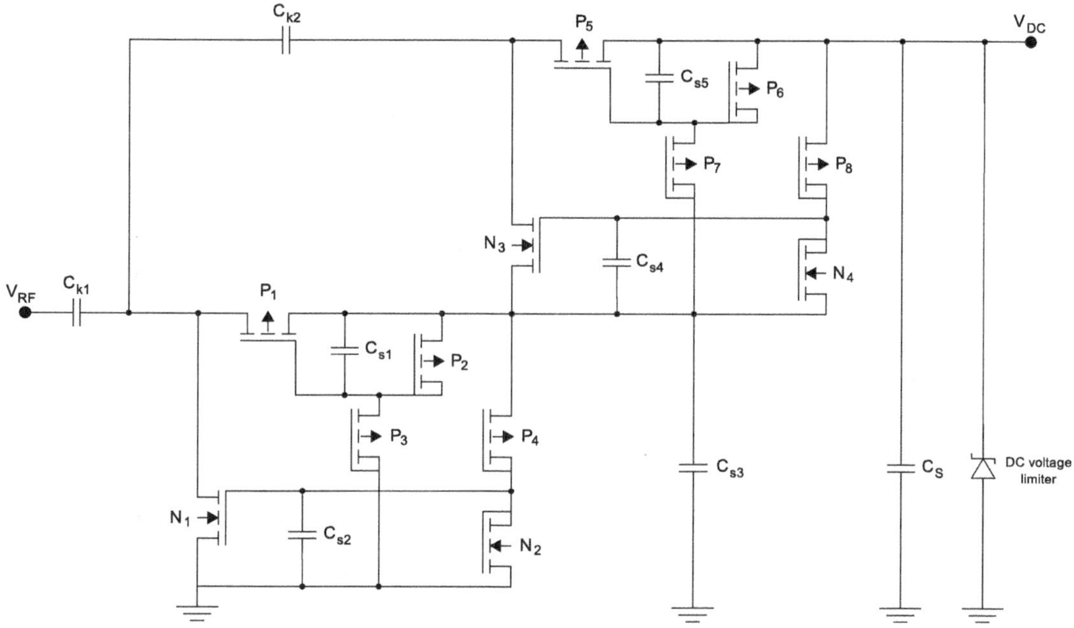

Figure 4.14 Single-ended two-stage rectifier with threshold cancelation circuit for UHF.

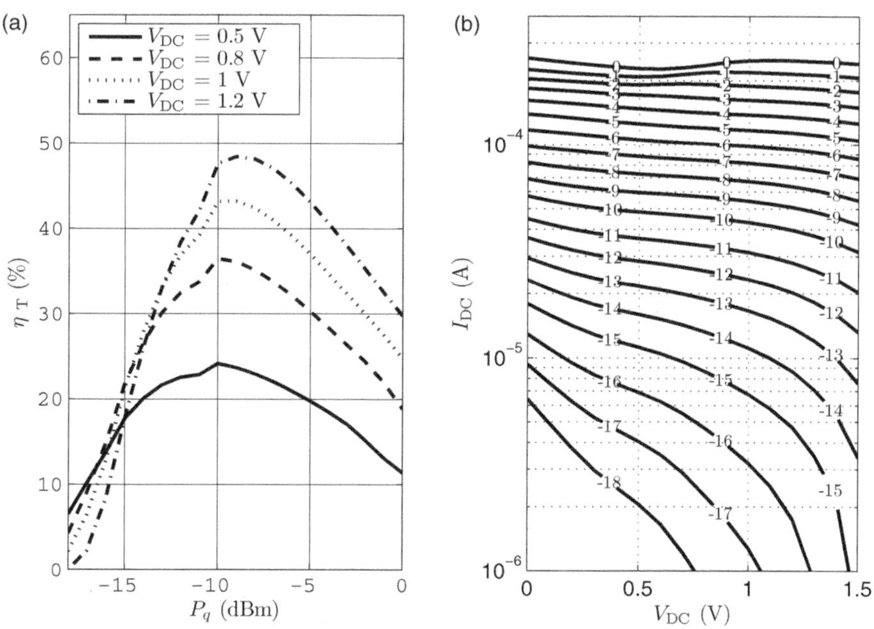

Figure 4.15 (a) Efficiency; and (b) IV-characteristic; of the integrated UHF rectifier.

10 μW (1 V and 10 μA) the efficiency is about 25%. The maximum efficiency for a DC output voltage of 1 V is about 43% and the efficiency of the rectifier at −15 dBm RF input power is still 20%.

4.3.2.2 Tag-to-reader communication

The communication principle which is used in passive UHF RFID systems is back-scattering of the incident RF power. In order to modulate the reflected power, the tag-to-reader signal is changed caused by the impedance of the RF node's analog frontend implementation. The EPC UHF Class1 Gen2 standard for RFID back-scattering communication permits two types of modulation: amplitude shift keying (ASK) and phase shift keying (PSK). Both modulate the complex impedance $Z_m = R_m + jX_m$ $(m = 1, 2)$ between two different states.

If the imaginary part of Z_m is zero, an ideal ASK modulation can be realized by a switchable resistance. Otherwise a more complex modulation circuit is required. Relating to the demonstrations in [2] and [33] the PSK modulation is more efficient than ASK, as the available power to the tag is kept almost constant during both modulation states. A PSK modulation can be achieved by a complex modulation circuit only, which is comparable to the ideal ASK modulation. The combination of PSK and ASK is used by common integrated back-scatter modulators in accordance with the fact, that most RFID transponders are using a switched resistance or a switched capacitance and that several parasitic effects get more important at high frequencies. Hence, the most reasonable modulation type depends on the available solid-state device technology.

For the tag-to-reader communication an integrated load modulator is used in passive HF RFID systems. The transformational coupling between the coil of the transponder and the coil of the reader causes a change of the transponder impedance, which can be detected by the reader.

4.3.2.3 Reader-to-tag communication

The bidirectional data transmission between the RFID transponder and the reader includes the ASK demodulator as an important circuit block. It recovers the information, which is ASK modulated on the carrier by the interrogator, and delivers the data to the digital logic. The more the reception of data is ensured at low power levels the higher is the reading range. In recognition of different modulation depths and different input power levels for UHF and HF, two demodulation blocks are used. A simplified block diagram of the ASK demodulator for the UHF mode is presented in Figure 4.16.

The input signal $Env_{In,UHF}$ of the demodulator represents the envelope of the carrier. The nonlinear resistors shift this signal and then the current source I_1 (upper path) or the current source I_2 and the capacitor C_1 (lower path) filter it. To receive the digital base band signal, a low-power comparator with hysteresis compares these two values with different time constants and slightly different voltage levels. The different dimensions of the current sources I_1 and I_2 cause the different voltage levels, which assure a robust demodulation.

Figure 4.17 shows a simplified schematic of the implemented HF demodulator, which looks very similar to the demodulator used for UHF. It is necessary to design the more

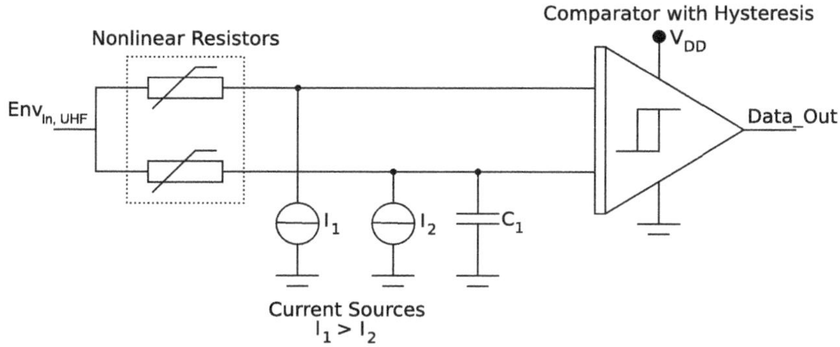

Figure 4.16 ASK data detector (UHF).

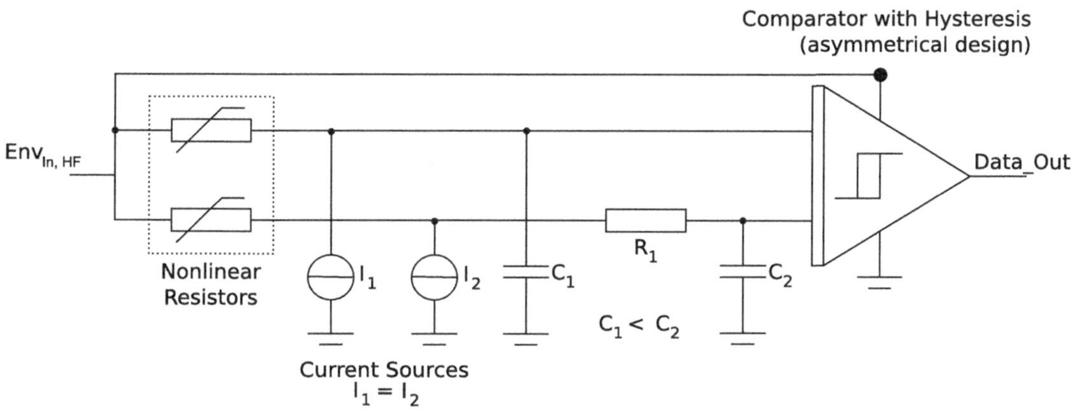

Figure 4.17 ASK data detector (HF).

complex filter structures, due to the lower modulation depth in HF (HF: [10–30%] ASK, UHF: [80–100%] ASK). The current sources are equally dimensioned to prevent different voltage levels in order to demodulate a very small change in the carrier's amplitude. Only the capacitors C_1, C_2, and the resistor R_1 adjust the time constant of the signals. To ensure a robust demodulation in HF the hysteresis behavior of the comparator is amplified by an unsymmetrical design and the use of the modulation state-dependent signal $Env_{In,HF}$ as supply voltage.

4.3.2.4 Clock generation and clock recovery

In the UHF range the RF frequency is between 860 MHz and 960 MHz. Here, clock recovery is inappropriate because the related power consumption would be too high. Therefore the implementation of a separate tag oscillator is required. Similarly to other transponder components, its architecture must meet low-voltage (typically 1.0 V to 1.5 V) and low-current ($\approx$ 1.5 μA) requirements. Different system parameters like temperature, supply voltage, phase noise, or process variations influence the related frequency value. According to the EPC protocol for UHF, frequency range variations of the clock signal must be within a $\pm$ 4% interval during the steady-state ($\pm$ 2.5% during

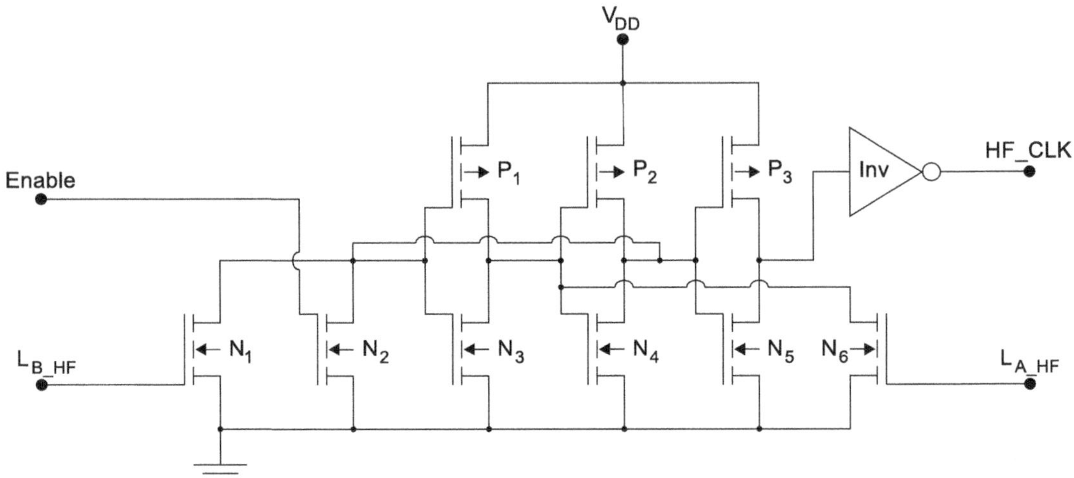

Figure 4.18 Integrated circuit which recovers the 13.56 MHz clock from the differential input signal.

the short period of back-scattering). These requirements are well below the demands of modern communication systems, so the use of simple components like RC oscillators [14] is possible. The implemented clock generator is a relaxation-type oscillator and works at 2.4 MHz.

A clock recovery circuit can be used at HF by reason of the lower carrier frequency. The integrated circuit, which recovers the 13.56 MHz clock from the differential input signal is illustrated in Figure 4.18.

Subsequently, the output signal *HF_CLK* needs to be divided by the factor 8 for use as the system clock of the transponder. There is a connection between the input pins L_{A_HF} and L_{B_HF}, and the corresponding HF input pins of the tag. To enable the clock recovery circuit the *Enable* pin is essential. The signal *Enable* is connected to $Short\_L_{B,HF}$ which specifies the ground potential and switches the clock recovery off at UHF. The transistor N_6 is switched on and the output *HF_CLK* is high during the positive half-wave on L_{A_HF} and in contrast, during the negative half-wave, N_6 is switched off and N_1 is switched on and therefore the output is low.

4.3.3 Sensor interface

4.3.3.1 Wireless sensing

Besides standard RFID signaling, wireless sensing is an important feature in the context of wireless sensor networks. Figure 4.19 shows typical application fields for this functionality. Data, such as the temperature or the pressure at the location of the transponder, can be acquired by sensor components included as an integral part of the transponder chip. These data must be digitized in the transponder by an analog-to-digital converter (ADC), in order to enable transmission back to the reader. The very limited available power is the main challenge for the ADC. As there is no need for very high conversion

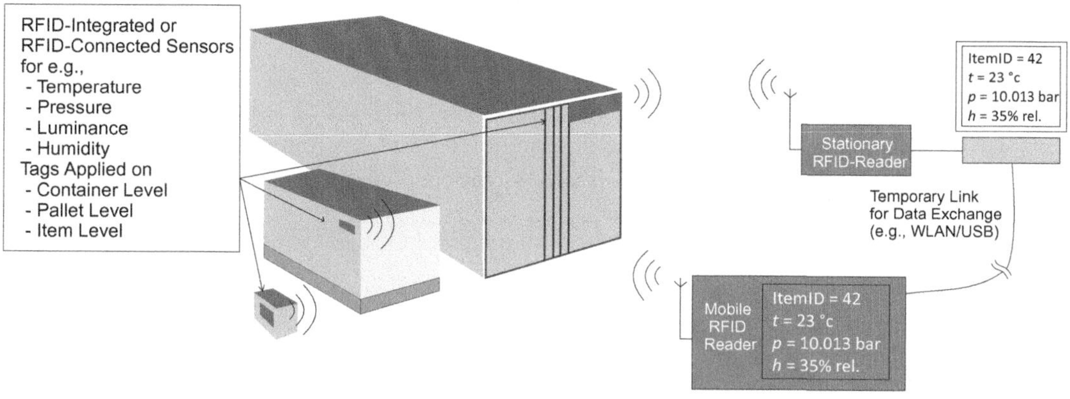

Figure 4.19 Typical application fields of wireless sensing (temperature and pressure surveillance).

rates and it consists of only a few components, a successive approximation architecture is best for data conversion. Low power techniques like time-domain comparison, charge scaling, and charge redistribution for the capacitive digital-to-analog conversion (DAC) and switching bias currents wherever possible decrease the power consumption of a successive approximation register ADC down to the microwatt-domain [43, 44]. This even decreases to several hundred nanowatts, when operating at low conversion rates. Hence, data may be collected by the sensors without needing an external power supply. The implementation of generic sensor interfaces in basic UHF transponders offers several advantages; in logistics, the current temperature of cargo goods can be observed. Another application area is facility surveillance where the humidity or luminance in storerooms must remain within fixed ranges, for example. There is no need to change the batteries after certain time periods, if all measurements can be achieved passively. Another result is the higher reliability because weak battery voltage and battery contact problems are eliminated. The combination of positioning and sensing for passive RFID tags might enable new uses. The creation of maps of sensor data becomes realizable and so, if many RFID-enabled objects are in the detection range of a reader, the objects that deviate from a specified tolerance might be located.

4.3.3.2 Timing constraint

The protocol chosen for the RFID communication in the HF and UHF mode was EPC HF draft2 [44] and EPC UHF Class1 Gen2 [16], respectively. Timeout conditions for a tag reply on a query sent by a reader are defined by these protocols. For UHF operation, this time is expressed as

$$T_1 = \max(RT\text{cal},\, 10T_{\text{pri}})T_{1\_\text{tolerance}}, \tag{4.7}$$

with

$$T_{1\_\text{tolerance}} = \begin{cases} (1-|FT|) - 2\ \mu\text{s}, & \text{minimal} \\ 1, & \text{nominal} \\ (1+|FT|) + 2\ \mu\text{s}, & \text{maximal} \end{cases} \tag{4.8}$$

where *RT*cal is the reader-to-tag calibration time,

$$T_{\text{ari}} = {}^{1}/_{BLF}, \tag{4.9}$$

is the minimum bit duration for the tag reply with *BLF* as back-scatter-link-frequency and *FT* as frequency tolerance. The maximum for *RT*cal determines the maximum timeout and *BLF* the minimum for *FT* and *BLF*. This concludes in a value of

$$T_{1\_\max} = \max\left(RT\text{cal}_{\max}, 10\frac{1}{BLF_{\min}}\right)(1 + |FT(BLF_{\min})|) + 2\ \mu\text{s}. \tag{4.10}$$

By using the minimum for *RT*cal and the maximum for *FT* and *BLF*, which results for $T_{1\_\min}$ in

$$T_{1\_\min} = \max\left(RT\text{cal}_{\min}, 10\frac{1}{BLF_{\max}}\right)(1 - |FT(BLF_{\max})|) - 2\ \mu\text{s}, \tag{4.11}$$

the minimum timeout can be calculated.

From the EPC standard, all minimum and maximum values can be determined to

$$\begin{aligned}
T\text{ari}_{\min} &= 6.25\ \mu\text{s},\\
T\text{ari}_{\max} &= 25\ \mu\text{s},\\
RT\text{cal}_{\min} &= 2.5 \cdot T\text{ari}_{\min} = 15.625\ \mu\text{s},\\
RT\text{cal}_{\max} &= 3 \cdot T\text{ari}_{\max} = 75\ \mu\text{s},\\
BLF_{\min} &= 40\ \text{kHz},\\
BLF_{\max} &= 640\ \text{kHz},\\
FT(BLF_{\min}) &= \pm 4\%,\\
FT(BLF_{\max}) &= \pm 15\%.
\end{aligned} \tag{4.12}$$

By applying these values to (4.8) and (4.9), the maximum and minimum timeouts result in

$$\begin{aligned}
T_{1_\max} &= \max\left(75\ \mu\text{s}, \frac{10}{40\ \text{kHz}}\right)(1 + 0.04) + 2\ \mu\text{s}\\
&= 250\ \mu\text{s} \cdot 1.04 + 2\ \mu\text{s} = 262\ \mu\text{s},
\end{aligned} \tag{4.13}$$

$$\begin{aligned}
T_{1_\min} &= \max\left(15.625\ \mu\text{s}, \frac{10}{640\ \text{kHz}}\right)(1 - 0.15) - 2\ \mu\text{s}\\
&= 15.625\ \mu\text{s} \cdot 0.85 - 2\ \mu\text{s} = 11.28\ \mu\text{s}.
\end{aligned} \tag{4.14}$$

The chosen data rate of the reader–tag–reader communication has an impact on the timing constraints for sensor data acquisition. The available time for one A/D conversion of the sensor interface including all gate delays of the RFID tag's control logic must be between 11.28 and 262 μs. Thus, the conversion rate of the ADC has to be approximately between 5 and 100 ksample/s. The faster the chosen conversion rate, the more sensor transponders of a sensor field can be interrogated within a certain time span. Hence, a conversion rate of around 100 ksample/s is chosen for the ADC.

Figure 4.20 shows the architecture of the sensor interface. On the left the ultra-low current temperature sensor is illustrated, which is connected to the first of four multiplexer

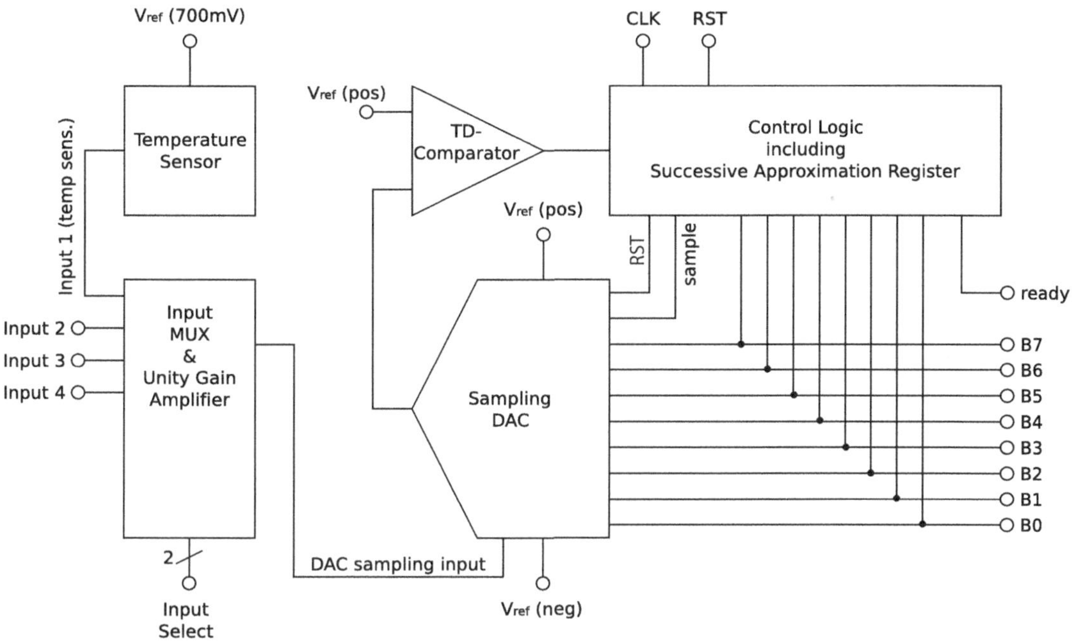

Figure 4.20 Sensor interface including an input multiplexer, a temperature sensor, and an eight-bit successive approximation ADC with DAC-integrated sampling, and a time-domain comparator.

inputs. The sensor is implemented using the temperature effects on a diode-connected NMOS transistor. This transistor is supplied with a constant current. This current is generated using the ZTC bias point technique mentioned in [45]. The unity gain buffer integrated into the multiplexer feeds the capacitive sampling DAC array of the ADC. The CMOS logic of the SAR-ADC controls the DAC and its analog output is linked to the time-domain comparator, representing the digitizing element of the sensor interface. Relating to [43], the current demand is kept low by the time-domain technique for the comparator. The control logic is designed with standard cells of the 0.13 μm CMOS process.

4.3.3.3 Time-domain comparator

A comparator is required in order to evaluate the DAC output signal in relation to the reference voltage and digitize each one-bit result before its storage in the successive approximation register. Figure 4.21 shows the time-domain comparator circuit (TDC).

Two identical and symmetrically laid out circuit parts convert the input voltage signals V_{in1} and V_{in2} to time-domain signals. It has been shown that the speed limit for the compare signal of the circuit is given by

$$f_{\text{compare}} < \frac{I_{\text{charge}}}{2CV_{\text{t,n}}}, \tag{4.15}$$

with I_{charge} and C as depicted in Figure 4.21 and $V_{\text{t,n}}$ as threshold voltage of NMOS transistor $N_{1.1}$ and $N_{2.1}$, respectively. By adding the dummy transistors $P_{1.1}$ and $P_{2.1}$

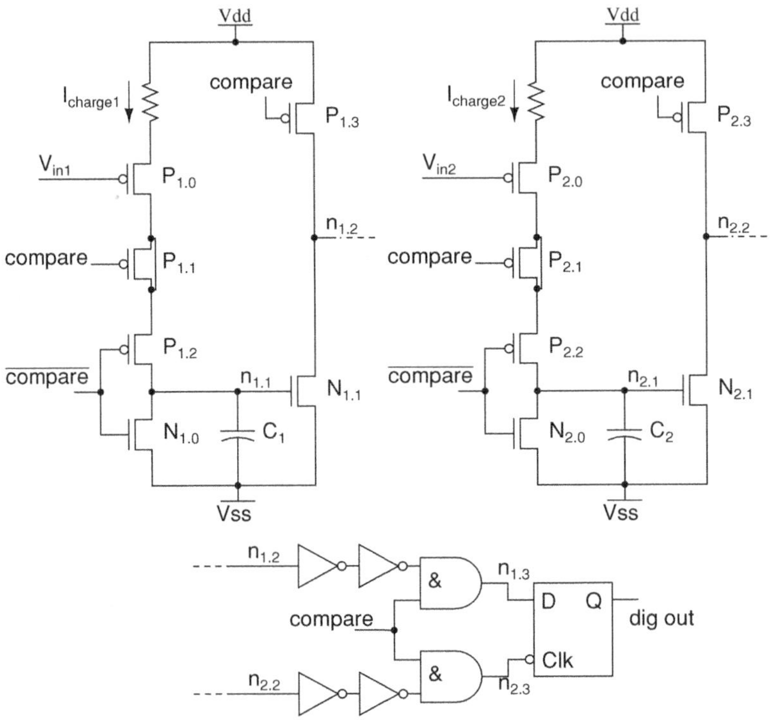

Figure 4.21 Time-domain comparator.

a further optimization is reached in order to minimize the clock feed through of the compare signal. This ensures a sensitivity of better than 1 mV which is sufficient for an LSB value of 1.96 mV.

4.3.3.4 Digital-to-analog converter

A capacitive DAC (CDAC) is used by the SAR-ADC for digital-to-analog conversion. The implementation is based on the concepts of charge redistribution [46] and charge scaling.

During one data conversion cycle the 8-bit DAC illustrated in Figure 4.22 performs the following steps: In phase φ_1 the rst- and sample-switches are closed and b0–b7 are turned to the V_{in}-path. This way, the nodes n_{LSB} and n_{MSB} are initialized to the potential V_{ref_pos} and all lower capacitance plates to V_{in}. The phase φ_2 is characterized by the open rst-switches, conserving a charge proportional to $(V_{ref\_pos} - V_{in})$ on the upper plates. If the sample-switch is set in phase φ_3, the potential at node n_{MSB} is calculated to $(2V_{ref\_pos} - V_{in})$. When switching sequentially b7–b0 between V_{ref_pos} and V_{ref_neg} in phase φ_4 (with V_{ref_neg} being V_{SS}), the output voltage V_{out} is described by the expression

$$V_{out} = 2V_{ref\_pos} - V_{in} - V_{ref\_pos}\left(\frac{1}{16}\frac{\Sigma_{i=0}^{3} b_i 2^i + 1}{2^4} + \frac{\Sigma_{i=0}^{3} b_4 + b_i 2^i}{2^4}\right), \quad (4.16)$$

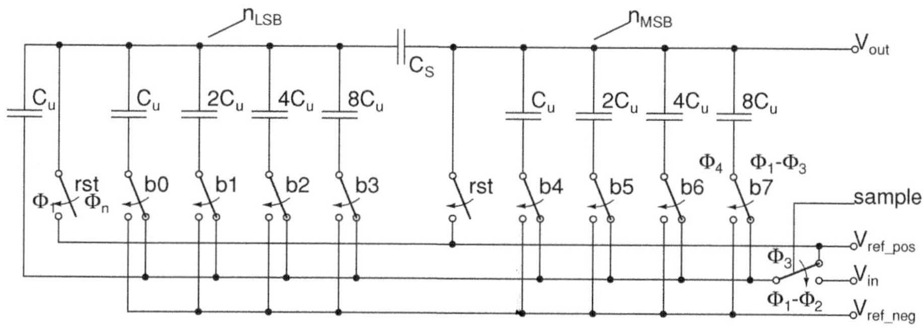

Figure 4.22 Capacitive sampling DAC using charge scaling.

where $(2V_{\text{ref\_pos}} - V_{\text{in}})$ represents the sample offset, the previously conserved charge on the top plates. Thus the CDAC is also used as a sample and hold circuit. The voltage step determines the minimum S&H capacity for an ADC corresponding to the LSB. This ADC is unipolar 8-bit, works with a reference voltage of 500 mV, and has an LSB voltage of 1.96 mV. The sampling noise voltage of a unipolar ADC is given by

$$v_{\text{n,sample}} = \sqrt{\frac{kT}{C_{\text{sample}}}}. \tag{4.17}$$

The complete DAC capacitance (depicted in Figure 4.22) is 31 C_{u} which equates to C_{sample}. As the unity capacitance C_{u} is 59.3 fF including all parasitic components, the resulting sampling noise voltage is 48 μV, which is well below the LSB value of 1.96 mV. This shows that the sampling noise for CDACs only has to be considered for ADC designs above 10-bit resolution. The reason for the unity capacitance being much bigger than needed with respect to the sampling noise is the matching requirements between the single unity capacitances themselves and related to the scaling capacitance. It is shown in [47] and [48] that the scaling capacitance in a capacitive charge-scaling DAC should be dimensioned according to

$$C_{\text{S}} = C_{\text{u}} \frac{2^{\frac{N}{2}}}{2^{\frac{N}{2}} - 1} = \frac{\text{sum}(C_{\text{u}}, \text{LSB})}{\text{sum}(C_{\text{u}}, \text{MSB})}, \tag{4.18}$$

where N is the resolution of the DAC. The numerator is the sum of all unity capacitances in the LSB-array and the denominator is the sum of all unity capacitances in the MSB-array. For a dual 4-bit charge-scaling CDAC, a C_{S} value of 16/15 C_{u} would be necessary.

The ADC consumes, depending on temperature and sample rate, 525 nA to 1.65 μA. High leakage conditions are shown by measurements at supply voltages above 1 V and at higher temperatures (>80 °C).

In order to verify the ADC, a measurement setup for both the static and dynamic characterization was applied. Figure 4.23 shows the ADC test chip. With a size of 0.042 mm^2 the homogeneous field of unity capacitances is most of the active area, the scaling and terminating capacitances (C_s and C_t) are a part of this field.

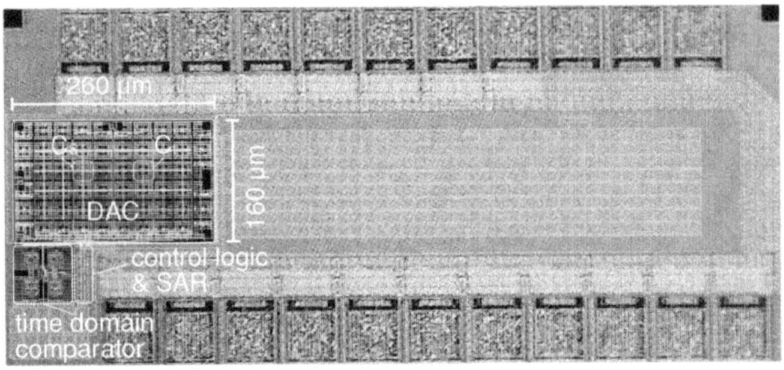

Figure 4.23 SAR-ADC test chip. The CDAC array is arranged in a dual 4-b common centroid layout.

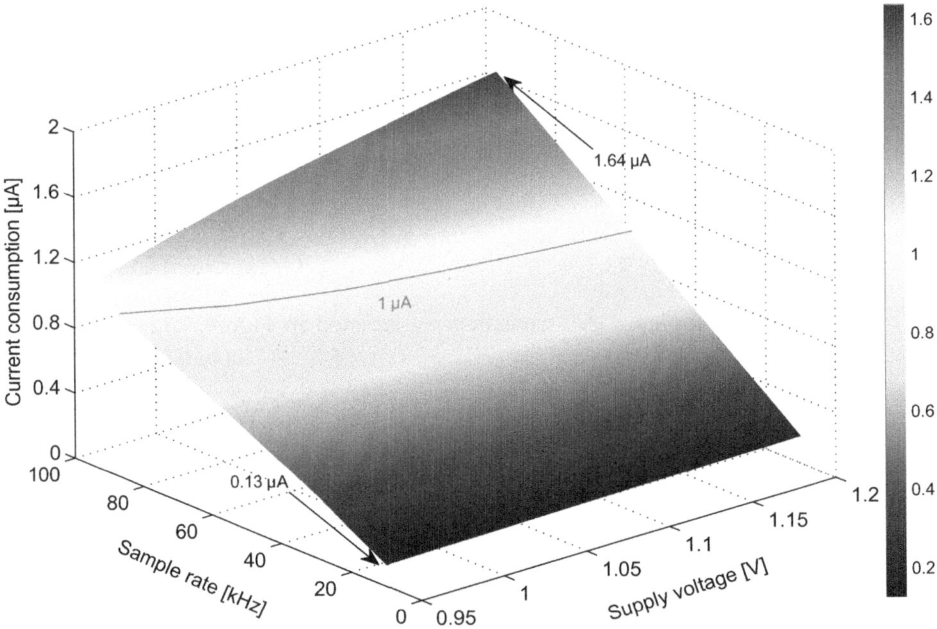

Figure 4.24 ADC current consumption over supply voltage and sample frequency. The 1 μA line marks the design target also showing the nonlinear relation to the supply voltage. The lowest current consumption at 10 kS/s is 130 nA and the highest recorded consumption at 100 kS/s is 1.64 μA.

In Figure 4.24, the current demand of the ADC is plotted against supply voltage and sample rate. Clearly, a linear dependency of the current from the clock frequency can be observed, whereas the dependency from the supply voltage is nonlinear. This identifies the SAR as the most current-consuming part of the ADC; the basic equation

$$P_{\text{dig}} = \alpha C_{\text{L}} V_{\text{DD}}^2 f_{\text{CLK}} \tag{4.19}$$

states these two dependencies for digital CMOS circuits. The 1 μA line marks the design target. The measured values spread from 130 nA to 1.64 μA. In particular, measurements

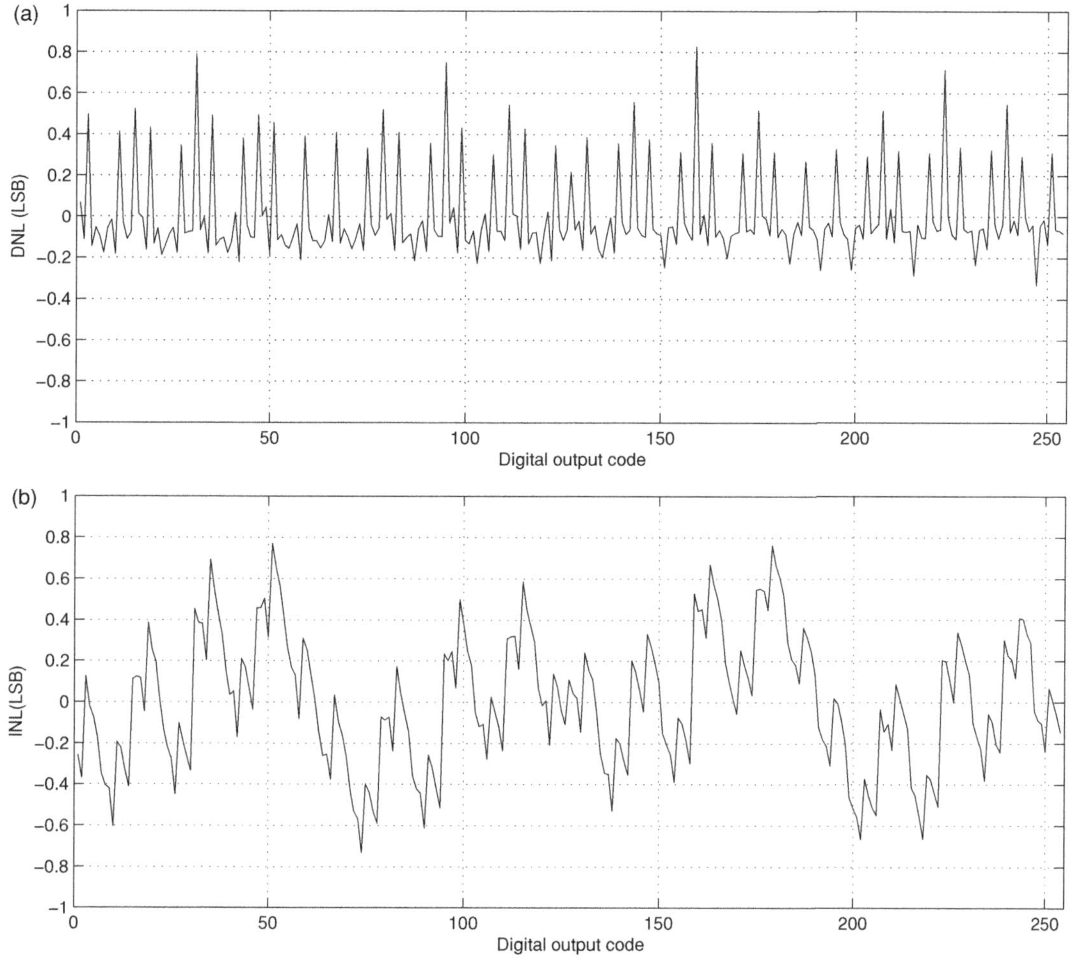

Figure 4.25 (a) DNL; and (b) INL; of the ADC acquired with the histogram method.

at supply voltages above 1 V and at higher temperatures (>80 °C) show high leakage conditions.

Using the histogram method at a very low input frequency ($f_{\text{in}} = 22$ Hz), the differential (DNL) and integral nonlinearity (INL) can be acquired. In Figure 4.25(a) the DNL over the digital output code is plotted. The values are between −0.36 and +0.81 LSB. Figure 4.25(b) plots the corresponding INL with values between −0.75 and +0.75 LSB.

Dynamic tests show a stable performance for different input- and sample-frequencies. Signal-to-noise-and-distortion ratios (SNDR) of 46 and 41.9 dB are reached at a sample rate of 40 kS/s for low- and Nyquist-frequency input signals, resulting in an effective resolution of 7.36 and 6.68 bit, respectively. In both cases, the current is less than 700 nA. The power demand diminishes but the harmonic distortions increase when lowering the supply voltage. At 0.9 V (leaving all other operating conditions as is) the

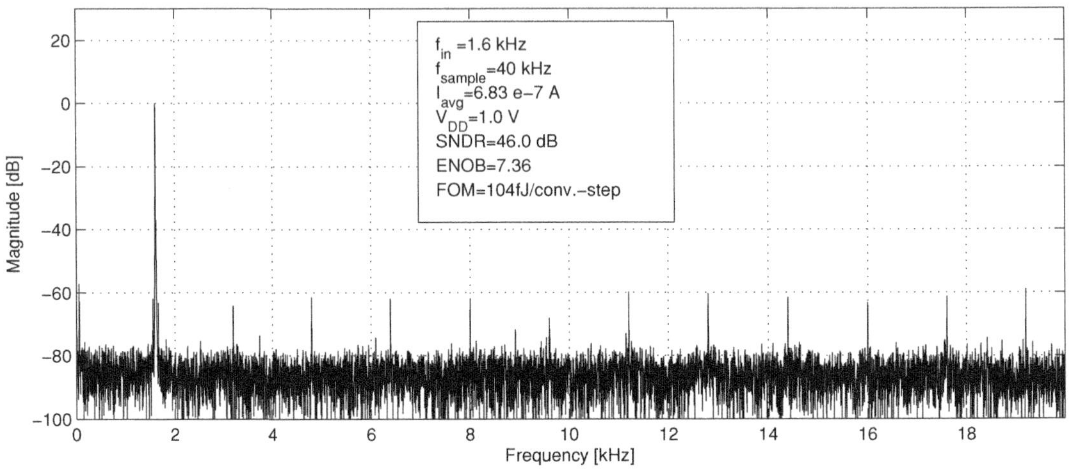

Figure 4.26 ADC output spectrum for a low-frequency input signal at 1.0 V supply.

current consumption drops to 525 nA and a still acceptable effective number of bits (ENOB) of 7.23 is reached. According to the commonly used ADC FOM definition

$$\mathrm{FOM} = \frac{P_{\mathrm{DC}}}{2^{\mathrm{ENOB}} f_{\mathrm{sample}}}, \tag{4.20}$$

a value of 79 fJ/conversion-step has been achieved with this design.

In Figure 4.26 the ADC output spectrum for a low-frequency input signal is shown.

Due to the lack of an anti-aliasing filter at the input of the ADC, the values for SNDR and ENOB decrease for higher input frequencies. An anti-aliasing filter is not necessarily needed, as only input signals near DC are expected for the different sensor inputs.

4.3.3.5 Temperature sensor

The integrated temperature sensor is based on the zero temperature coefficient (ZTC) bias point of MOS transistors, as described by Filanovsky and Lim in [49] and later shown by Zhai *et al.* in [50]. At this bias point, temperature effects on electron mobility μ and threshold voltage V_{T} compensate for each other. This results in a constant drain current I_{D} over temperature, as illustrated in Figure 4.27.

The circuit shown in Figure 4.27 requires a reference voltage V_{ref} to generate a reference current I_{ref} through N_0. By using a current mirror (P_0,P_1), an equivalent constant current I_{const} is generated. As the current is stable over temperature, the voltage drop V_{GS} over N_1 is given by [49]

$$V_{\mathrm{temp}} = V_{\mathrm{GS}} = V_{\mathrm{GSF}} + \alpha_{\mathrm{VT}} T \left(1 - \sqrt{\frac{I_{\mathrm{D}}}{I_{\mathrm{DF}}}}\right), \tag{4.21}$$

where V_{GSF} and I_{DF} ($= I_{\mathrm{const}}$) are the parameters of the common intercept point of both temperature-dependent characteristics (μ & V_{T}) and α_{VT} is the threshold voltage temperature coefficient. V_{temp} is a linear function of the temperature T.

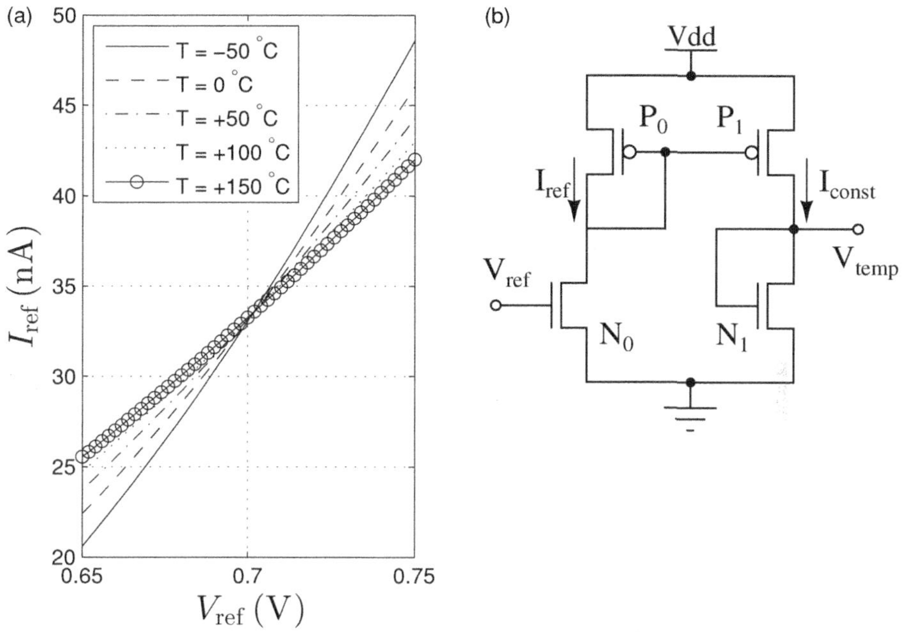

Figure 4.27 Temperature sensor: (a) illustration of the ZTC bias point of the transistor; (b) schematic of the sensor.

The solution presented in [50] needs an additional voltage stabilizing circuit for V_{dd} (e.g. a low drop out regulator), due to the fact that the current through a MOS FET mainly depends on the gate source voltage. Within a passively supplied RFID, the supply voltage V_{dd} is highly fluctuating. Hence, instead of using the ZTC bias point of a PMOS transistor, the reference voltage is generated independently by a current through an NMOS transistor.

4.3.4 Distance measurement

4.3.4.1 Local positioning

Positioning systems can measure the distance to a desired target based on different principles, for example received signal strength indication (RSSI) and round-trip time of flight (RToF). With the first one, the received field strength of the related target is evaluated and with the latter, the run-time of a signal from the reader to the target and back is used. The distance between reader and transponder information could be evaluated using a frequency-modulated continuous wave (FMCW) radar by using the technique of modulated back-scattering [51–53]. This is illustrated in Figure 4.28. Standard FMCW systems calculate the distance between the radar station and the related (passive) target based on the run-time of the FMCW signal. With passive RFID transponders as the desired target, the transmitted FMCW signal from the reader is back-scattered by the reflection coefficient of the transponder antenna, as well as by any other (undesired)

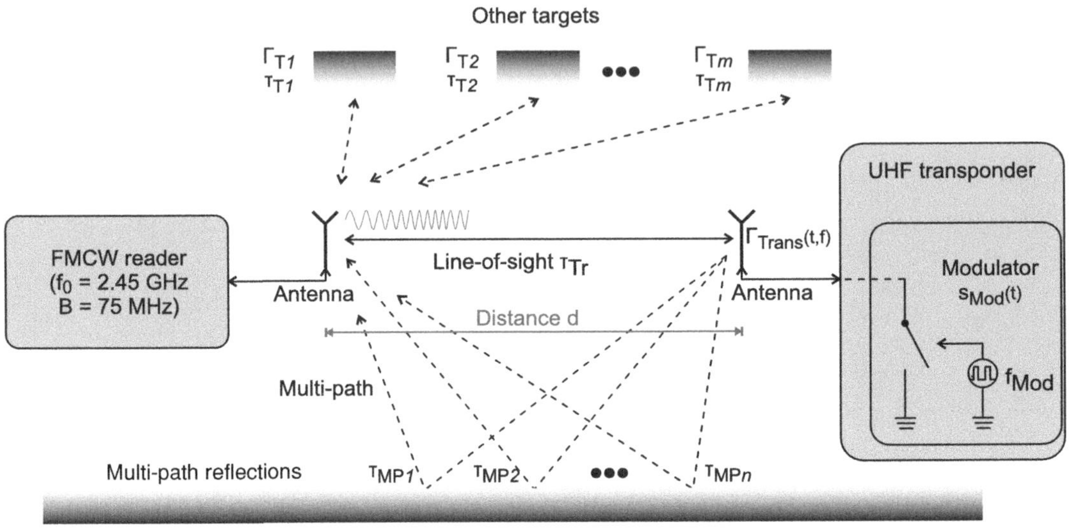

Figure 4.28 Distance measurement with frequency-modulated continuous wave (FMCW) radar.

target within its operating range. Therefore, the back-scatter modulator of the transponder is continuously switched on and off by a constant modulation signal f_{mod} and the reflection coefficient is varied between two states. This causes a frequency shift of the back-scattered FMCW signal (modulated back-scattering). This way, it is possible to distinguish between the back-scattered transponder signal and the signal of any other target close to the FMCW reader that is not modulated. Here, the design of the reader antenna is a trade-off between its detectable operating range and the handling of multipath effects. However, the possible resolution of an FMCW radar is dependent on the bandwidth of the FMCW signal. Because of the smaller allowed bandwidth at UHF frequencies given by regulation, higher frequency ranges at 2.45 GHz or 5.8 GHz should be used for this technique.

Positioning and distance measurement can be used for enabling the tracking of persons and objects as well. For example, a container with a UHF transponder affixed can easily be detected in storerooms. Access control with RFID is often used in security-relevant departments of a company.

Using a tag with the additional possibility for distance measurement, the current whereabouts of a person who stays in that area could be tracked. Furthermore, position tracking can be applied for comfort functionalities in modern automotive applications: this feature can activate several functions of a car when the driver is moving to it. Examples of other positioning systems are given in [54, 55].

The localization principle of an RFID tag is based on a FMCW radar system with modulated reflection. A continuous wave signal with a linear frequency modulation is transmitted by the FMCW reader. The signal starts at a lower frequency f_0 and, within the modulation interval, is linearly increased in frequency to an upper level of $f_0 + B$ (Figure 4.29).

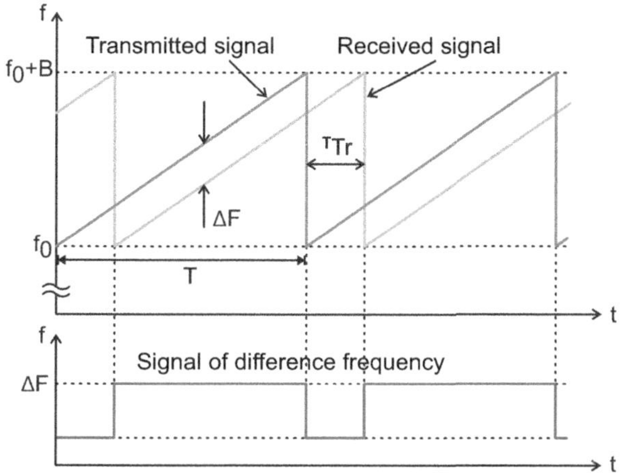

Figure 4.29 Frequency curve of transmitted and received FMCW signal.

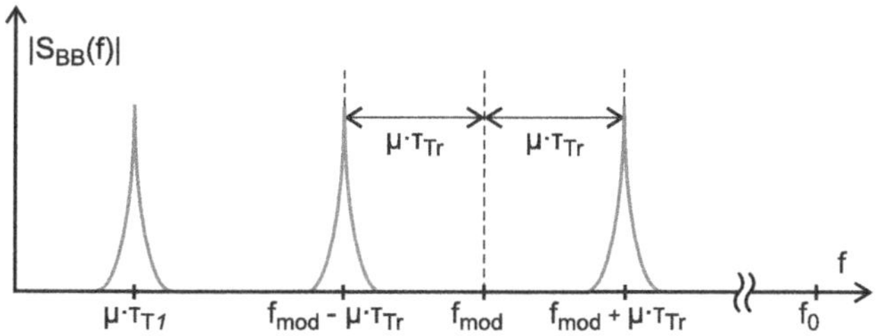

Figure 4.30 Base band spectrum with modulation.

The signal from the FMCW reader is reflected at the antenna of the RFID transponder at the distance d and at any other target close to the reader. Therefore, after activating the distance measurement with the related custom command, the reflection coefficient of the transponder antenna is characteristically modulated. Thereby it can be distinguished from false targets in the operating range of the reader.

A divide-by-8 frequency divider derives the modulation frequency of about 300 kHz from the internal clock generator of the RFID tag. In order to use its whole bandwidth B, this modulation should be performed for the duration of a modulation interval T of the FMCW ramp. For a good resolution of the distance measurement [56], this is very important. When the back-scattered signal of the transponder arrives at the antenna of the reader, it is delayed by the time of flight τ_{Tr}. Then a mixing of this signal with the transmitted signal [57] takes place.

At the output of the mixer for the duration of one ramp interval T the signal is then digitized and the further data processing is done in the digital domain. It consists of two mainlobes around a center frequency f_{mod} (Figure 4.30). Its value is determined by the modulation frequency f_{mod} of the transponder. The distance d between reader and transponder and the known ramp frequency define the distance Δf of the two mainlobes.

4.3.4.2 Modulated back-scattering with passive transponders

An alternative approach for implementing a distance measurement with the basic back-scatter modulator and the EPC standard is shown in [58]. Here, the tag preamble or a user memory filled with zeros is used to generate the rectangular modulation signal for continuous modulation instead of deriving it from the transponder oscillator.

4.3.4.3 Measurement setup for distance measurement

Finally, distance measurement using the technique of modulated back-scattering is evaluated. The whole algorithm for the distance measurement works as already described in [59, 60]. The accuracy of the distance measurement depends on the available power of the transponder, the transponder antenna used, and the reflector used [59].

Within this setup, distance measurement is intended as an optional feature for passive low-cost transponders without the need for additional high-performance components for localization.

When using the back-scatter modulator of the passive transponder also for distance measurement, only a small extension in the digital part of the transponder circuit (clock divider, implementation of the related custom command) is necessary. It does not noticeably increase the whole power consumption and chip area. Here, the distance information can be used for comfort functionalities.

Figure 4.31 shows the whole measurement setup for distance measurement using this multi-frequency RFID system. The two readers and the transponder are arranged on a positioning stage. The transponder chip includes all necessary components to be compliant to the EPC protocol. The design of the digital part is described briefly in [60, 36]. The reader antennas [61, 62] are attached close to each other.

4.3.4.4 User defined EPC custom command for distance measurement

The initial bit sequence 11100000 (0xE0) is fixed by the EPC protocol to recognize a transmitted command as a defined custom command [16]. The following four bits determine the “Locate” command. The command parameter sets the intended duration of the continuous modulation. Within this setup, it can be adapted by the user from 0.6 ms to 1.4 ms in 16 steps.

After the custom command for the distance measurement is sent by the UHF reader, the transponder performs a continuous modulation with its back-scatter modulator. The distance measurement is then carried out with the FMCW reader. The algorithm for distance calculation is described in [60]. The reflection coefficient of the transponder antenna has different values at 866 MHz and 2.45 GHz. At UHF, most of the power is absorbed by the transponder chip, while at 2.45 GHz most of the power is reflected back to the FMCW reader to increase the possible operating range for the distance measurement.

Therefore, performing only a PSK modulation between the two modulation states of the back-scatter modulator would be best at 2.45 GHz. The design of the antenna is described in [63, 64]. The antenna structure is fabricated on a PCB substrate in the

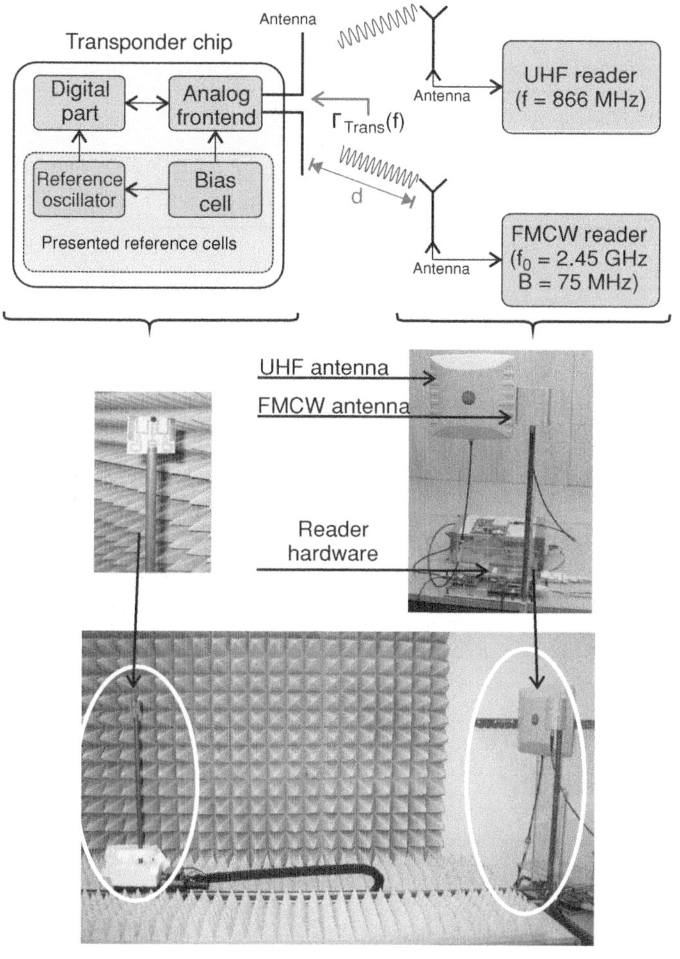

Figure 4.31 Distance measurement with whole RFID system.

ID-1 format (85.60 mm 53.98 mm) [65]. Its impedance is sufficiently matched to the RF port of the transponder chip so a matching network is not used between the transponder antenna and the chip.

The EPC code of our multi-standard transponder could be read with a commercial reader setup [61, 66] at up to about 8 m with a radiated power of 33 dB. Unfortunately, commercial UHF readers cannot be used for the distance measurement because up to now the custom command option is still not implemented in any available reader. Therefore, a custom-made UHF reader based on an FPGA board was used on which the EPC protocol is programmed. It allows convenient implementation of the user-defined custom command. Measurements have shown a lower sensitivity of the RF frontend included in the FPGA solution than the commercial setup. It leads to a limited detection range of about 2.0 m, but the comparative measurement with the commercial reader has shown an extended operating range of the transponder.

Table 4.2 Custom commands for the sensor extension functionality.

Code	Parameter 1 (source)	Parameter 2 (function)
0xE0EX	X = 0b00xx → src 1	X = 0bxx00 → single
	X = 0b01xx → src 2	X = 0bxx01 → continuous
	X = 0b10xx → src 3	X = 0bxx10 → reserved
	X = 0b11xx → src 4	X = 0bxx11 → reserved

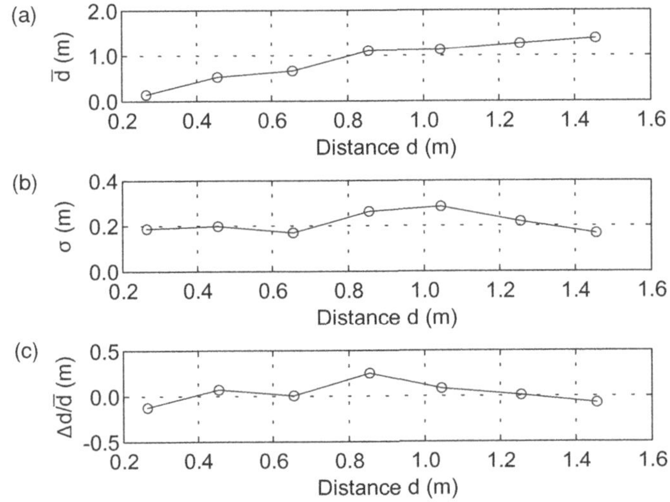

Figure 4.32 Measurement results with RFID system: (a) Mean distance $\bar{d}$; (b) Standard deviation σ of distance; (c) Absolute error Δ d/$\bar{d}$ of distance calculation.

Figure 4.32 demonstrates the measurement results with the RFID system. Ten measurements were performed for every distance value. Figure 4.32(a) shows the measured mean distance value, Figure 4.32(b) shows the related standard deviation, and Figure 4.32(c) shows the absolute error.

By using custom commands declared in the RFID protocol standard *EPC UHF class1 gen2*, the data conversion is obtained along with the localization. The range from 0xE000 to 0xE0FF is reserved for these commands, resulting in 256 possibilities to trigger actions not described by the standard.

Table 4.2 illustrates the custom commands used for sensor data acquisition within the UHF RFID test system. They were added to the finite-state machine of the transponder, which already contains all mandatory EPC commands. If a single sensor reading is taken from source3, the corresponding custom command, sent to the RFID Transponder, is 0xE0E8. After transmission of the command code within an established EPC session (see Table 4.3), the ADC starts a conversion on the specified input source and the acquired sensor data are transmitted back to the UHF RFID reader as shown in Figure 4.33.

The actual sensor reply consists of four bytes: ADC data, ADC calibration data, sensor source information, and a counter. The ADC calibration data byte is reserved for

Table 4.3 User-defined EPC custom command for distance measurement.

Command	Code	Parameter
Locate	0*xE*0*FX*	$X = 0b0000 .. 0b1111$ ($X = 0b1000$: 1ms)

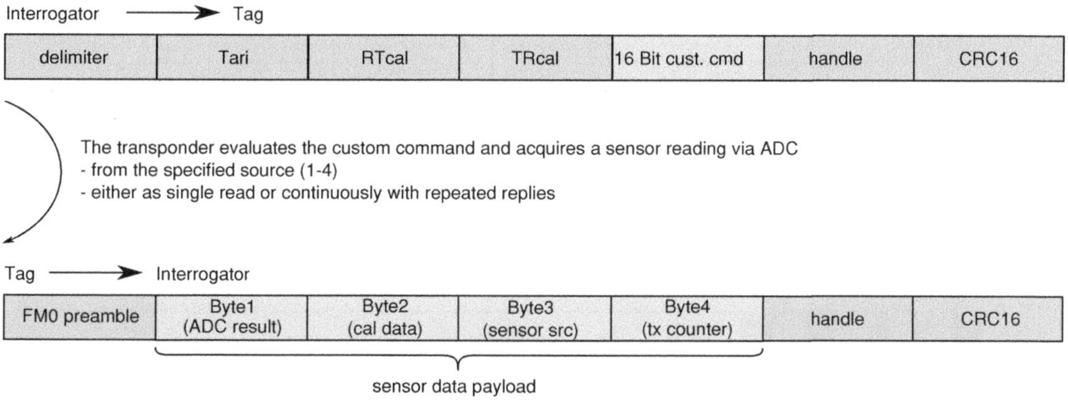

Figure 4.33 A sensor data acquisition is triggered with a standard compliant custom command and the result is transmitted in a compatible encoding but with a nonstandard frame structure.

future use and was set to a constant (0x55) at the current state. The sensor information links the ADC data received by the reader to a specific sensor source. Depending on the command mode (single or continuous), the acquisition and data transmission take place only once or infinitely until the supplying field is switched off. Using the latter mode requires no further interaction from the reader device. In this mode the counter increases with each conversion, providing a timestamp-like identification of the ADC data. After 256 acquisitions the counter overflows and starts at 0x00 again.

4.3.5 CMOS RFID system tests

UHF range measurements with a commercial UHF RFID reader at 2 W transmit power and a sensor-transponder showed operational distances of up to 8 m. Additional sensor measurements using our custom command-capable UHF reader at the lower transmit power level resulted in similar distances for sensor and non-sensor operation. This shows that the ADC – turned off for non-sensor operation – consumes less energy as measured in the stand-alone setup.

In Figure 4.34 a trace of a sensor challenge and reply recorded with the Agilent VSA Software using the RFID demod option is depicted. The sequence up to the *Handle* and an inventory round of the EPC protocol are absolutely compliant. Requesting a single data conversion of sensor source “0” the custom command is sent in data block 7. An

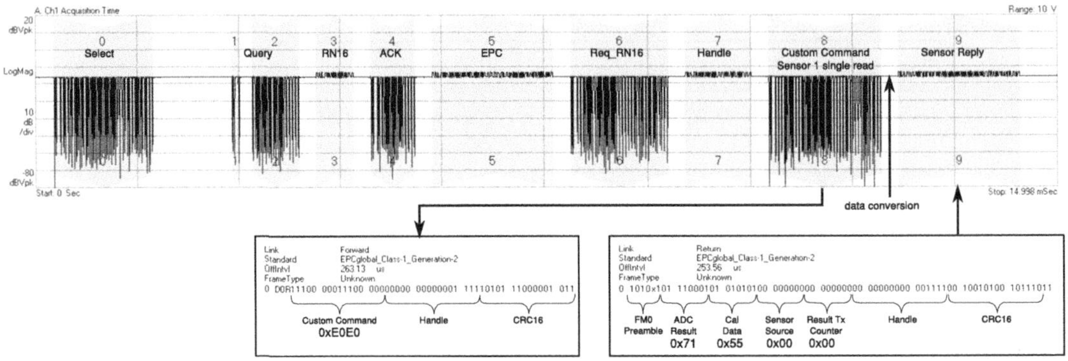

Figure 4.34 Trace of a complete EPC-compliant communication with sensor custom command. A single sensor value is requested.

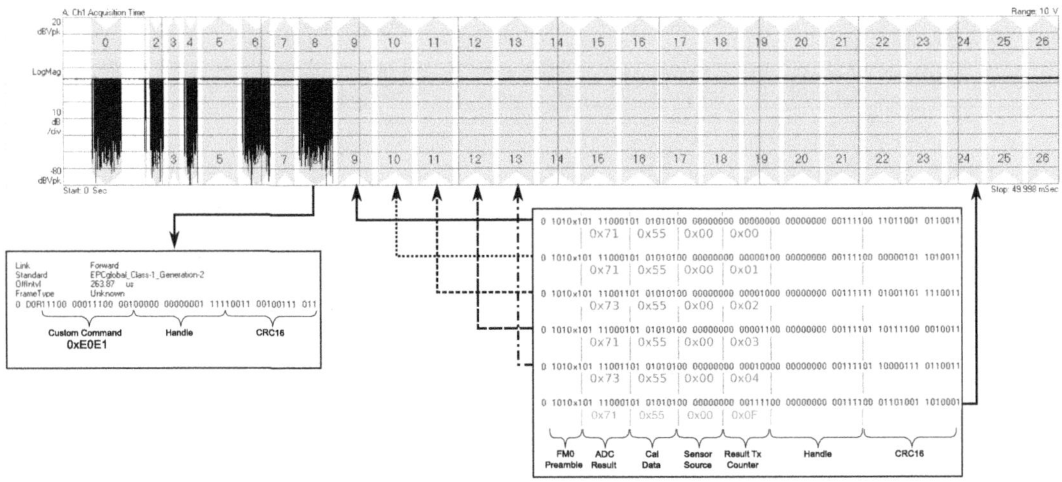

Figure 4.35 Trace of complete EPC-compliant communication with continuous sensor value request and replies.

ADC result of 0x71, calibration data (0x55), and the sensor source information (0x00) are the content of the reply, which is sent directly after conversion (data block 8).

The value 0x71 corresponds to an environment temperature of approximately 22 °C, as sensor source one (0x00) is the internal temperature sensor. Figure 4.35 presents an example of a continuous-mode request. The EPC sequence is the same as explained for Figure 4.34. A continuous data conversion and transmission (0xE0E1) is now requested by the sensor custom command. Immediately after command reception, the transponder starts transmitting back sensor replies with short A/D conversion phases in between. Valid ADC results are decoded in combination with the (previously-mentioned) transmission counter, by using the VSA software. Even if data transmissions take longer, the ADC result fluctuates only by two LSB (0x71–0x73). Reply number 16 (block 24) still shows the same stable result (0x71).

4.4 Comparison of SAW and CMOS RFID sensors

The state-of-the-art of both CMOS- and SAW-based passive RFID sensors has been discussed. The principal difference between passive CMOS and SAW RFID-based systems is the higher dynamic range of SAW-based systems. A SAW transponder is a linear device and so the system dynamic range is determined by the interrogating transceiver. Otherwise the transponder, which behaves like a bandpass nonlinearity, limits the dynamic range of CMOS-based RFID systems. Therefore, when strong interferers hit the transponder, CMOS-based RFID systems suffer from desensitization effects. Additionally, the required minimum DC voltage gives the lower end of the dynamic range of CMOS systems. The required signal-to-noise ratio at the interrogating reader's receiver determines the sensitivity of SAW systems. Passive SAW-based RFID systems are therefore noise-limited, whereas passive CMOS-based RFID systems are "voltage-limited." Considering the existing worldwide frequency and power regulations, a higher read range than CMOS systems is shown by practical SAW systems. The fact that the programming of the SAW transponders has to be done during fabrication gives one major drawback of SAW-based RFID systems. The crucial difference between SAW transponders and CMOS transponders is the programming: the SAW transponders are one-time programmable devices and in contrast CMOS transponders can be programmed during operation. Due to the possibility of making CMOS transponders smart, another advantage of them is the possibility to run communication protocols or to implement multi-transponder systems by using subcarrier techniques. Moreover, the encryption of CMOS transponders is possible. Of course, these are just a few examples of differences between SAW- and CMOS-based passive RFID systems focusing on the different nature of the transponder type. However, both types of transponder are able to perform the sensing of physical, chemical, and biological quantities either directly on-chip or by incorporating interfaces to external sensors.

References

[1] R. Martinez-Catala and J. Barrett, "A modular wireless sensor platform with fully integrated battery," *IEEE Transactions on Components and Packaging Technologies*, **32**, (*3*), 617–626, 2009.

[2] U. Karthaus and M. Fischer, "Fully integrated passive UHF RFID transponder IC with 16.7-μW minimum RF input power," *IEEE Journal of Solid State Circuits*, **38**, (*10*), 1602–1608, 2003.

[3] K. Seemann and R. Weigel, "The system design of integrated passive transponder devices," in *Microwave Conference, APMC 2006*, Asia-Pacific, Yokohama, pp. 1114–1117, Dec. 2006.

[4] "ERC recommendation 70–03 (relating to the use of short range devices (SRD))," European Communications Office, October 2010. [Online]. Available: http://www.erodocdb.dk/Docs/doc98/o_cial/pdf/REC7003E.PDF.

[5] NXP, "UCODE G2iL and G2iL+ (SL3S1203 1213 product short data sheet)," Oct. 2010. [Online]. Available: http://www.nxp.com.

[6] J. Essel, D. Brenk, J. Heidrich, G. Holweg, and R. Weigel, "Design of a highly efficient CMOS rectifier for passive communication systems," in *Proc. Asia Pacific Microwave Conf. APMC 2010*, pp. 896–899, 2010.

[7] J. Essel, D. Brenk, J. Heidrich, R. Weigel, G. Hofer, and G. Holweg, "A low-power multi-standard frontend for wireless sensor networks," in *Proc. IEEE Sensors*, pp. 1124–1128, 2010.

[8] N. Cho, S.-J. Song, S. Kim, S. Kim, and H.-J. Yoo, "A 5.1 μW UHF RFID tag chip integrated with sensors for wireless environmental monitoring," in *Proc. 31st European Solid-State Circuits Conf. ESSCIRC*, pp. 279–282, 2005.

[9] M. van Elzakker, E. van Tuijl, P. Geraedts, *et al.*, "A 10-bit charge-redistribution ADC consuming 1.9 μW at 1 MS/s," *IEEE Journal of Solid-State Circuits*, **45**, (*5*), 1007–1015, 2010.

[10] J.-P. Curti, M. Declercq, C. Dehollain, and N. Joehl, *Design and Optimization of Passive UHF RFID Systems*, Springer, 2007.

[11] S. A. Bhalerao, A. V. Chaudhary, and R. M. Patrikar, "A CMOS low voltage charge pump," in *20th International Conference on VLSI Design, 2007*, held jointly with *6th International Conference on Embedded Systems*, IEEE, pp. 941–946, 2007.

[12] J. Essel, D. Brenk, J. Heidrich, and R. Weigel, "A highly efficient UHF RFID frontend approach," in *Proc. IEEE MTT-S International Microwave Workshop on Wireless Sensing, Local Positioning, and RFID, IMWS 2009*, pp. 1–4, 2009.

[13] J. Essel, D. Brenk, J. Heidrich, *et al.*, "Highly efficient multistandard RFIDS enabling passive wireless sensing," in *Proc. Asia Pacific Microwave Conference APMC*, pp. 2228–2231, 7–10 Dec. 2009.

[14] J. Heidrich, D. Brenk, J. Essel, *et al.*, "Low voltage reference cells for UHF transponders with advanced features," in *Proc. IEEE International Conference on Microwaves, Communications, Antennas and Electronics Systems, COMCAS 2009*, pp. 1–8, 9–11 Nov. 2009.

[15] Z. Wang, X. Sun, C. Zhang, and Y. Li, "Issues in integrated circuit design for UHF RFID," in *IEEE International Workshop on Radio-Frequency Integration Technology, RFIT 007*, pp. 322–328, 2007.

[16] EPCglobal, "Radio-frequency identity protocols – class-1 generation-2 RFID v1.2.0," EPC-global Standard, EPCglobal Inc., Oct. 2008, http://www.epcglobalinc.org.

[17] B. Warneke, M. Scott, B. Leibowitz, *et al.*, "An autonomous 16 mm^3 solar-powered node for distributed wireless sensor networks," in *Proc. IEEE Sensors*, vol. 2, pp. 1510–1515, 2002.

[18] P. V. Nikitin and K. Rao, "Antennas and propagation in UHF RFID systems," in *IEEE International Conference on RFID*, pp. 277–288, 2008.

[19] V. Plessky and L. Reindl, "Review on SAW RFID tags," *IEEE Transactions on Ultrasonics, Ferroelectrics and Frequency Control*, **57**, (*3*), 654–668, 2010.

[20] L. Reindl, "Wireless passive sensors: Basic principles and performances," in *IEEE Sensors*, pp. 1607–1610, Oct. 2008.

[21] V. Kalinin, R. Lohr, A. Leigh, and G. Bown, "Application of passive SAW resonant sensors to contactless measurement of the output engine torque in passenger cars," in *IEEE International Frequency Control Symposium*, pp. 499–504, Jun. 2007.

[22] W. Buff, F. Plath, O. Schmeckebier, *et al.*, "Remote sensor system using passive SAW sensors," in *Proceedings of Ultrasonics Symposium*, vol. 1, pp. 585–588, 1994.

[23] A. Canabal, P. Davulis, E. Dudzik, and M. Pereira da Cunha, "CDMA and FSCW surface acoustic wave temperature sensors for wireless operation at high temperatures," in *IEEE International Ultrasonics Symposium*, pp. 807–810, Sept. 2009.

[24] L. Reindl, G. Scholl, T. Ostertag, *et al.*, "SAW devices as wireless passive sensors," in *IEEE Ultrasonics Symposium*, vol. 1, pp. 363–367, 1996.

[25] R. Fachberger, G. Bruckner, G. Knoll, *et al.*, "Applicability of LiNbO3, langasite and GaPO4 in high temperature SAW sensors operating at radio frequencies," *IEEE Transactions on Ultrasonics, Ferroelectrics and Frequency Control*, **51**, (*11*), 1427–1431, 2004.

[26] E. A. Pereira da Cunha, "Wireless acoustic wave sensors and systems for harsh environment applications," in *IEEE Topical Conference on Wireless Sensors and Sensor Networks (WiSNet)*, pp. 41–44, Jan. 2011.

[27] A. Pohl, G. Ostermayer, L. Reindl, and F. Seifert, "Monitoring the tyre pressure of cars using passive SAW sensors," in *IEEE Ultrasonics Symposium*, vol. 1, pp. 471–474, 1997.

[28] B. Dixon, V. Kalinin, J. Beckley, and R. Lohr, "A second generation in-car tyre pressure monitoring system based on wireless passive SAW sensors," in *IEEE International Frequency Control Symposium and Exposition*, pp. 374–380, 2006.

[29] A. Pohl, R. Steindl, and L. Reindl, "Measurements of vibration and acceleration utilizing SAW sensors," in *Proc. Sensor'99*, 1999.

[30] V. Kalinin, R. Lohr, A. Leigh, J. Beckley, and G. Bown, "High-speed high dynamic range resonant SAW torque sensor for kinetic energy recovery system," in *European Frequency and Time Forum*, pp. 1–8, 2010.

[31] L. Reindl, C. Ruppel, A. Kirmayr, *et al.*, "Radio-requestable passive SAW water-content sensor," *IEEE Transactions on Microwave Theory and Techniques*, **49**, (*4*), 803–808, 2001.

[32] J. Kraus and R. Marhefka, *Antennas for all Applications*, 3rd Edn., McGraw-Hill College, 2002.

[33] D. Pardo, A. Vaz, S. Gil, *et al.*, "Design criteria for full passive long range UHF RFID sensor for human body temperature monitoring," in *IEEE International Conference on RFID*, pp. 141–148, Mar. 2007.

[34] U. Karthaus and M. Fischer, "Fully integrated passive UHF R_D transponder IC with 16.7-μW minimum RF input power," *IEEE Journal of Solid-State Circuits*, **38**, (*10*), 1602–1608, 2003.

[35] G. De Vita and G. Iannaccone, "Design criteria for the RF section of UHF and microwave passive RFID transponders," *IEEE Transactions on Microwave Theory and Techniques*, **53**, (*9*), 2978–2990, 2005.

[36] D. Brenk, J. Essel, J. Heidrich, *et al.*, "Wireless sensing by means of passive multistandard RFID tags," in *Proc. IEEE Sensors*, pp. 61–64, 25–28 Oct. 2009.

[37] D. Brenk, J. Essel, J. Heidrich, and R. Weigel, "Advances in wireless sensors and sensor networks," *Lecture Notes in Electrical Engineering*, pp. 125–150, Springer, 2010.

[38] K. Seemann and R. Weigel, "The system design of integrated passive transponder devices," in *Proc. Asia-Pacific Microwave Conference APMC 2006*, pp. 1114–1117, 12–15 Dec. 2006.

[39] N. Tran, B. Lee, and J.-W. Lee, "Development of long-range UHF-band RFID tag chip using Schottky diodes in standard CMOS technology," in *Radio Frequency Integrated Circuits (RFIC) Symposium*, IEEE, pp. 281–284, Jun. 2007.

[40] J. O. McSpadden, L. Fan, and K. Chang, "Design and experiments of a high-conversion-efficiency 5.8-GHz rectenna," *IEEE Transactions on Microwave Theory and Techniques*, **46**, 2053–2060, (Dec.) 1998.

[41] T. Salter, G. Metze, and N. Goldsman, "Parasitic aware optimization of an RF power scavenging circuit with applications to smartdust sensor networks," in *Radio and Wireless Symposium, RWS'09*, IEEE, pp. 332–335, 2009.

[42] K. Kotani, A. Sasaki, and T. Ito, "High-efficiency differential-drive CMOS rectifier for UHF RFIDs," *IEEE Journal of Solid State Circuits*, **44**, (*11*), 3011–3018, 2009.

[43] D. Brenk, J. Essel, J. Heidrich, and R. Weigel, "Ultra low-power techniques for sensor-enhanced RFID tags," in *Proc. IEEE MTT-S International Microwave Workshop on Wireless Sensing, Local Positioning, and RFID IMWS*, pp. 1–4, 2009.

[44] EPCglobal, "Radio-frequency identity protocols – HF version 2 RFID – draft version 0.3," EPCglobal Standard, EPCglobal Inc., Jul. 2007, http://www.epcglobalinc.org.

[45] I. Filanovsky and S. T. Lim, "Interaction of threshold voltage and mobility temperature dependencies applied to stabilization of current and voltage," in *Proc. 43rd IEEE Midwest Symposium on Circuits and Systems*, vol. 3, pp. 1022–1025, 8–11 Aug. 2000.

[46] J. McCreary and P. Gray, "All-MOS charge redistribution analog-to-digital conversion techniques. I," *IEEE Journal of Solid State Circuits*, **10**, (*6*), 371–379, 1975.

[47] P. E. Allen and D. R. Holberg, *CMOS Analog Circuit Design*, 2nd Edn., Oxford University Press, 2002.

[48] F. Maloberti, *Data Converters*, Springer, 2007.

[49] I. Filanovsky and S. T. Lim, "Temperature sensor applications of diode-connected MOS transistors," in *Proc. IEEE International Symposium on Circuits and Systems ISCAS* 2002, vol. 2, pp. II-149–II-152, 26–29 May 2002.

[50] Y. Zhai, S. Prakash, M. Cohen, and P. Abshire, "Detection of on-chip temperature gradient using a 1.5 V low power CMOS temperature sensor," in *Proc. IEEE International Symposium on Circuits and Systems, ISCAS*, pp. 1171–1174, 2006.

[51] P. Heide, J. Ilg, R. Roskosch, *et al.*, Patent US000006946949b2, Sep. 2005.

[52] M. Vossiek, R. Roskosch, and P. Heide, "Precise 3-D object position tracking using FMCW radar," in *Proc. 29th European Microwave Conference*, vol. 1, pp. 234–237, 1999.

[53] J. Heidrich, D. Brenk, J. Essel, *et al.*, "Local positioning with passive UHF RFID transponders," in *Proc. IEEE MTT-S International Microwave Workshop on Wireless Sensing, Local Positioning, and RFID, IMWS*, pp. 1–4, 2009.

[54] Y. Park, J. W. Lee, and S. Kim, "Improving position estimation on RFID tag for localization using RFID reader transmission power control," in *IEEE International Conference on Robotics and Biomimetics, ROBIO 2008*, pp. 1716–1721, Feb. 2009.

[55] U. Hatthasin, S. Thainimit, K. Vibhatavanij, N. Premasathian, and D. Worasawate, "An improvement of an RFID indoor positioning system using one base station," in *6th International Conference on Electrical Engineering/Electronics, Computer, Telecommunications and Information Technology, ECTI-CON 2009*, vol. 2, IEEE, pp. 862–865, 2009.

[56] D. K. Barton, *Radar Systems Analysis and Modeling*, Artech House, 2005.

[57] L. Wiebking, *Entwicklung eines zentimetergenauen mehrdimensionalen Nahbereichs-Navigationssystems*, VDI Verlag GmbH, 2003.

[58] A. Povalac and J. Sebesta, "Phase of arrival ranging method for UHF RFID tags using instantaneous frequency measurement," in *Proc. ICECom*, vol. 1, pp. 1–4, 2010.

[59] J. Heidrich, D. Brenk, J. Essel, *et al.*, "Distance measurement with UHF transponders by means of modulated backscattering (invited paper)," in *ICECOM 2010*, Sept. 2010.

[60] D. Brenk, J. Essel, J. Heidrich, and R. Weigel, "Advances in wireless sensors and sensor networks," *Lecture Notes in Electrical Engineering*, vol. 64, Springer-Verlag GmbH, 2010.

[61] Poynting Antennas (Pty) Ltd, RFID Patch Antenna Operating Frequency – 860–960 MHz circular polarization, Product code: PATCH-A0025.

[62] Huber+Suhner, "Planar antenna for wireless communication CP," 2004 www.hubersuhner.com.
[63] J. Meyer, R. Herschmann, and B. Geck, "A testbed for precision impedance measurements of planar symmetrical RFID antennas," in *European Microwave Conference (EuMC)*, pp. 1754–1757, 2010.
[64] J. Meyer, R. Herschmann, and B. Geck, "Antenna design for multistandard transponder ICS," in *RFID-Systec*, June 2010.
[65] ISO/IEC 7810, "ISO/IEC 7810 identification cards – physical characteristics," Nov. 2003.
[66] Feig Electronic, OBID i-scan ID ISC.LR2000.

5 Unconventional RFID systems

Alírio Boaventura, Ricardo Fernandes, João Nuno Matos, and
Nuno Borges Carvalho

5.1 Introduction

Non-conventional RFID systems include those that are not exclusively designed for identification purposes, but which also implement other advanced tasks such as location or sensing.

The most widespread, yet simple, RFID system is the one-bit EAS (electronic article surveillance) system whose only purpose is to detect the presence/absence of a tagged object in the vicinity of a reader. On the other hand, current state-of-the-art *N*-bit RFID systems serve a broader range of purposes, and the technology is experiencing tremendous advances. RFID readers are more robust, effective, and energy-efficient. Tags can now be made thinner, cheaper, often physically flexible, and more energy-efficient. Sensors, only used in wireless sensor nodes before, are currently making their way towards RFID tags. Some examples of these sensor-enabled tags can readily be found in the literature [1]. Multi-sensor tags including processing capabilities can also be found [2].

In order to be more autonomous and effective, beyond the energy harvesting subsystem, advanced tags also incorporate energy storage and power management units (Figure 5.1). Moreover, some tags integrate more than one method to harvest and store energy, as in Figure 5.1.

As the complexity of passive tags increases (due to sensor, analog-to-digital data conversion, and processing needs) the energy demand also increases, and consequently the system coverage range and the overall energy efficiency become a major concern. Some interesting approaches have been proposed, among them the use of specific signal shapes to improve the RF-DC efficiency of the harvester receivers. One of these examples unconventionally uses multi-sine signals with high PAPR (peak to average power ratio) waveforms (Figure 5.2) that benefit the rectification process on the harvester side [3]. Other kinds of power optimized waveforms, such as chaotic signals, are now being used to improve the RF-DC conversion efficiency [4]. RFID technology is starting to be viewed as a potential low cost localization technology, and a few interesting solutions already exist [2]. Some of them only make use of standard tags and readers, while others brings UWB (ultrawideband) approaches into the equation.

The integration of sensors into passive RFID tags noticeably extends their functionalities and brings a new cluster of opportunities. However, such integration also brings

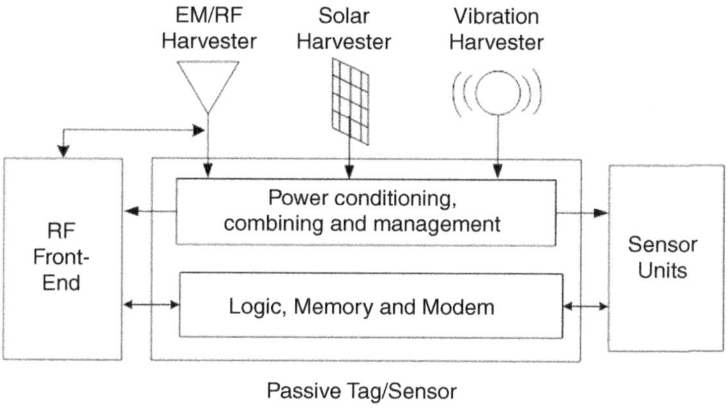

Figure 5.1 An advanced RFID/wireless sensor tag with multi-harvesting capability.

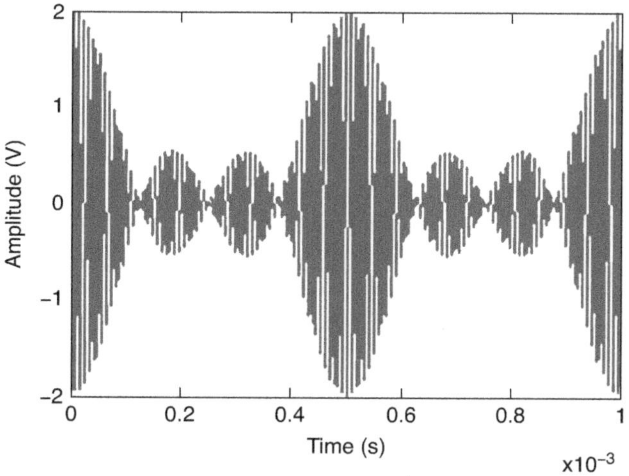

Figure 5.2 High PAPR multi-sine signal for RFID coverage extension.

several challenges due to the increased complexity and energy consumption. Some of these challenges are well described in [5] through the implementation of what the authors called a WISP (wireless identification and sensing platform). Unlike a conventional RFID tag, a passive RFID sensor tag or WISP requires a sensor integration, analog-to-digital converter, and increased processing capabilities. All these needs result in increased energy consumption.

The choice of suitable sensors for these applications is a major concern. Reference [6] and particularly [7] provide extensive lists of sensors. In addition, [7] also presents the power consumption of several types of sensors, such as temperature and acceleration. It is important to note that most of these sensors only require (one/few/some) microwatts to operate properly. This is the reason why their integration in RFID indeed makes sense.

Other important information in [7] is a list of commercially available sensor-enabled tags. The tags in this list are interesting because they are quite distinct from each other, not only in terms of target application but also in terms of how the sensed data are encoded and transmitted.

In addition to the list of sensors mentioned before, [6] also contains additional useful information including a good and up to date introduction to RFID, just like [8]. It also describes very well the current state of the art in the usage of paper as a very low cost substrate for RFID tags.

Although all the previous information refers to sensors that require a power source to operate (a remote power source in the case of passive RFID), it should be pointed out that other alternatives exist in which the sensor does not need any power supply at all. Such approaches rely on the use of a passive component (e.g. a capacitive, resistive, inductive, or ferroelectric sensor) whose electrical property (e.g. capacitance) is sensitive to an external physical quantity (e.g. pressure, temperature). Examples can be found in [9–12].

Energy harvesting and/or wireless power transmission are fundamental for powering up passive RFIDs and sensors. Energy harvesting is not new to RFID. In fact, passive tags have always fully relied on some kind of dedicated RF energy harvesting. Additional harvesting mechanisms, however, are something new. An example of a tag with multi-harvesting capability is depicted in Figure 5.1. Passive tags with multiple harvesting systems make sense, because if there are more harvesters, the available power increases. If the available power increases the range also goes up. Since coverage range is one of the most delicate issues in RFID, any improvement is naturally more than welcome. A tag with a dual harvesting system is proposed in [13]. This tag successfully combines the traditional RF method with a solar energy harvester. RF plus solar represents one of the combinations most often found in the literature. The tag prototype in [13] consists of a modified WISP, a planar inverted-F antenna, and a solar cell mounted on top of the antenna. Under 150 lux (typical of a warehouse), the authors measured a range of about 24 meters, significantly more than the 4 meters obtained with the same tag with the solar harvester turned off. It should be noted that the authors were not able to achieve ranges higher than 24 meters, even with better lighting conditions (or using active tags), because of the sensitivity of the reader.

This amazing coverage range increase opens door for a variety of new RFID applications, for instance location scenarios, where a tagged person or object is expected to be found in closed and indoor environments. This location scenario is one of the hardest in terms of RF system design since it imposes not only an identification requirement, but also positioning based first on the received signal, but also on signal time-of-flight, which allows higher location resolution in indoor scenarios. We can say that RFID is already a localization system, albeit a quite limited one.

This chapter brings to the table some new ideas regarding energy efficiency and coverage range improvement using multi-sine signals, location based on UWB RFID systems and multi-RFID schemes used for household applications. The conceptual ideas are followed by hardware implementation details and also test and measurement results. The content of each topic discussed is summarized in the following subsections.

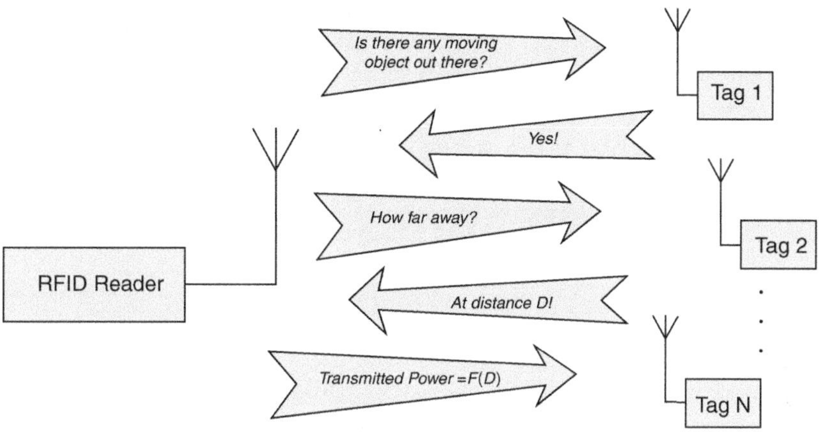

Figure 5.3 Using feedback information to save energy in RFID systems.

5.1.1 Efficient and energy-aware approaches

In passive RFID systems, the RFID tag is remotely powered by the reader through an RF carrier. On the other hand, the reader also accesses the tag memory wirelessly to read/write data. Such a reader is thus simultaneously a communication and an energy supplier device. In typical application scenarios the reader is always on and continuously broadcasting RF energy to the medium, searching for RFID tags in its field. Therefore, a huge amount of energy is wasted in this process. This becomes especially critical when the tag reading rate is low, because energy is wasted even without tags in the field of the reader. In this sense, improved energy-aware systems should be designed. In this chapter, two aspects are addressed: energy transfer efficiency and efficient power management. The minimum transmitted power (and the maximum reading range) of passive RFID systems is primarily imposed by the activation power level of RFID tags. Hence, it is important to optimize the efficiency of RFID tags in order to lower their activation power level. This can be achieved by designing improved circuits with improved harvesting capabilities combined with low power electronics, or alternatively by using efficient waveforms that boost the energy transfer efficiency [3, 14]. Concerning the power management aspect, a proposal for energy reduction in RFID readers will also be presented. The proposed scheme requires feedback information which provides an estimate of the distance between the tag and the reader so that the transmitted power can be adjusted according to the distance. In order to be even more effective the feedback mechanism can also include information about the presence/absence of tagged objects or even the number of tags in the field of the reader (Figure 5.3). It should be noted that such feedback information must be gathered using an energy-inexpensive method. For instance, low power motion sensors can be used to detect the presence of tagged objects in the field and an RSSI (received signal strength indication) method can be applied to obtain an estimation of the distance between reader and tag.

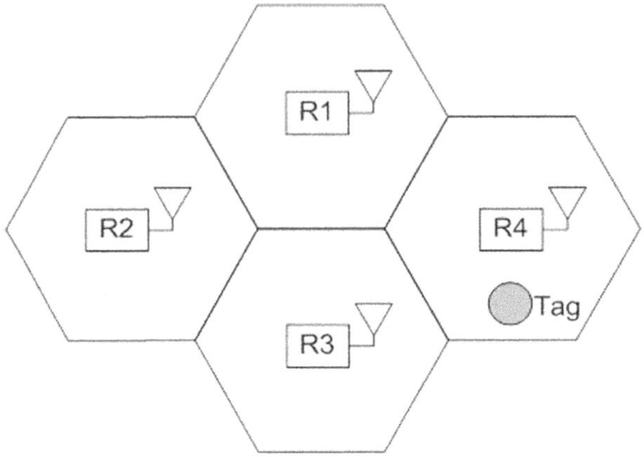

Figure 5.4 Indoor location using RFID systems.

5.1.2 RFID for location

In an RFID location configuration, if there are multiple readers with non-overlapping zones of coverage as in Figure 5.4, the position of a particular tag can be determined by simply instructing the readers to search for it. The position of the reader that returns a positive answer is the estimate of where the tag is. The accuracy of this method is certainly not the best, but it can be improved by taking the received signal strength (RSS) into account. Furthermore, some RFID readers already have the capability to measure RSS values, so there is no additional cost. However, it is well known that the RSS is strongly affected by environmental factors. This technique and others, based on the same principles, are explained in [15]. One of them corresponds to the LANDMARC system, which is worth mentioning. In this system, reference tags are deployed prior to localization. The improvement in accuracy because of these reference tags can be very noticeable. The RSS of each normal tag is read, just as before, but it is also compared with those corresponding to nearby reference tags. The position of the tag is computed using a weighted average of the reference positions. The more similar the RSS of a reference is to the RSS of the unknown tag, the higher the weight. This method is interesting because interference and multipath effects that affect the unknown tag are very likely to affect nearby tags in the same way. All the localization techniques mentioned so far are tag-based, due to the fact that the target objects carry tags.

Passive UWB RFID tags such as the one described in [2] represent another approach. Combining RFID with UWB has its advantages. The accuracy tends to increase in comparison to the RSS-based systems. The immunity to multipath-related problems increases. The low power characteristic is kept and the cost does not change too much. Passive UWB RFID tags include an IR-UWB transmitter (where the IR means impulse radio) instead of the usual back-scatter modulator but, except for that, there are (essentially) no differences. Regarding [16], the authors report an accuracy of 16.8 cm with line of sight, using a TDoA (time difference of arrival) algorithm. The tag is able to generate 620 ps long UWB pulses, with 220 mV (of peak-to-peak amplitude).

References [17–19] describe a UWB-based RFID system worth mentioning. The system introduces the concept of a power node, namely a device designed to radiate energy to tags, but only that. Power nodes can be turned on and off by the reader. The reader in this case is only a UWB receiver, essentially. Once within range of a power node each tag waits for a short (random) amount of time and then sends its identification number to the reader. It is very important to note that the UWB signals generated by the tags are only used for communication purposes (meaning that no TDoA or any other localization technique is utilized). Localization is performed in a different way. For any given unknown tag, the reader first identifies the corresponding power node (by using the on–off capability mentioned above) and then instructs the power node to reduce its output power gradually, until the tag runs out of power. One of the most interesting features of this system is the separation between power and data. Splitting the two makes sense because this way the coverage can be easily enhanced by increasing the number of power modules, which are inexpensive compared to readers. This system has a data read range of 10 meters, a power-up range of 1 meter and an accuracy in the order of 0.25 meters. In Section 5.3 we explain this concept further and present a UWB-RFID combined system.

5.1.3 RFID for household applications

RFID is a very desirable technology due to its extremely low power consumption, which can sometimes even be zero power consumption, one example being passive sensors that harvest energy from the environment. In home appliances, sometimes termed household applications or home automation, the use of RFID-based systems will imply a significant reduction in chemical battery use. Most of the systems could see their energy life extended and, sometimes, could even eliminate use of batteries for operation of the sensor itself. Some examples include the alarm sensors inside a house, the remote controllers used for TVs and Hi-Fi systems, or also some thermal or humidity sensors.

This will create a new approach to household systems and specially will force new approaches to the wireless power transmission or energy harvesting that is needed to power up these sensors. As an illustrative example of application of RFID in home automation, a multi-RFID system especially conceived for application in a battery-less remote control system is presented in Section 5.4. In such an approach the remote controller is completely energy independent and does not need any type of chemical battery. It is expected that this technique or a similar one may allow saving millions or even billions of batteries in the future, having a huge impact through reduction of chemical battery pollution.

5.2 Efficient and energy-aware approaches

In this section an approach to improve energy transfer efficiency in RFID systems is presented first. High PAPR multi-sine signals are used instead of the conventional single carrier. High PAPR signals excite the envelope detectors and charge pumps in a more effective way, forcing them to produce much more DC power. When compared with a single carrier with the same average power, this approach will result in increased coverage range or equivalently less transmitted power to achieve the same communication distance.

In conventional single reader configurations the RFID reader transmits the same power independently of the distance between tag and reader, no matter whether or not there are tags in the field and independently of the number of tags. The introduction of feedback information could save a considerable amount of energy. This is the second problem to resolve in this section, where an energy-efficient scheme is proposed. The introduction of feedback information in the system provides an estimate of the distance between the tag and the reader. This allows adjustment of the transmitted power according to the distance.

5.2.1 Extending the coverage range of RFID systems

RFID tags are normally excited using a modulated carrier; this carrier allows power-up of the TAG and also communication with it. Nevertheless, a more robust signal can be used to power up the TAG in order to increase the coverage range using the same average power. One such signal is the multi-carrier or multi-sine approach.

Multi-carrier signals are commonly used in OFDM systems to achieve high spectral efficiency and robustness in environments affected by fading and multipath. Recently the benefits of multi-carrier signals for wireless energy transfer have been explored. Multi-sine signals have been brought to the RFID arena by the work presented in [20], where a multi-sine scheme was used to extend the reading range of UHF tags. In [21] a survey was presented where several commercial UHF RFID tags were tested with multi-sine signals. In [3] the nonlinear behavior of RF-DC converters (the RF-DC circuit is the one responsible for converting RF signals into DC to power up the RFID TAG), and the efficiency enhancement when using multi-sine signals have been investigated. To do so, a Schottky diode detector was excited both with a single carrier and with several multi-sine signals.

The results have been compared and the conclusion is that multi-sine signals can provide higher RF-DC efficiency than a single carrier in diode detectors and charge pumps. Basically, the inherent high PAPR (peak-to-average power ratio) nature of multi-sine signals excites the nonlinearity of diode detectors and charge pumps in a more efficient way, forcing the devices to produce much more DC power [3, 20]. This can potentially improve the sensitivity of RFID tags and consequently increase the system communication range. In [3] a mathematical description of the rectification process has been presented. Also, comment on the nonlinear phenomena behind the improvement of RF-DC conversion efficiency in Schottky diode detectors when using multi-sine excitations has been provided.

High PAPR signals have been used in some work to increase the RF-DC conversion efficiency and ultimately to increase the system coverage range. A first use of this concept can be found in [22]. In that work the authors used a high PAPR signal to reduce the average power transmitted by a UHF RFID reader, while keeping the same communication distance. Instead of keeping the RF carrier always turned on during the entire bit period of a high level ASK symbol, the RF carrier is intermittently switched on and off with a given duty cycle. Therefore, the transmitted average power is reduced according to the duty cycle. A power reduction of about 75% has been achieved for a duty cycle of 40% [22].

However, the main drawback of this approach is related to the method itself. Switching the carrier on and off with a square shape introduces new frequency components and degrades the spectral efficiency of the system. Quite recently, as reported in [23, 24], ultrawideband (UWB) signals have been proposed for efficient low power transmission. A train of short duration pulses with relatively high amplitude exhibits very low average power (and consequently high PAPR). In [24] a conversion efficiency of 50% has been achieved in a Schottky diode voltage doubler fed with a relatively low power level of −3dBm. The obvious disadvantage of the UWB approach is the increased spectral bandwidth and the need for wideband antennas and components. Furthermore, contrary to the multi-sine approach, the UWB scheme cannot be applied to conventional RFID systems which inherently have a narrowband.

In the work presented in [14], a multi-sine front-end scheme was integrated with a commercial RFID reader in order to extend the reading range. In [14] only the downlink path was implemented and the response from the tag was visualized in an oscilloscope. In this chapter we revisit the work presented in [14] and we present the corresponding extension that addresses the uplink path. Thus, both the downlink and the uplink are studied and a multi-sine design is presented to guarantee that the reader can actually access the tag ID with minimal hardware changes. Furthermore, it is proved that the multi-sine signal can be designed in such a way that no additional changes are needed in the traditional receiver hardware architecture to operate with it. Practical consequences of this proof are that a commercial RFID reader can demodulate and decode a properly designed multi-sine signal reflected back from the tag. Thus only a few hardware changes are required, so as to generate the multi-sine signal. Two experiments are conducted to evaluate the reading range improvement with respect to the conventional approach (single carrier) with the same average power. In the first scenario only the downlink path is implemented by the reader combined with the front-end and an oscilloscope is used to visualize the tag response in order to decide whether the tag does or does not respond [14]. In the second measurement scenario, the downlink path is implemented by the reader combined with the front-end and the uplink is implemented solely by the reader. In the second case the decision on successful tag response is taken when the reader actually accesses the tag ID.

5.2.1.1 The radio link using multi-sine signals

The radio link of passive RFID systems exhibits very peculiar characteristics. Because a passive tag has no battery and is entirely powered by the RF power radiated by the reader, the downlink operation has a twofold goal: data communication and energy transference. Moreover, since the passive tag has no local oscillator, the uplink communication is realized by power reflection (the so-called back-scattering communication). In this section we analyze the downlink and uplink paths when using multi-sine excitations instead of

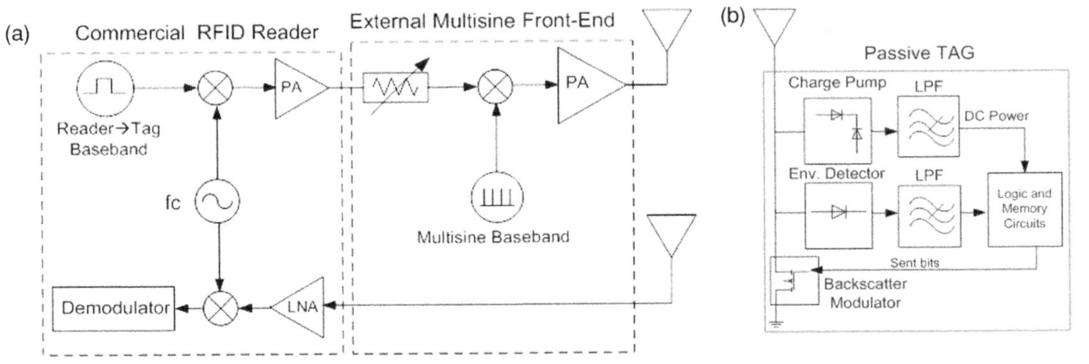

Figure 5.5 a) RFID reader and external multi-sine front-end. b) Passive RFID tag. If the multi-sine signal is properly designed a conventional reader is able to demodulate and decode the tag response without any hardware change. Hence, the tag response is directly forwarded to the reader receiver. The attenuator in the forward path is used to find the activation power level of the tag.

a conventional single carrier. This includes the three link operations: power-link, data downlink, and data uplink.

For the following mathematical description, consider Figure 5.5(a) (basic configuration of the RFID reader and multi-sine front-end) and Figure 5.5(b) (block diagram of a typical passive tag). Also consider the baseband multi-sine signal defined as a sum of N subcarriers harmonically spaced by $\Delta\omega$:

$$x_{\mathrm{BB}}(t) = \sum_{n=1}^{N} A_n \cos(n\Delta\omega t + \varphi_n), \tag{5.1}$$

where N is the number of tones, $\Delta\omega$ is the frequency separation between tones and φ_n is the phase of each tone.

In order to analyze the multi-sine power transfer and data communication the tag's diode detector and the charge pump (Figure 5.5b) are modeled by an even order Taylor series, as done previously in [3]:

$$S[x(t)] = k_2 x(t)^2 + k_4 x(t)^4 + k_6 x(t)^6 + \cdots, \tag{5.2}$$

where k_2, k_4, and k_6 are the model coefficients.

5.2.1.2 Multi-sine power-link

As proved before in [3] and [20], multi-sine signals provide higher RF-DC conversion efficiency in diode detectors and charge pumps (commonly used in RFID tags) when compared with a single carrier with the same average power. This leads to a sensitivity gain and reading range extension in RFID systems.

In order to explain the efficiency enhancement, the system described by Equation (5.2) (truncated to order four) is fed with a single carrier and with a multi-sine signal. In both cases the same average power is considered. Afterwards the DC component for the two cases (obtained after low-pass filtering) is compared. Accordingly, if the system is excited with a single carrier $x(t) = B\cos(\omega_{ct} + \varphi_1)$ then the DC component after low-pass filtering is given by [3]:

$$S_{\mathrm{DC}_{1\mathrm{Tone}}} = \frac{B^2 k_2}{2} + \frac{3B^4 k_4}{8}. \tag{5.3}$$

Now, consider a multi-sine signal, for instance a bandpass multi-sine centered at frequency ω_c with four tones equally spaced and with the same amplitude (Equation (5.4)):

$$x(t) = A\cos(\omega_{1t} + \varphi_1) + A\cos(\omega_{2t} + \varphi_2) + A\cos(\omega_{3t} + \varphi_3) + A\cos(\omega_{4t} + \varphi_4). \tag{5.4}$$

After a symbolic computation we end up with the following output DC component:

$$\begin{aligned} S_{\mathrm{DC}_{N\mathrm{Tone}}} &= \frac{4A^2k_2}{2} + \frac{21A^4k_4}{2} + \frac{3A^4k_4}{2}\cos(2\varphi_3 - \varphi_2 - \varphi_4) \\ &+ \frac{3A^4k_4}{2}\cos(-2\varphi_2 + \varphi_1 + \varphi_3) + 3A^4k_4\cos(\varphi_1 - \varphi_2 - \varphi_3 + \varphi_4). \end{aligned} \tag{5.5}$$

By taking $A = B/2$ the average power of the multi-sine signal equals the average power of the single carrier. The conclusion is that the DC provided by the multi-sine is higher than the one provided by the single carrier ($S_{\mathrm{DC}_{N\mathrm{Tone}}} > S_{\mathrm{DC}_{1\mathrm{Tone}}}$). Moreover and more importantly, $S_{\mathrm{DC}_{N\mathrm{Tone}}}$ is a function of tone phase and by aligning all the tones in phase zero ($\varphi_1 = \varphi_2 = \varphi_3 = \varphi_4 = 0°$) the DC is maximized.

5.2.1.3 Multi-sine data downlink

As proved before in [3] and as revised in the previous subsection, from the viewpoint of power transmission, multi-sine signals are beneficial since they provide increased RF-DC efficiency in RF-DC converters such as the charge pump circuits commonly used in passive RFID tags. This can potentially improve the communication range of RFID systems. However, the multi-sine signal must be properly designed in order to still guarantee data communication. In this subsection we analyze the downlink data communication and we present some design rules to guarantee communication with multi-sines. For simplicity, in the following analysis all link factors are ignored. Additionally, the tag's envelope detector and charge pump (Figure 5.5b) will be modeled by an even order polynomial series as done previously.

From now on, in order to facilitate the analysis, all signals will be represented by complex exponentials. The multi-sine modulated signal at the transmitter antenna (Figure 5.5a), is given by the mixture of the base band signal $m(t)$ with an RF carrier at frequency ω_c, followed by a mixture with a zero phase[1] base band multi-sine signal:

$$x(t) = m(t)\left\{\sum_{n=1}^{N} A_n \frac{\mathrm{e}^{\mathrm{j}n\Delta\omega t} + \mathrm{e}^{-\mathrm{j}n\Delta\omega t}}{2}\right\}\left\{A_C \frac{\mathrm{e}^{\mathrm{j}\omega_c t} + \mathrm{e}^{-\mathrm{j}\omega_c t}}{2}\right\}. \tag{5.6}$$

For instance, if a two-tone base band multi-sine is considered ($N = 2$) then the bandpass modulated multi-sine $x(t)$ (after being shifted in frequency) will have spectral

[1] Zero phase arrangement is proven to be the most efficient [3].

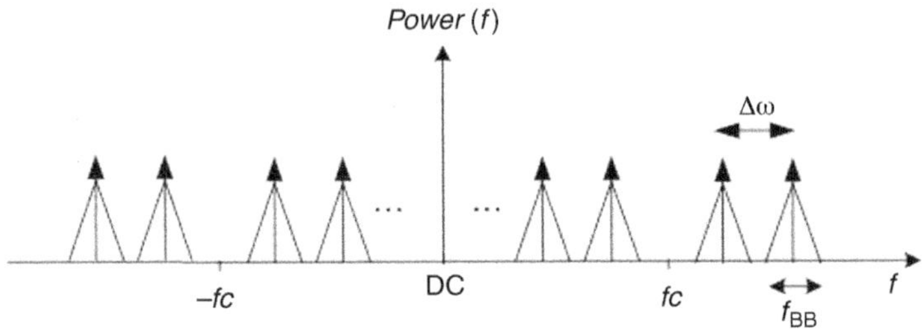

Figure 5.6 Spectrum of a modulated multi-sine signal. Tone separation in the multi-sine must be higher than the bandwidth of the reader-to-tag base band signal in order to avoid spectrum overlapping. If overlapping occurs, the tag will not be able to recover the reader's base band information.

components at $-(\omega c + 2\Delta\omega)$, $-(\omega c + \Delta\omega)$, $-(\omega c - \Delta\omega)$, $-(\omega c - 2\Delta\omega)$, $(\omega c - 2\Delta\omega)$, $(\omega c - \Delta\omega)$, $(\omega c + \Delta\omega)$, $(\omega c + 2\Delta\omega)$:

$$x(t) = m(t)\left\{\begin{array}{l} \frac{A_1 A_C}{4} \mathrm{e}^{\mathrm{j}(\omega_c+\Delta\omega)t} + \frac{A_1 A_C}{4} \mathrm{e}^{\mathrm{j}(\omega_c-\Delta\omega)t} \\ + \frac{A_1 A_C}{4} \mathrm{e}^{-\mathrm{j}(\omega_c-\Delta\omega)t} + \frac{A_1 A_C}{4} \mathrm{e}^{-\mathrm{j}(\omega_c+\Delta\omega)t} \\ + \frac{A_2 A_C}{4} \mathrm{e}^{\mathrm{j}(\omega_c+2\Delta\omega)t} + \frac{A_2 A_C}{4} \mathrm{e}^{\mathrm{j}(\omega_c-2\Delta\omega)t} \\ + \frac{A_2 A_C}{4} \mathrm{e}^{-\mathrm{j}(\omega_c+2\Delta\omega)t} + \frac{A_2 A_C}{4} \mathrm{e}^{-\mathrm{j}(\omega_c-2\Delta\omega)t} \end{array}\right\}. \tag{5.7}$$

The frequency spectrum of $x(t)$ is depicted in Figure 5.6. As can be seen the frequency separation between tones $\Delta\omega$ must be higher than the bandwidth of the data signal f_{BB}, so that the information can be successfully recovered at the tag side. Ideally, $\Delta\omega$ must be much higher than f_{BB} ($\Delta\omega \gg f_{\mathrm{BB}}$). As an example, the EPC standard imposes a base band data rate ($f_{\mathrm{BB}}/2$) of 26.7 to 128 kbps [25] which requires a bandwidth f_{BB} of 53.4 to 256 kHz. In our experiments the minimum tone separation used is 500 kHz which is higher than the maximum bandwidth required by the standard. The increased bandwidth is, however, the main drawback of the multi-sine scheme. Nevertheless, the bandwidth values used in this work (maximum 3 MHz) are quite far from the maximum imposed by ISM UHF regulations.

In order to understand the multi-sine downlink, assume a unitary channel response (no attenuation or phase shift) and consider that the tag's envelope detector behaves as a pure squarer, $S[x(t)] = k_2 x(t)^2$ (a second order model is enough to describe the basic envelope detection operation). Thus, the square of the signal (5.3) will provide DC/baseband components coming from the product of symmetric frequency components, $-(\omega_c + 2\Delta\omega)^* (\omega_c + 2\Delta\omega)$, $-(\omega_c - \Delta\omega)^* (\omega_c - \Delta\omega)$, $-(\omega_c - \Delta\omega)^* (\omega_c - \Delta\omega) - (\omega_c + 2\Delta\omega)^* (\omega_c + 2\Delta\omega)$. Accordingly, the base band component at the output of the envelope detector (after low-pass filtering) is given by:

$$y_{\mathrm{BB}}(t) = \frac{k_2}{4}[m(t)]^2 \left(A_1^2 A_C^2 + A_2^2 A_C^2\right). \tag{5.8}$$

From Equation 5.8 it can be seen that the tag's detector is able to recover the base band information $m(t)$ sent by the reader through the "multi-sine carrier," provided that the condition, $\Delta\omega > f_{BB}$, is verified. To be precise, if the base band information is an ASK signal [25] with period T, simply composed by two levels ($a_k = 0$, $a_k = 1$), $m(t) = \sum_{k=-\infty}^{+\infty} a_k \delta(t - kT)$, then $[m(t)]^2$ will also be between two levels providing essentially the same information as $m(t)$. The recovered information is proportional to the square of the base band signal, and so proportional to its power.

More importantly, no changes are needed in the architecture of a typical tag to demodulate these new kinds of signals. In the transmission path, changes are only required on the reader side, namely to incorporate the multi-sine signal. In [14] this is done by using the external front-end as in Figure 5.5a. However it could be efficiently implemented by an SDR (software defined radio) approach simply by updating the RFID reader's firmware.

In the next subsection it is shown that if the multi-sine is properly designed then no changes are needed in the conventional architecture of the reader receiver. This design requirement imposes a multi-sine with an odd number of tones, where the base band multi-sine has a DC component and the central component of the bandpass multi-sine coincides with the reader's local oscillator.

5.2.1.4 Multi-sine data uplink

The tag-to-reader data communication is made by power reflection, by the so-called back-scattering method. This communication is performed by a two-step operation: first the reader illuminates the tag with an unmodulated carrier (in this case a "multi-sine carrier"), and second, the tag modulates its antenna reflection coefficient according to the information to be sent to the reader. This information can be represented by a time-varying reflection coefficient $\Gamma(t)$. Consider now that the tag is illuminated by an unmodulated three-tones multi-sine signal where the amplitudes of the subcarriers are respectively A_1, A_0, and A_2 and the phase relation between the subcarriers is 0°. This multi-sine is designed in such a way that the central tone with amplitude A_0 is placed at the same frequency as the reader local oscillator. Thus, unlike the signal in Equation (5.6) that used an even number of tones, with no subcarrier at the local oscillator frequency, the bandpass version of this signal will have additional spectral components at ω_c and $-\omega_c$ as can be seen in Equation (5.8). Actually, this is the necessary condition to allow a conventional receiver to demodulate the reflected signal from the tag without any hardware change. Hence, the reflected signal from the tag is given by:

$$r(t) = \left\{ \begin{array}{l} \Gamma_0(t)\dfrac{A_0 A_C}{2}e^{j\omega_c t} + \Gamma_0(t)\dfrac{A_0 A_C}{2}e^{-j\omega_c t} + \Gamma_1(t)\dfrac{A_1 A_C}{4}e^{j(\omega_c+\Delta\omega)t} \\ +\Gamma_2(t)\dfrac{A_1 A_C}{4}e^{j(\omega_c-\Delta\omega)t} + \Gamma_2(t)\dfrac{A_1 A_C}{4}e^{-j(\omega_c-\Delta\omega)t} \\ +\Gamma_1(t)\dfrac{A_1 A_C}{4}e^{-j(\omega_c+\Delta\omega)t} + \Gamma_3(t)\dfrac{A_2 A_C}{4}e^{j(\omega_c+2\Delta\omega)t} \\ +\Gamma_4(t)\dfrac{A_2 A_C}{4}e^{j(\omega_c-2\Delta\omega)t} + \Gamma_3(t)\dfrac{A_2 A_C}{4}e^{-j(\omega_c+2\Delta\omega)t} \\ +\Gamma_4(t)\dfrac{A_2 A_C}{4}e^{-j(\omega_c-2\Delta\omega)t} \end{array} \right\}, \quad (5.9)$$

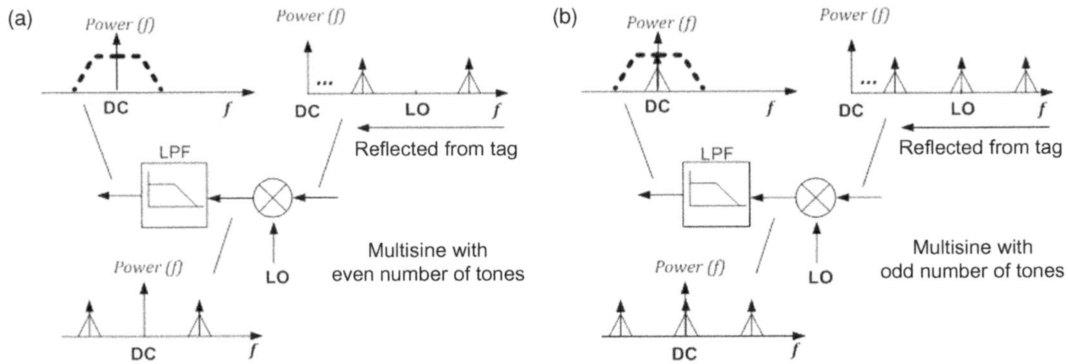

Figure 5.7 Even (a) and odd (b) multi-sine schemes. Low pass filter is represented by a dashed line. In the first case (even multi-sine), a conventional reader is not able to recover the tag baseband information, while in the second case (odd multi-sine), the reader is allowed to access and process tag baseband information. Multi-sine is required to have an odd number of tones and the central tone must coincide with the reader's local oscillator.

where $\Gamma_0(t)$, $\Gamma_1(t)$, $\Gamma_2(t)$, $\Gamma_3(t)$, and $\Gamma_4(t)$ are the time-varying reflection coefficients seen by the subcarrier components. Assuming again no link factors, the signal $r(t)$ reaches the reader receiver antenna and is forwarded to the down-conversion mixer (Figure 5.5a) where it is mixed with the local oscillator at frequency ω_c. The output of the down-conversion mixer is given by:

$$z(t) = r(t)\left\{A_C \frac{e^{j\omega_c t} + e^{-j\omega_c t}}{2}\right\}. \tag{5.10}$$

The only components of $z(t)$ falling on base band frequencies result from the product of the components at ω_c and $-\omega_c$. After passing the signal $z(t)$ through a low-pass filter (present in traditional receiver architectures) with a cut-frequency of f_{BB} a sample of the tag's base band information $\Gamma_0(t) = \Gamma(t)$ – reflection coefficient of the central frequency – can be obtained:

$$z_{BB}(t) = \frac{1}{2}\Gamma(t) A_0 A_C^2. \tag{5.11}$$

The previous mathematical proof is illustrated in Figure 5.7: in the first case a multi-sine with an even number of tones and having no component at the local oscillator frequency is used (Figure 5.7a). In this case a conventional reader is not able to recover the tag base band information. On the other hand, in the second case (Figure 5.7b), the tag is illuminated with a multi-sine with an odd number of tones where the central tone coincides with the reader's local oscillator. Thus, the tag modulates its base band response in all the multi-sine subcarriers including the central one. This allows a conventional reader to access and process the tag base band information. In fact the necessary condition is the frequency positioning of the multi-sine central component. The odd number of tones requirement is not mandatory, but it is convenient since it guarantees the symmetry of the signal spectrum and it simplifies the multi-sine signal generation and hardware design. Moreover, the bandwidth of the low-pass filter of the reader receiver has to be

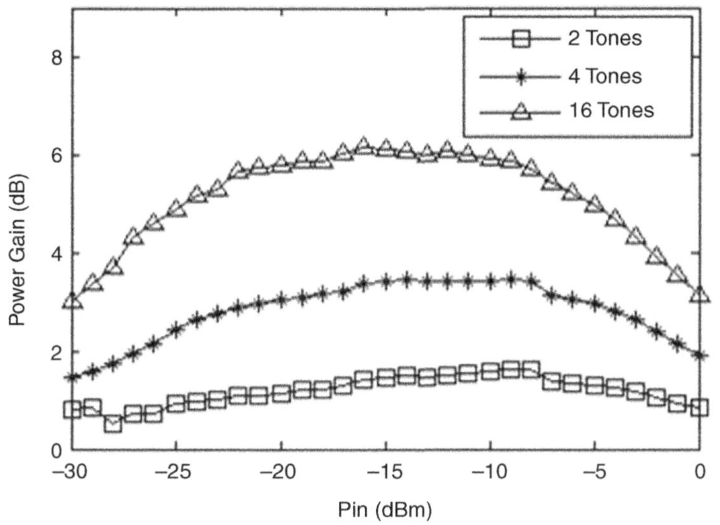

Figure 5.8 Measured DC power gain (single diode detector) as function of input power, for different multi-sine signals.

tight enough to reject all the lateral multi-sine components, leaving only the central one. While these considerations were not important in the previous downlink analysis (since the tag receiver uses envelope detection rather than coherent detection), they are crucial in the uplink case.

5.2.2 Laboratory test beds and measurements

In this subsection we validate the previous theory and we evaluate the reading range improvement provided by multi-sine signals in RFID systems. To do so three experiments are conducted: in the first experiment: (Setup 1) a simple Schottky diode detector is directly fed with a single carrier and with three different multi-sines with zero phase arrangement. In the other two experiments over-the-air measurements have been performed. In the first one an oscilloscope is used to visualize the tag response and to decide whether the tag does or does not respond (Setup 2). In the last measurement scenario (Setup 3), the downlink path is implemented by the reader combined with the front-end and the uplink is implemented solely by the reader. In this case, the decision on successful tag response is taken when the reader actually accesses the tag ID.

5.2.2.1 Measurement setup 1

First of all some bench measurements are performed. A simple Schottky diode detector is directly fed with a single carrier and three different multi-sines with zero phase arrangement. A metric of DC power gain is defined (Equation (5.9)) relating the DC power collected with a multi-sine and DC power obtained with a single carrier. The measured results are depicted in Figure 5.8 where we can observe a power gain as high as 6 dB for a 16-tones multi-sine.

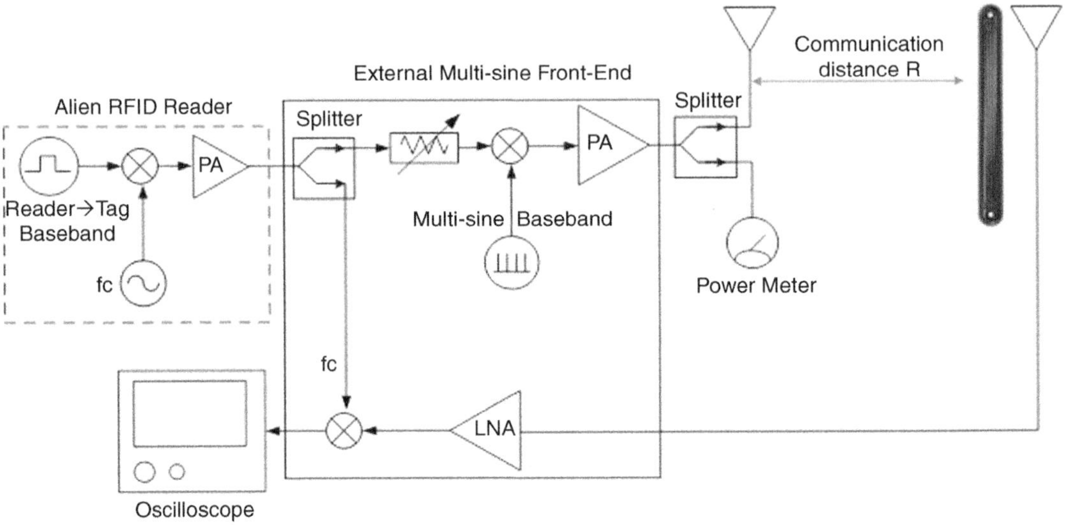

Figure 5.9 Laboratory setup used to measure tag sensitivity gain.

5.2.2.2 Measurement setup 2

In this measurement scenario a commercial RFID reader [26, 27] (compliant with ISO/IEC 18000–6 and with the EPCglobal Class-1 Generation-2 protocol [25]) is equipped with an external front-end. The front-end incorporates the multi-sine waveform in the original reader signal. Figure 5.9 depicts the laboratory test bed used in the measurements, similar to the one used in [20].

The reader's transmitter output is mixed with a baseband multi-sine and then amplified and transmitted. Two power splitters are used, the first one to have a sample of the local oscillator to be used as local oscillator in the receiving path and a second one used to provide a measure of the radiated average power. The signal reflected from the tag is down-converted (mixed with f_c) so that the tag baseband response can be visualized in an oscilloscope. This will provide us with information on whether the tag is or is not activated by identifying the tag response after each reader command (Figure 5.10). A power attenuator is used to control the radiated power. The receiving antenna and the tag (Class1 Gen2 compliant) are placed at a fixed distance R from the transmitter antenna. Figure 5.11 shows the experimental setup used.

In order to evaluate the improvements achieved by the multi-sine scheme, a power gain metric, $G_P = P_{CW} - P_{min}$ (dB), is defined as the difference between the minimum average power needed to activate the tag (and get a response) using a single carrier signal (P_{CW}) and the minimum average power required to activate the tag using a multi-sine signal (P_{min}) for the same distance R. G_P can be seen as a sensitivity improvement. To measure such a gain, first the tag is illuminated with a single carrier, and by varying the attenuation of the transmission path the minimum average power (P_{CW}) to activate the tag at a distance R is found. In the second experiment we repeat the previous procedure but now using a multi-sine excitation.

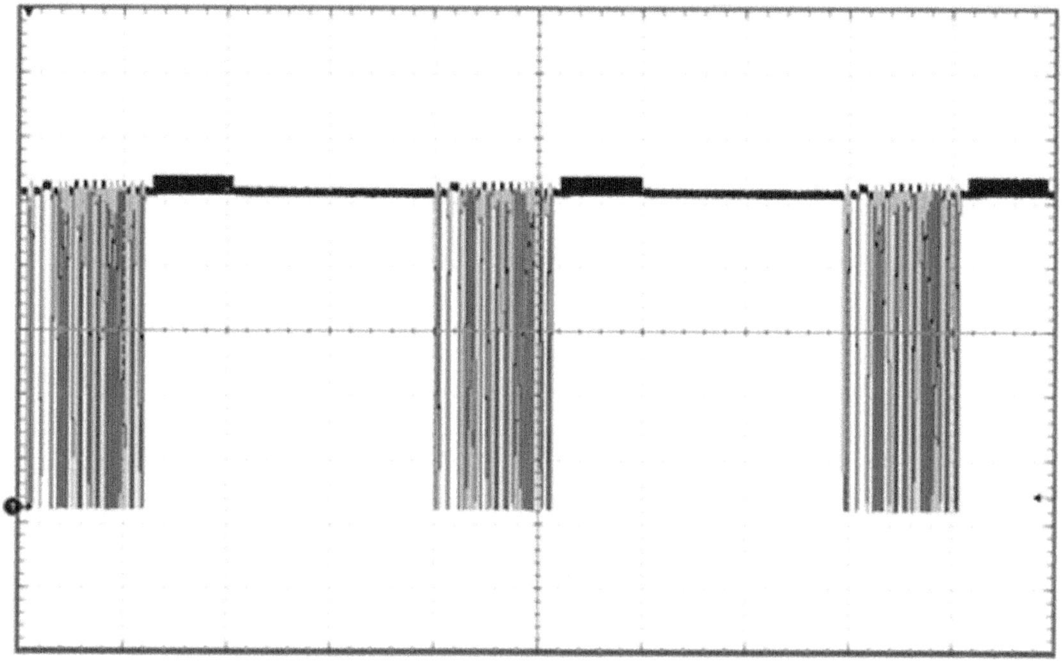

Figure 5.10 Baseband signal visualized in the oscilloscope: reader commands followed by tag responses.

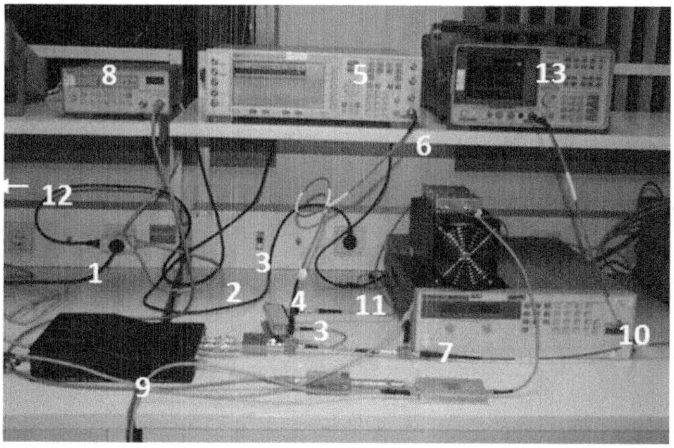

Figure 5.11 RFID reader and multi-sine front-end setup: 1- RFID reader, 2- power splitter, 3- variable attenuator, 4- mixer, 5- signal generator, 6- power amplifier, 7- power splitter, 8- power meter, 9- cable to the transmitting antenna, 10- cable from the receiving antenna, 11- mixer, 12- oscilloscope, 13- spectrum analyzer.

Table 5.1 Expected sensitivity gains and reading range gains obtained in the first measurement scenario.

Signal	$\Delta\omega$ (MHz)	P_{min} (dBm)	G_P (dB)	r (m)	r (%)
Single-tone		19.3	0.0	0.0	
2-tones	3.0	17.4	+1.9	+0.46	24.2
	2.0	17.5	+1.8	+0.44	23.2
	1.0	17.8	+1.5	+0.36	18.9
4-tones	3.0	16.5	+2.8	+0.72	37.9
	2.0	17.2	+2.1	+0.52	27.4
	1.0	17.7	+1.6	+0.38	20.0
8-tones	2.0	16.2	+3.1	+0.81	42.6
	1.0	16.4	+2.9	+0.75	39.4
	0.5	no response	–	–	–

By using the measured gain G_P and the propagation law we can estimate the expected communication range gain r (in meters). The minimum DC power needed to activate the tag at a distance R when using a single carrier, is $P_{DC_{min}} = \eta_{RD} P_t G_t G_r \left(\frac{\lambda}{4\pi R}\right)^2$, where η_{RD} is the RF-DC conversion efficiency, P_t is the transmitted power, G_t and G_r are the transmitter and receiver antenna gains respectively, λ is the wavelength, and R is the distance between reader and tag antennas. On the other hand, when using a multi-sine with the same transmitted power P_t, a power gain G_P, and an extra reading range r will be obtained. So, the Friis formula can be rewritten as follows:

$$P_{DC_{min}} = G_P \eta_{RD} P_t G_t G_r \left(\frac{\lambda}{4\pi(R+r)}\right)^2.$$

From the two previous Friis equations an estimate for the reading range gain can be obtained, as in [1]: $r = R(\sqrt{G_P} - 1)$ (m).

Table 5.1 shows the measurement results. Four different signals were used in the measurements, a single carrier and three multi-sines with two, four, and eight tones. Each multi-sine was also tested with three different tone separations. The reference signal (single carrier) required a minimum transmitted power $P_{CW} = 19.3$ dBm for a reference fixed distance of $R = 1.9$ m. For the fixed distance R and based on the reference power P_{CW} and on the minimum power for each multi-sine P_{min} we are able to obtain the tag sensitivity gain G_P. Finally, we can estimate the expected reading range gain r. The best result is obtained for the 8-tones multi-sine with an optimal tone separation of 2 MHz, indicating a communication distance improvement of 43%.

The gains are degraded with decrease of the tone separation and we can also observe that in some cases the tag does not respond. This is due to the ripple effect on the tag's detector and charge pump imposed by the very low frequency separation between tones in the multi-sine.

5.2.2.3 Measurement setup 3

In this measurement setup, instead of evaluating the tag response using an oscilloscope, the tag response is directly forwarded to the reader receiver as in Figure 5.5a, and the

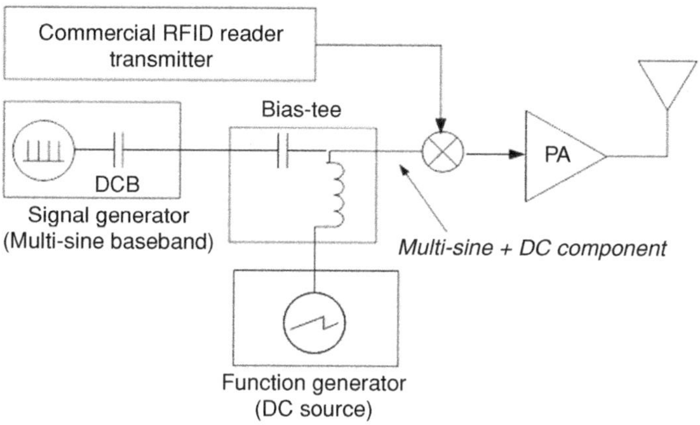

Figure 5.12 Setup used to impose a DC component in the baseband multi-sine. Hence, the band-pass multi-sine will have a frequency component at LO and will fulfill the odd number of tones and central frequency positioning requirements. The DC voltage is varied in order to level the amplitude of the central tone and to match it to the other subcarriers.

reader actually accesses the tag ID. In this case, the downlink path is implemented by the reader combined with the front-end and the uplink is implemented solely by the reader. The decision on successful tag response is taken when the reader actually accesses the tag ID. Contrary to the very complex receiver configuration (a matched filter multi-sine demodulator) that has been proposed in [21], we propose a very simple scheme that only implies a proper multi-sine design and does not require any additional hardware to be deployed. Nevertheless, to implement this scheme the multi-sine requirements imposed in the previous section must be fulfilled, namely the multi-sine nature (odd number of tones), the central tone positioning (coinciding with the local oscillator), and tone separation and bandwidth requirements. The latter requirements should avoid spectrum overlapping and guarantee compliance with the internal low-pass filter of the reader.

As stated before, the bandpass multi-sine signal must have a component at the local oscillator frequency and consequently the baseband multi-sine signal must have a DC component. Since the baseband multi-sine is being provided by a signal generator (that employs a DC block at the output), it has no DC component. For this reason the setup presented in Figure 5.12 is used to force a DC component in the baseband multi-sine. A function generator combined with a bias-tee as in Figure 5.12 is used for this purpose.

Table 5.2 shows the measurement results obtained in the second scenario, where the RFID reader is actually accessing the tag ID. The signal description x-tones+1 stands for a multi-sine with x lateral subcarriers plus a central one (actually a multi-sine with an odd number of tones). In this case the distance between reader and tag was $R = 1.7$ m and the minimum power to activate the tag with the single carrier signal was 16.5 dBm. Although the gains are lower than the first case setup, a good improvement of almost 25% is achieved with an 8-tone multi-sine with 2 MHz tone separation.

Table 5.2 Sensitivity gains and reading range gains.

Signal	$\Delta\omega$ (MHz)	P_{min} (dBm)	G_P (dB)	r (m)	r (%)
Single-tone		16.5	0.0	0.0	
2-tones+1	2.0	15.4	+1.1	+0.23	13.5
	1.0	16.9 (–)	–0.4	–0.08	–4.7
4-tones+1	2.0	14.9	+1.6	+0.34	20.0
	1.0	16.2	+0.3	+0.06	3.5
8-tones+1	2.0	14.6	+1.9	+0.42	24.7
	1.0	16.2	+0.3	+0.06	3.5
	0.5	no response	–	–	–

5.2.3 Results discussion

In the envelope detector experiment the use of multi-sine signals has shown an effective enhancement in the RF-DC conversion efficiency. In this case a 16-tones multi-sine has provided a DC power gain of 6 dB over the single carrier. Over-the-air measurements have been also carried out. An RF front-end was integrated with a commercial RFID reader in order to increase the reading distance using multi-sine signals. Both the forward link (reader–tag) and the reverse link (tag-to-reader) have been implemented. No additional changes were needed in conventional receiver architecture of commercial RFID readers in order to receive and demodulate the multi-sine signal backscattered from the tag. Good coverage range improvement has been achieved.

5.2.4 Protocols for reduction of energy consumption

The transmitted power level of RFID readers and passive sensor interrogators is critical. While the minimum power level is imposed by tag/sensor activation level (or sensitivity), the maximum transmitted power is also important for several reasons. First, the maximum transmitted power is limited by regulations. Second, RFID reader emissions can cause interference in neighboring systems. Finally, optimization of the power level is important for energy saving purposes.

Dynamic power control is a well-known strategy applied in cellular mobile systems to save energy both in the mobile station (MS) and in the base station (BS). Recently, some researches have addressed power control algorithms in dense RFID networks (where several RFID readers operate in adjacent cellules) [28, 29]. The objective is to reduce interference caused by reader emissions and to improve the network performance as well as the overall coverage range. This is achieved by dynamically adjusting the RFID reader output power according to the interference level detected during tag interrogation.

In reference [30], a method for managing the power consumption of an RFID reader is presented that includes the use of proximity and motion sensors to detect an RFID-tagged object in the field of the reader. The work conducted in [31] presents an energy-efficient RFID inventory algorithm, which automatically estimates the number of tags to be read and incrementally adjusts the power level. This way no excessive power is transmitted and energy is saved (the authors claimed more than 60% energy reduction).

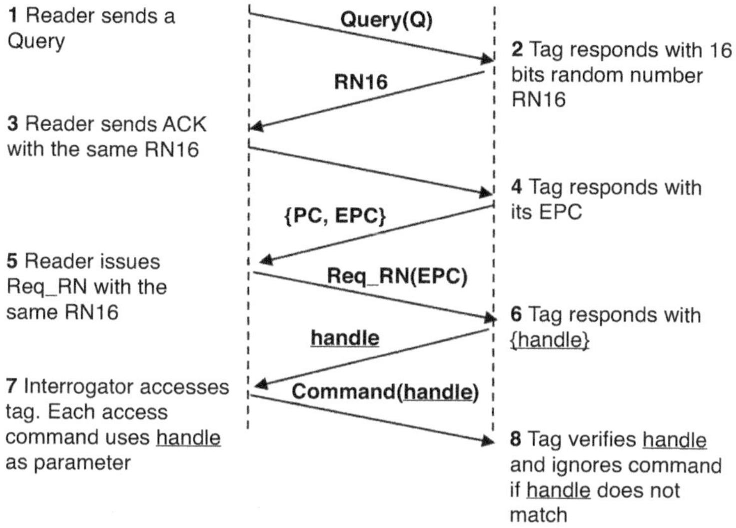

Figure 5.13 Example of tag inventory and access after ISO18000–6C protocol.

However, at the reader–tag communication level, conventional single RFID reader configurations do not employ power control, the reader transmits a fixed power level independently of the number of tags in the field and distance between the tag and the reader. In this subsection we propose a dynamic power control architecture and algorithm based on the received signal strength indication (RSSI) received from the tag. Such a system employs an RSSI receiver in order to measure the strength of the received signal. Based on the tag RSSI, an estimate of the distance from the reader can also be obtained and the transmitted power can be re-adjusted.

5.2.4.1 Proposed architecture and algorithm

This scheme is intended to be applied in passive RFID UHF systems (ISO18000–6C, EPCglobal Class1 Gen2 [25]). The basic protocol interaction between an EPCglobal compliant interrogator and a tag is depicted in Figure 5.13 [25].

After the first tag response (RN16) to the first reader command (Query), the reader measures the RSSI from the tag response and re-adjusts its output power according to the estimated distance from the tag (received RSSI provides information about distance between reader and tag). An adequate algorithm to estimate distance from the RSSI should be implemented. Figure 5.14 shows a simplified flowchart of the proposed power control algorithm. The architecture of the proposed hardware is presented in Figure 5.15.

5.2.4.2 Signal strength measurement

Typical RFID receivers use in-phase and quadrature (IQ) receivers (Figure 5.16) to demodulate the signal reflected back from the tag [32]. The IQ configuration is used to overcome the problems of simple ASK/OOK demodulators [33]: the signal reflected back from the tag is a phase-shifted version of the carrier transmitted by the reader and the phase shift depends on the distance between the reader and the tag. If the distance is

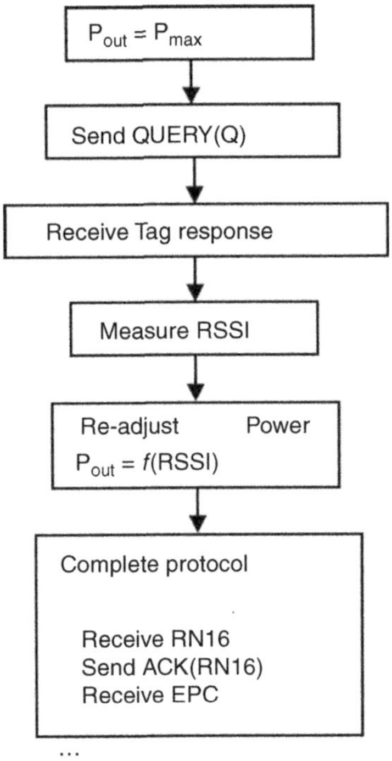

Figure 5.14 Flowchart of the proposed power control algorithm.

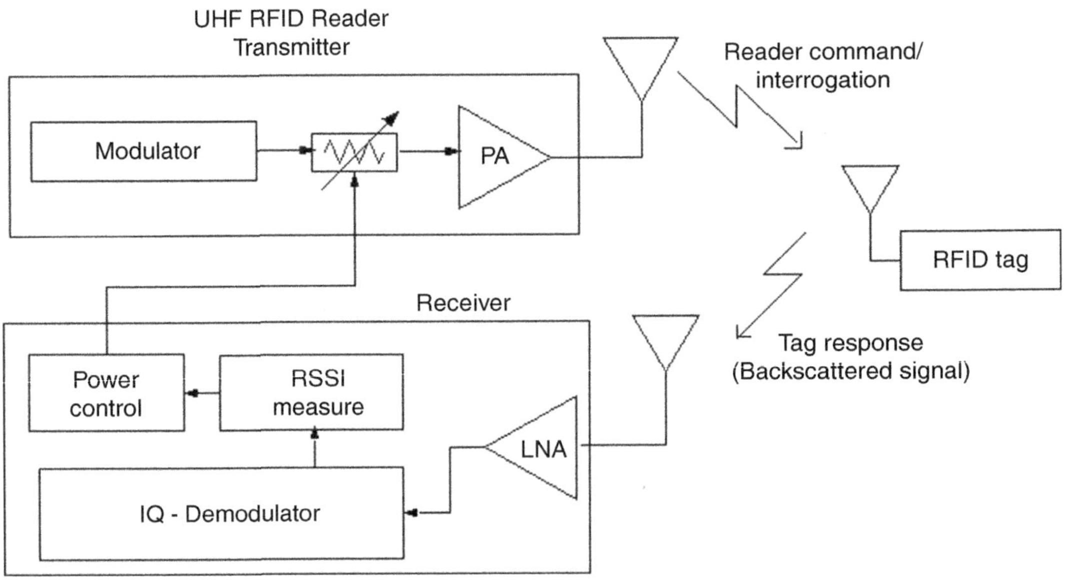

Figure 5.15 Hardware implementation including an RSSI receiver and output power control.

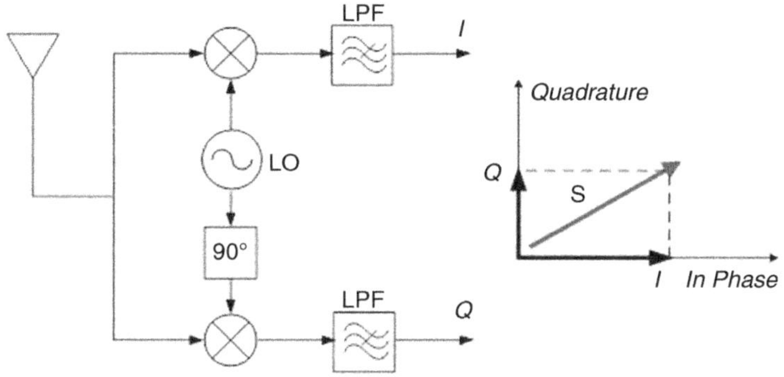

Figure 5.16 IQ demodulator typically used in RFID receivers (left) and received constellation diagram (right).

such that the local oscillator and the received signal are in quadrature (90° phase shift), then the output of an ASK/OOK demodulator is zero and the information from the tag is lost. By using the IQ architecture with two complementary branches the information loss is avoided. The same IQ receiver hardware can be used to measure the RSSI. An indication of the signal strength can be given by:

$$S^2 = I^2 + Q^2. \tag{5.12}$$

5.2.4.3 Preliminary measurements

In this work the first part of the power control scheme (basically, the first four steps of the flowchart of Figure 5.14) is implemented with a dedicated measurement setup (similar to the one used in [20]), that is composed of an RFID reader, some external components (e.g. mixers), an oscilloscope (to capture the signals), and Matlab scripts (to analyze the waveforms and compute the RSSI). We intend to obtain an indication of the signal strength received from the tag as a function of the distance between the reader and the tag. Based on this indication we will evaluate whether it can or cannot provide useful information about the distance. To do so we interrogate the tag with a commercial RFID reader and use a dedicated measurement setup to measure the air interface waveforms which will be stored for further analysis and calculation of a signal strength indication using Matlab. The baseband waveforms of several reader commands and tag responses were captured for the same transmitted power and for different distances between the reader and tag. As an example, Figure 5.17 depicts the baseband captured at the reader location for the tag at 2 meters away from the reader. Figure 5.18 is a zoom of the tag response received at the reader location. Table 5.3 shows the amplitudes of tag response, A. Based on these amplitudes it is possible to calculate an RSSI metric and have an estimation of the distance.

A dynamic power control scheme for application in RFID has been presented and, as a first step, a setup has been implemented to study the suitability of such a scheme. For a fixed transmitted power level we conduct static measurements for different distances in order to evaluate the amplitude of the tag response. Based on the received amplitudes

Table 5.3 Amplitude of the tag baseband signal at the receiver for different distances between reader and tag.

Distance (m)	0.9	1.3	1.75	2	2.9	4.4
A (mV)	1.6	0.63	0.53	0.50	0.44	0.28

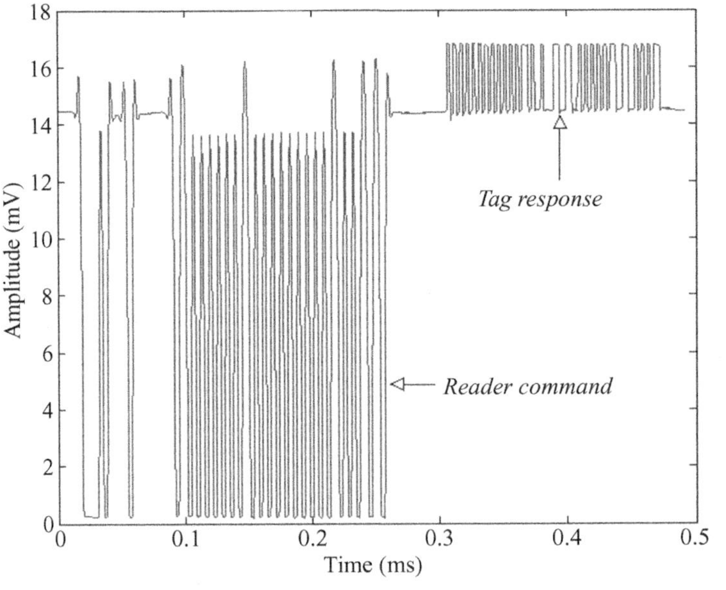

Figure 5.17 Baseband waveform of reader commands followed by tag responses for a reader–tag distance of 2 meters.

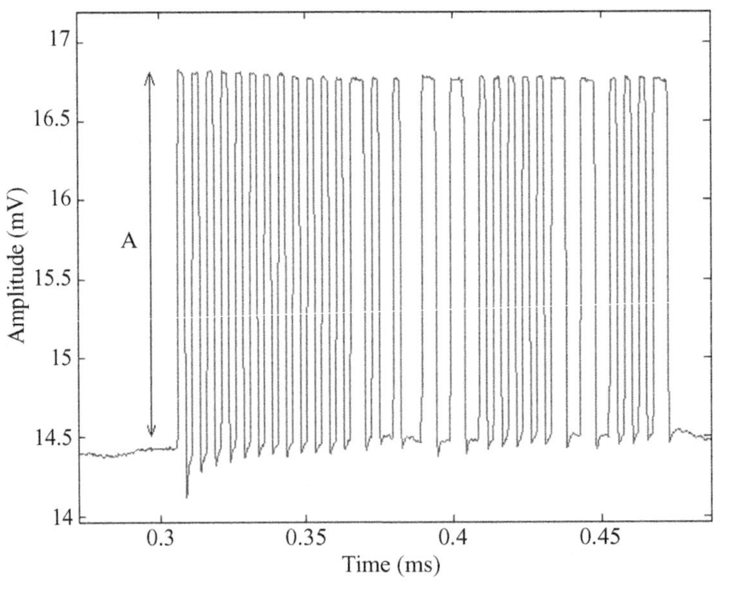

Figure 5.18 Zoomed version of tag response.

an RSSI metric and an estimation of distance can be calculated in order to dynamically re-adjust the transmitted power level.

5.3 UWB location based on passive sensors

In the past the acronym RFID meant wireless identification (pure and simple), today it means not only that but also wireless sensing. In fact, sensor-enabled RFID tags are becoming more and more common. In the future RFID will probably mean identification, sensing and high-resolution localization. It is already possible to find in the literature some very interesting ways to transform typical RFID systems into something much more suitable for localization applications. In general, these techniques do not require any kind of hardware changes and focus on increasing localization accuracy by deploying additional tags and readers, usually at known locations. These modified RFID systems can be thought of as the first generation localization-enabled RFID systems.

Next generation commercial RFID readers and tags will include built-in localization-related features. It is also likely that these localization-related features will be based on ultrawideband (UWB) pulses. For instance, an interesting example of a UWB-based RFID system is presented in [2]. In this case each tag consists simply of a UWB pulse generator, a microcontroller unit (MCU), and a CR2032 battery. Even though there is no battery-free operation and the tags are not able to receive any data (only transmit), this example clearly shows that mixing RFID and UWB can yield some very interesting results. Each tag periodically generates pulses with peak-to-peak voltages of approximately 2.8 V (which gives around 15.5 dBm peak considering a 50 ohm load) and duration of 1.5 ns. The pulses move away from the tag and towards several readers placed at fixed, well known locations. The differences between the pulse arrival times reported by each one of the receivers are then used to figure out where the tags are located. This localization technique is very popular and is known as time difference of arrival (TDoA). The reported accuracy is 10 cm, measured in a 6 by 6 meter room, and using a total of four non-coherent receivers with a sensitivity of –71 dBm. This kind of accuracy is, in general, not achievable using the typical RSSI-based techniques. In order to integrate UWB into passive RFID tags, the challenge is to design pulse generators able to operate with little available power, but capable of producing pulses strong enough to be easily detected by readers and short enough so that the localization is accurate.

To be considered UWB a signal must have a bandwidth of at least 500 MHz absolute (measured at 10 dB below the peak) or 0.2 fractional (which is given by the ratio between the absolute bandwidth and the center frequency). This UWB definition was initially proposed by the Federal Communications Commission (known as FCC), and is currently the most often found in the literature. In terms of regulations, for the United States the FCC established the core UWB frequency band as being from 3.1 to 10.6 GHz. Within this band the transmitted power spectral density (PSD) limit is higher, fixed, and equal to –41.3 dBm/MHz. In the time domain there are basically no restrictions to be taken into account. In Europe there are also no restrictions on the time domain, but the maximum

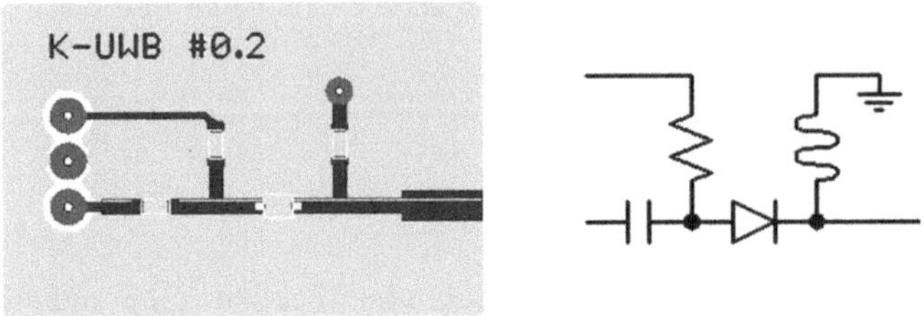

Figure 5.19 Physical layout of the PIN diode-based UWB pulse generator and corresponding schematic.

PSD is equal to –41.3 dBm/MHz only from 6 to 8.5 GHz. Outside the previous band it is always either –65 dBm/MHz or lower. Additional details about regulations can be found in [34]. In general, UWB pulse generators are based on high speed logic gates, transistors driven in the avalanche region, or one of several types of diodes, which can be avalanche, tunnel, or step recovery (SRDs). Pulse generators made with logic gates usually only need two of these (NOT plus AND is a popular example). The circuit is designed to force a race condition at the output gate, causing a short glitch. Creating UWB pulses using this method is, however, not straightforward. The reason is that generating shorter pulses requires faster logic, and the faster logic families tend to consume more power. A comparison between several types of pulse generators (logic-based pulse generator included) is presented in [35]. Transistors driven in the avalanche region can be used to generate extremely short pulses but usually require high supply voltages, as mentioned in [36]. For instance, the pulse generator described in [35] uses a supply voltage of 300 V. High voltages are also often linked to avalanche diodes. In fact, the latter are normally only used to generate high power pulses (needed for applications such as long range radar). Tunnel diodes are very suitable for use in extremely short (sub-nanosecond) pulse generators. This is true because of their small transition time (the smallest of all the diode types mentioned). The disadvantage of this kind of diodes is their relatively low output voltage [36]. Another point worth noting is that tunnel diodes are not easy to obtain, due to low volume manufacturing. Lastly, SRDs are not as good as tunnel diodes in terms of transition times but do not suffer from the output voltage problem. SRDs are commonly found in literature related to UWB pulse generation (much more than any other kind of diode). These diodes are currently seen as very promising devices for low power UWB applications, yet (like tunnel diodes) they are very hard to obtain. Some authors even call it a "dying art." A table with the most relevant parameters of SRDs is provided in [36]. PIN diodes usually do not appear in the UWB-related literature, but can also be used to produce UWB pulses. They are not as good as SRDs as pulse generators, but are cheap and more abundant.

In [37] a UWB pulse generator based on a PIN diode is presented. Its circuit layout and schematic are both shown in Figure 5.19.

This pulse generator consists of a PIN diode (model BA592), a 1 nF capacitor, an 820 Ω resistor (for biasing purposes) and a 3.9 nH inductor mounted on a 0.813 mm thickness board reference RO4003C (a low loss substrate designed for high frequencies).

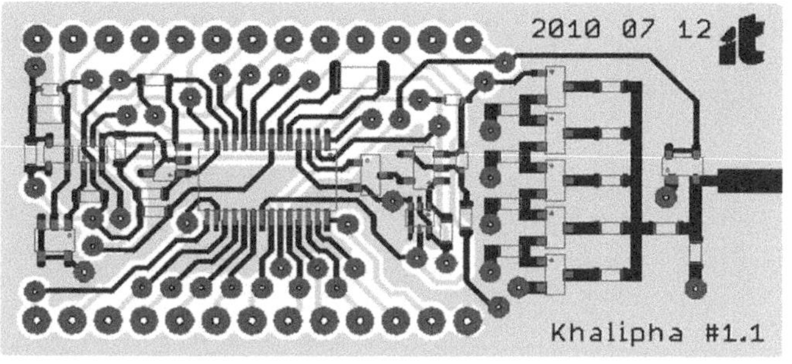

Figure 5.20 Physical layout of the passive RFID tag (shown on a 1:1 scale) used to test the UWB pulse generator [1].

Before describing how the pulse generator works it should be noted that the bottom and top pins of the three-pin connector seen in the layout correspond to input and bias, respectively. They were not short-circuited directly on the board only for testing purposes, but represent only one input. If the input voltage is set to 0 V the pulse generator is turned off and the power consumed is virtually zero. Once the voltage is increased to 1.8 V the diode becomes forward biased and begins to accumulate charge. As the input voltage is decreased again the current in the capacitor increases (from right to left in the figure) quite sharply. Some of this current is supplied by the load (a 50 Ω load in this case), but the largest part comes from the inductor. The diode loses charge continuously during this process. Once it fully discharges, its impedance changes from a low value to a very high value in a very short time, the transition time mentioned earlier. The result is that the current passing through the inductor is redirected to the 50 Ω load, producing a positive pulse. This UWB pulse generator was designed to be used in conjunction with the passive RFID shown in Figure 5.20.

This tag includes a multi-level voltage multiplier (Dickson configuration) composed of Schottky diodes and tuned to harvest as much RF energy as possible around 866.6 MHz. The harvested energy is then converted to DC, using a 1.8 V low-dropout voltage regulator. In terms of processing capabilities, the tag includes a 16-bit flash MCU from the MSP430 series, model F2132. The UWB pulses are triggered by this MCU, using a free digital output pin. Each time the selected pin is switched from high to low (1.8 V to 0) a pulse like the one shown in Figure 5.21 is produced.

Obviously, in order to build a passive RFID localization system, generating UWB pulses without using batteries represents only one of the challenges to overcome. For instance, in the example mentioned previously the tags were battery-powered, meaning that the range of the system was basically limited only by the sensitivity of the UWB receivers. In passive RFID systems the tags typically run out of power long before the sensitivity of the readers can even cause any problems. Since deploying the UWB receivers very close to one another in order to avoid gaps in the energy coverage would be a waste of resources, a concept like the one illustrated in Figure 5.22 may be very useful.

The system would consist of power and control units. Power units would be designed (as their name suggests) to radiate power to the tags. They would also be optionally

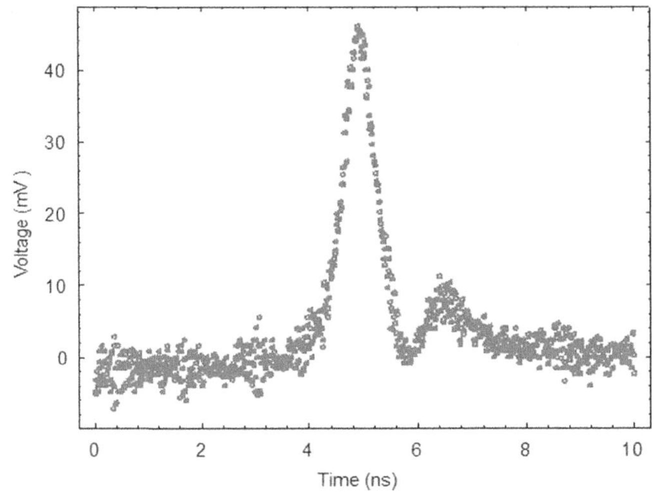

Figure 5.21 Pulse generated by the UWB pulse generator, in response to a high to low transition (1.8 V to 0) produced by the RFID tag.

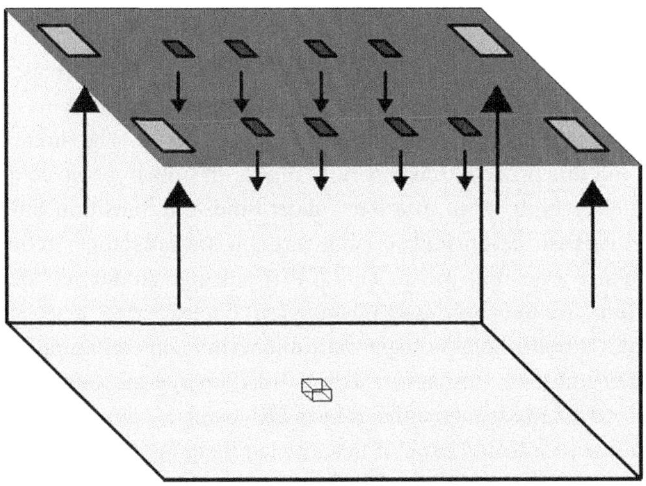

Figure 5.22 Potential structure for an RFID-UWB passive localization system.

capable of receiving simple instructions (like turn on, turn off, increase power output, or decrease power output) via an integrated wireless receiver. Power units would obtain their own energy from a nearby AC mains socket or by making using of a renewable energy source (such as a solar panel). Control units would communicate wirelessly with the tags and control the power units. Similarly to the power units, control units would be powered either by a nearby AC mains socket or by a renewable energy source. Separate devices for communication and power make sense, not only because the communication range is in general higher than the power range (as previously mentioned), but also because UWB receivers would in principle be more expensive than the power units due to the signal processing they have to do. Using this strategy, improving coverage would simply mean increasing the number of power units as necessary and perhaps installing a couple of new UWB receivers.

5.4 RFID for household applications

As an example of RFID application in home automation, this section presents a battery-less remote control system using multiple passive RFID chips. RFID tags were originally conceived to operate separately/independently and their applications are typically limited to the unique identification of people, animals, and objects. The arrangement of several RFID tags working together and sharing the same antenna fits well into the category of unconventional RFID systems. However, an unplanned attempt to interconnect several tags working together and sharing the same antenna would fail. In this section we disclose a multi-RFID scheme for application in a battery-free remote control system. We propose and design a novel N-port network in order to include controllability of ordinary RFID chips and in order to properly interconnect them in a common structure with a shared antenna.

5.4.1 A battery-less remote control system based on a multi-RFID scheme

Remote control systems are used to wirelessly control TVs, doors, game consoles, air-conditioners, and many other devices. Conventionally, remote control systems are based on infrared (IR) technology and the controller requires batteries to operate. Such IR battery-powered systems present two major drawbacks. First, optical communication using IR signals requires line-of-sight, otherwise the communication fails. More importantly, the remote controller uses disposable batteries which leads to several problems: first, there is a cost associated with battery maintenance. Second, the limited lifetime of batteries is a constant source of frustration. More seriously, disposable batteries generate huge amounts of toxic waste and release heavy metals at the end of their lifecycle. So, beyond monetary cost, there is a tremendous environmental price to pay. Disposable batteries take hundreds of years to decompose in nature, and if not properly treated they can present a serious risk to public health and a threat to the environment.

In order to overcome the communication problems associated with IR technology – short range and need for line-of-sight – recently, radio frequency remote controls have been adopted by manufacturers for TV and audio handsets instead of the traditional IR-based technology. Due to the inherent advantages of wireless radio communication over IR for low data rates, namely long range and non-line-of-sight communication, manufacturers have started to provide some RF-based remote control solutions. While in the past IR technology was preferred and became widespread thanks to its cost effectiveness and low complexity implementation compared to past RF solutions, nowadays, as the RF/microwave industry is growing and reaching large volume production, the price per unit is being considerably lowered and it is likely that in the coming years IR technology will be substituted by radio. Nevertheless, both in conventional IR technology and in the radio frequency approaches based on conventional super-heterodyne architecture, there is a need for disposable batteries; that, as mentioned before, has many drawbacks. Our proposal is prompted by and based on a very peculiar class of radio technology (passive RFID), where the mobile devices have no self-battery to power up their electronics, but instead all the energy needed is wirelessly delivered by a fixed device (the RFID reader) [32, 38]. In order to provide a cost-effective solution a widespread and commercially available low cost passive technology is selected, the UHF EPCglobal Gen2 Class1 [25].

Accordingly, in this section we describe a novel eco-friendly battery-less remote control system based on wireless power transmission and passive UHF RFID. The proposed remote controller is a fully passive device that does not require the use of batteries or other installed power source to operate. This approach can save a huge amount of disposable batteries that otherwise would generate toxic substances at the end of their lifecycle.

The controlled device (e.g. a TV, a game console, or a door) incorporates an RFID reader system that remotely powers-up the remote controller device and communicates with it using UHF radio waves. The battery-less remote controller incorporates a plurality of N passive RFID chips/tags, N keys/switches, and a novel N-port microstrip network specially conceived to interconnect the various RFID chips. Each remote controller's key is associated with an RFID chip by a unique identifier (UID) which allows the device to be controlled to identify the key pressed by the user.

The microstrip network combined with the switches ensures proper interconnection between the chips and allows them to share the same antenna in such a way that only the chip associated with the depressed key is read. In the following, a brief description of the system is presented followed by an analytical model of the novel N-port microstrip network. The proper arrangement of this network along with the tags and the switches is also proposed. Finally a functional demonstration prototype is successfully tested and validated, in which the remote control system is integrated in a TV set and the basic control functionalities, CH +, CH −, Vol +, and Vol −, are implemented as a proof of concept.

5.4.2 Proposed system

The device to be controlled (DC), e.g. TV, a game device, a garage door, or other equipment, wirelessly energizes the battery-less remote controller using radio waves. For this purpose, the DC incorporates internally or externally an RFID reader that remotely energizes the RC and accesses the memory of the passive RFID tags in order to identify the key pressed by the user.

Compared to conventional infrared (IR) remote control systems this system presents several advantages, namely the use of a clean and eco-friendly form of energy transference using radio waves, no need for batteries, elimination of costs associated with battery maintenance and treatment of toxic waste, long range, and no need for line-of-sight communication thanks to the use of radio waves. It is also a very low cost solution thanks to the use of a mature passive RFID technology (UHF EPC technology [25]).

The system includes a remote controller (Figure 5.23(a)) that is made up with passive UHF RFID chips/tags, the device to be controlled, which can be a TV, a cable TV Box,

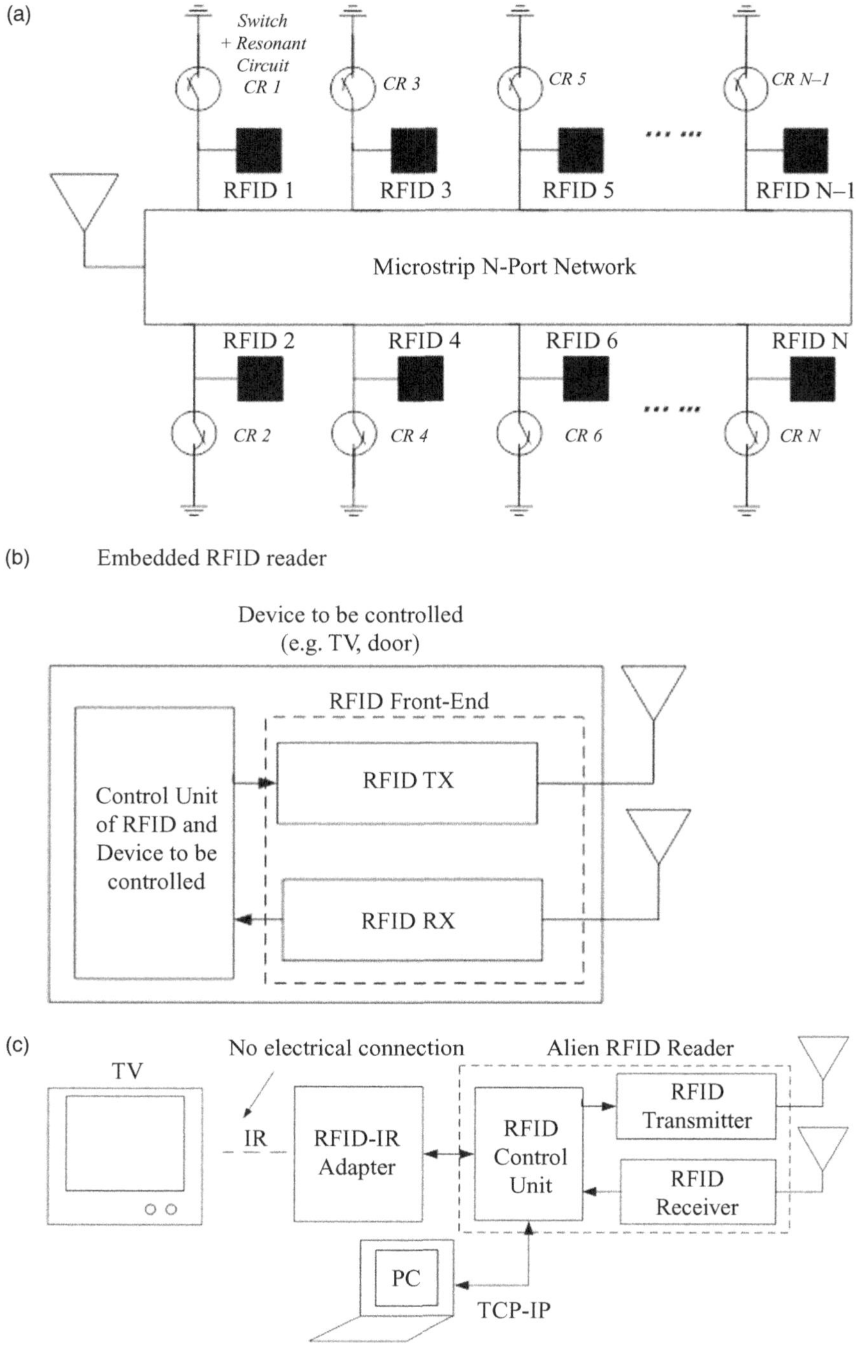

Figure 5.23 Proposed battery-less remote control system: a) The multi-tag passive remote controller; b) Device to be controlled with embedded RFID reader; c) Device to be controlled with external RFID reader plus RFID-IR adapter to guarantee compliance with already installed IR equipment.

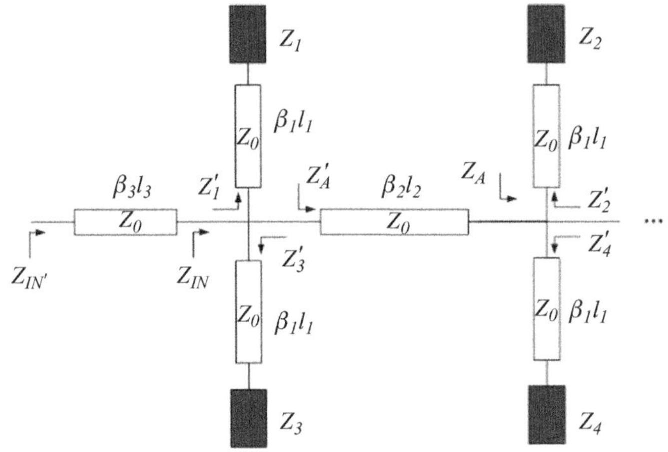

Figure 5.24 Model of the N-port microstrip network. The port Z_{IN} corresponds to the common antenna port and the ports Z_i correspond to the N keys of the remote control.

or a garage door. To access the information on the tags' memory, an RFID reader is internally (as in Figure 5.23(b)) or externally (as in Figure 5.23(c)) mounted on the DC. Additionally, in order to guarantee compliance with already installed systems (e.g. old TVs) an RFID-IR adapter/interface can be used (Figure 5.23(c)). The controller consists of N passive RFID tags associated with N keys/switches. A novel N-port microstrip network (Figure 5.24), specially designed for this purpose, guarantees that: by default all the tags are inactive (silent); once a key is pressed by the user the respective tag goes to active mode and is read by the RFID reader to identify the depressed key; and inactive tags don't interfere with the active one. The remote controller communicates with the reader and receives power to operate the tags through an antenna via RF waves, precisely in the UHF band (860–960 MHz).

5.4.3 The novel *N*-port microstrip network

A novel N-port microstrip network (Figure 5.24) has been proposed to interconnect the N RFID tags and the respective N switch-controlled resonant circuits. This network combined with the switch-controlled resonant circuits must guarantee that only one RFID tag, the active one, is connected to the antenna (port Z_{IN}') at the time. Moreover, all other inactive tags should not interfere with the active one. That is, the port Z_{IN}' must see at each instant only the active tag port $Z_{i=n}$ while all other ports remain invisible. The following requirements must be met:

- Only one impedance $Z_{i=n}$, corresponding to the active port n should be routed to the input port (antenna port).
- Active port (active RFID tag) is selected by the user using a switch.
- All other ports remain invisible to the input/antenna port.
- All other ports do not interfere with port n.

- Insertion loss between antenna port and active port n is ideally 0 dB.
- Cross-talk between input port and all inactive ports is ideally null.
- Cross-talk between active port and all others is ideally null.
- Each impedance Z_i attached to port i assumes two values: $Z_i = 0$ or $Z_i = Z_{\text{TAG\_MATCHED}}$.
- By default the impedance of all ports is $Z_i = 0$, superimposed by a switch-controlled resonant circuit resonating at the operating frequency.
- By acting on the switch n the user sets $Z_{i=n} = Z_{\text{TAG\_MATCHED}} = Z_0$.

The model of the proposed N-port microstrip network is shown in Figure 5.24 and is analytically described in the following, using the lossless model of a microstrip transmission line.

Considering the lossless model of a microstrip transmission line, the input impedance seen looking into each branch i of the network is given by:

$$Z'_i = Z_0 \frac{Z_i + jZ_0 \tan(\beta_1 l_1)}{Z_0 + jZ_i \tan(\beta_1 l_1)}, \tag{5.13}$$

where $i = 1, 2, \ldots n \ldots N$, N is the total number of ports in the network, corresponding to the total number of keys in the remote controller, Z_i is the load attached to branch i, Z_0 is the characteristic impedance of the microstrip line, β_1 is the phase propagation constant, and l_1 is the physical length of the line.

Assuming that the central transmission line ($\beta_2 l_2$) is forced to produce a delay equal to a multiple of a half wavelength ($\beta_2 l_2 = k\cdot 180°$), then the input impedance of the network is given by:

$$Z_{\text{IN}} = Z'_1 // Z'_2 // Z'_3 // Z'_4 \ldots // Z'_N = \left[\frac{1}{Z'_1} + \frac{1}{Z'_2} + \frac{1}{Z'_3} + \frac{1}{Z'_4} + \cdots + \frac{1}{Z'_N}\right]^{-1}. \tag{5.14}$$

Additionally, consider that the impedance termination at each branch, Z_i, can assume two different impedance values corresponding to the cases where the tag is activated and deactivated:

$$\begin{cases} Z_n = Z_{\text{TAG\_MATCHED}} = Z_0, \text{ if the key}(n) \text{ is pressed,} \\ Z_i = 0, \text{ by default, if no action is performed.} \end{cases} \tag{5.15}$$

Finally, assuming that the phase shift of each branch is set to be a quarter wavelength plus a multiple of a half wavelength ($\beta_1 l_1 = 90° + k\cdot 180°$), and that only one tag is activated each time ($Z_{i=n} = Z_{\text{TAG\_MATCHED}} = Z_0$) while the rest remain deactivated ($Z_{i\neq n} = 0$, superimposed by a resonant circuit in parallel with the tag as in Figure 5.25), then according to Equations (5.13–5.15), the input impedance is as follows:

$$\begin{aligned} Z_{\text{IN}} &= \left[\begin{array}{l} \dfrac{Z_0 + jZ_1 \tan(90°)}{Z_0(Z_1 + jZ_0 \tan(90°))} + \dfrac{Z_0 + jZ_2 \tan(90°)}{Z_0(Z_2 + jZ_0 \tan(90°))} \\ \cdots + \dfrac{Z_0 + jZ_n \tan(90°)}{Z_0(Z_n + jZ_0 \tan(90°))} \cdots + \dfrac{Z_0 + jZ_N \tan(90°)}{Z_0(Z_N + jZ_0 \tan(90°))} \end{array}\right]^{-1}. \\ &= Z_n = Z_0 \end{aligned} \tag{5.16}$$

The conclusion of Equation (5.16) is that if the N-port network is properly designed and only one key is pressed each time (only one tag active each time) then the antenna

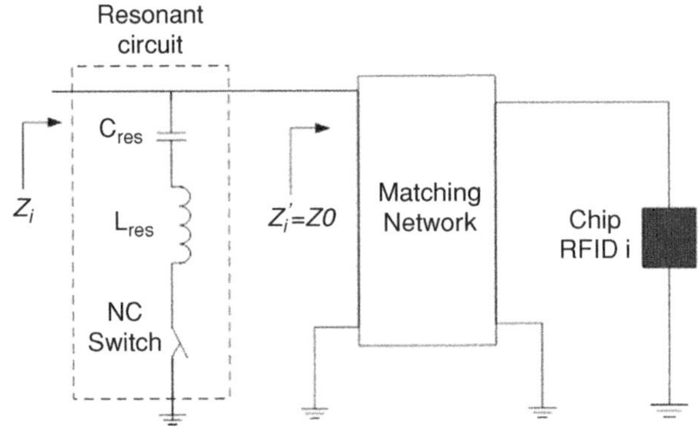

Figure 5.25 Port termination consisting of a parallel association of a switch-controlled series resonant circuit and an RFID chip matched to the characteristic impedance of the lines.

port will see only one RFID chip while the other will be deactivated and will not interfere with the active one.

Figure 5.25 depicts the termination of each network port. This termination is formed by a parallel association of a switch-controlled series resonant circuit and an RFID chip matched to the characteristic impedance of the lines. This configuration approximates to the condition expressed by Equation (5.15) and allows the user to activate the desired RFID chip by acting on the switch. The RFID chip is matched to the characteristic impedance of the system (Z_0) by the matching network block. A switch-controlled series resonant circuit formed by C_{res}, L_{res}, and a switch is put in parallel with the matched chip.

When the switch is closed it presents inductive behavior and it closes the series circuit. A switch that is normally closed is used, therefore this is the default state. In this state the capacitor C_{res} and the inductor L_{res} are dimensioned in such way that the series circuit (capacitor, inductor, and inductive-switch) resonates and presents a null impedance at the operating frequency (866.6 MHz). In this state the respective RFID chip is inactive and cannot be read by the reader. Otherwise, if the switch is open, it exhibits a capacitive behavior and the series circuit (capacitor, inductor, and capacitive-switch) presents very high impedance. In this state the RFID reader can access the respective chip. This state is reached when the user presses the respective key opening the switch.

The system is based on a commercially available low cost RFID, the remote controller unit can be manufactured at very low cost and can be integrated into old systems that use conventional infrared technology. This makes this solution quite affordable and cost-effective.

A major drawback of the proposed system is related to the need to have the RFID reader always turned on, which can degrade the system efficiency. Future work will try to solve this problem, for instance, by combining passive RFID with other energy harvesting technologies (e.g. harvest energy from the user finger pressure). A strategy can be developed where the RFID reader is turned off by default and the user

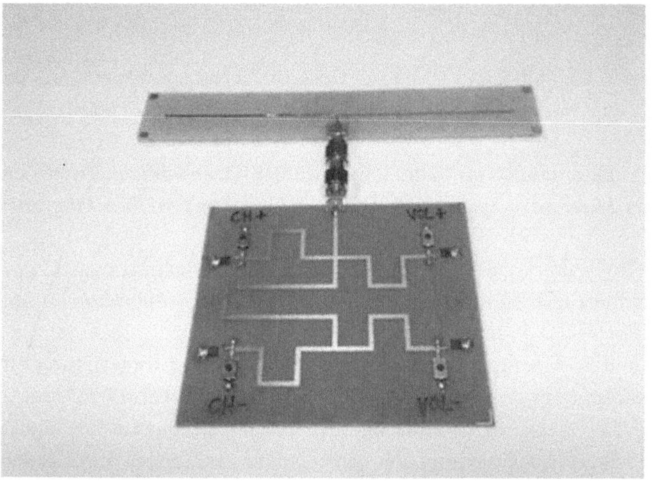

Figure 5.26 A prototype of a four-keys remote controller fabricated in low cost FR4 substrate. The concept can be extended to any number of keys. Keys are in the top and RFID chips are in the bottom.

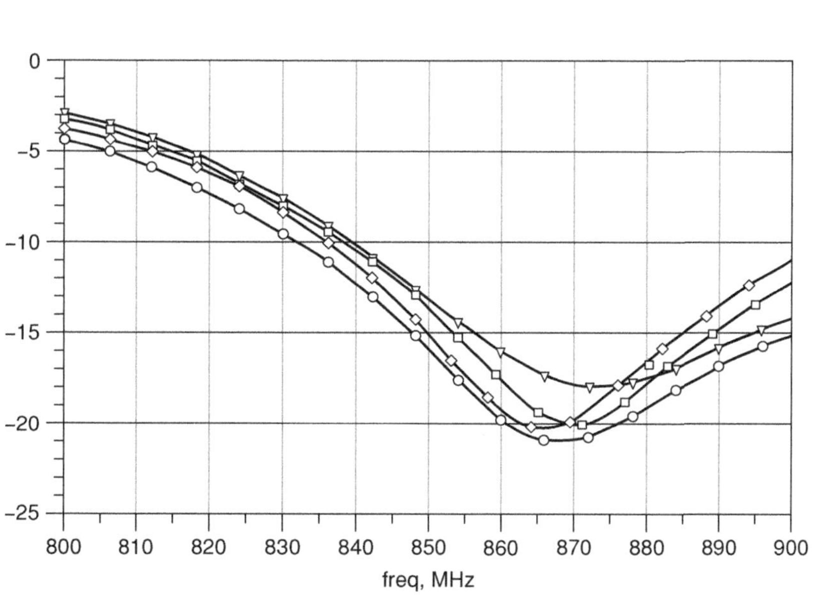

Figure 5.27 Measured return loss of a four-keys remote controller seen from the antenna port, when a key is pressed by the user: key 1 pressed (circles), key 2 pressed (diamonds), key 3 pressed (squares), key 4 pressed (triangles).

mechanical energy is used to generate a beacon pulse solely to wake-up the RFID reader, while further communication is made by the backscattering (RFID) mechanism. Figure 5.26 shows a prototype remote controller and Figure 5.27 measured return loss results.

References

[1] R. Fernandes, N. Carvalho, and J. Matos, "Design of a battery-free wireless sensor node," in *EUROCON – International Conference on Computer as a Tool* (*EUROCON*), IEEE, pp. 1–4, April 2011.

[2] Y. Zhou, C. L. Law, and J. Xia, "Ultra low-power RFID tag with precision localization using IR-UWB," in *Microwave Symposium Digest (MTT), IEEE MTT-S International*, pp. 1–4, June 2011.

[3] Alírio J. S. Boaventura and Nuno Borges Carvalho, "Maximizing DC power in energy harvesting circuits using multi-sine excitation," *IMS2011 – International Microwave Symposium*, Baltimore, USA, June 2011.

[4] A. Collado and A. Georgiadis, "Improving wireless power transmission efficiency using chaotic waveforms," in *Proc. IEEE MTT-S IMS 2012*, Montreal, 17–22 June 2012.

[5] A. P. Sample, D. J. Yeager, P. S. Powledge, A. V. Mamishev, and J. R. Smith, "Design of an RFID-based battery-free programmable sensing platform," *IEEE Trans. Instrum. Meas.*, **57**, (*11*), 2608–2615, 2008.

[6] A. Rida, L. Yang, and M. Tentzeris, "RFID-enabled sensor design and applications," in *Integrated Microsystems*, Artech House, 2010.

[7] A. Ruhanen, M. Hanhikorpi, F. Bertuccelli, *et al.*, *Sensor-enabled RFID Tag Handbook, BRIDGE*, ist-2005–033546, January 2008.

[8] M. Bolic, D. Simplot-Ryl, and I. Stojmenovic, *RFID Systems: Research Trends and Challenges*, in Wiley online library, 2010. http://eu.wiley.com/WileyCDA/wileyTitle/productCd-0470746025.html.

[9] G. Scholl, F. Schmidt, T. Ostertag, *et al.*, "Wireless passive SAW systems for industrial and domestic applications," in *IEEE International Frequency Control Symposium* 1998, pp. 595–601, May 1998.

[10] J. C. Butler, A. J. Vigliotti, F. W. Verdi, and S. M. Walsh, "Wireless, passive, resonant-circuit, inductively coupled, inductive strain sensor," *Sensors and Actuators A: Physical*, **102**, (*12*), 61–66, 2002.

[11] V. Viikari, H. Seppa, T. Mattila, and A. Alastalo, "Wireless ferroelectric resonating sensor," *IEEE Transactions on Ultrasonics, Ferroelectrics and Frequency Control*, **57**, (*4*), 785–791, 2010.

[12] V. Viikari, H. Seppa, and D.-W. Kim, "Intermodulation read-out principle for passive wireless sensors," *IEEE Transactions on Microwave Theory and Techniques*, **59**, (*4*), 1025–1031, 2011.

[13] A. Sample, J. Braun, A. Parks, and J. Smith, "Photovoltaic enhanced UHF RFID tag antennas for dual purpose energy harvesting," in *IEEE International Conference on RFID*, pp. 146–153. April 2011. *Also see*: A. Georgiadis, A. Collado, S. Via, and C. Meneses, "Flexible hybrid solar/EM energy harvester for autonomous sensors," in *Proc. 2011 IEEE MTT-S Intl. Microwave Symposium (IMS)*, Baltimore, USA, 5–10 June, 2011.

[14] Alírio Soares Boaventura and Nuno Borges Carvalho, "Enhanced front-end to extend reading range of commercial RFID readers using efficient multi-sine signals," *IMS2012 – International Microwave Symposium*, Montréal, Canada, June 2012.

[15] L. Ni, D. Zhang, and M. Souryal, "RFID-based localization and tracking technologies," *Wireless Communications, IEEE*, **18**, (*2*), 45–51, (April) 2011.

[16] M. Baghaei-Nejad and L.-R. Zheng, "Low cost and precise localization in a remote-powered wireless sensor and identification system," in *19th Iranian Conference on Electrical Engineering (ICEE)*, p. 1, May 2011.

[17] J. G. Bolander, P. A. Lovoi, and G. A. Zawolkow, "System for powering and reading RFID tags," *Tagent*, October 2010, uS7812725B2.

[18] *Talon Integrated RTLS Tag*, Tagent, tc1022, http://www.tagent.com.

[19] *Talon UWB Reader/Controller*, Tagent, July 2009, tr10, http://www.tagent.com.

[20] M. S. Trotter, J. D. Griffin, and G. D. Durgin "Power-optimized waveforms for improving the range and reliability of RFID systems," in *IEEE International Conference on RFID*, 2009.

[21] M. S. Trotter and G. D. Durgin "Survey of range improvement of commercial RFID tags with power optimized waveforms," *IEEE RFID Conference*, pp. 196–202, 2010.

[22] Hisanori Matsumoto and Ken Takei, "An experimental study of passive UHF RFID system with longer communication range," *Proceedings of Asia-Pacific Microwave Conference*, 2007.

[23] Chun-Chih Lo, Yu-Lin Yang, Chi-Lin Tsai, Chieh-Sen Lee, and Chin-Lung Yang, "Novel wireless impulsive power transmission to enhance the conversion efficiency for low input power," *Microwave Workshop Series on Innovative Wireless Power Transmission*, IEEE, pp. 55–58, 2011.

[24] Yu-Lin Yang, Chin-Lung Yang, Chi-Lin Tsai, and Chieh-Sen Lee, "Efficiency improvement of the impulsive wireless power transmission", *Microwave Workshop Series on Innovative Wireless Power Transmission, IEEE*, pp. 175–178, 2010.

[25] EPC Radio-frequency identity protocols Class-1 Generation-2 UHF RFID, Protocol for communications at 860 MHz – 960 MHz, Version 1.2.0, 2008.

[26] Alien Technology, *Hardware Setup Guide ALR-8800*, 2006.

[27] Alien Technology, *Reader Interface Guide*, September 2007.

[28] K. Cha, A. Ramachandran, and S. Jagannathan, "Adaptive and probabilistic power control algorithms for dense RFID reader network," *Proceedings of the 2006 IEEE International Conference on Networking, Sensing and Control*, ICNSC '06, 2006.

[29] Kainan Cha, S. Jagannathan, and D. Pommerenke, "Adaptive power control protocol with hardware implementation for wireless sensor and RFID reader networks," *IEEE System Journal*, **1**, (*2*), 145–149, 2007.

[30] Tong-Hsiao Chang, Karl Edward Keppeler, and Charles Rinkes, US Patent 20090309704 A1, *Methods and Systems for RFID Reader Power Management*, 2009.

[31] Xunteng Xu, Lin Guy, Jianping Wang, and Guoliang Xing, "Negotiate power and performance in the reality of RFID systems," *IEEE International Conference on Pervasive Computing and Communications (PerCom)*, 2010.

[32] Daniel M. Dobkin, *The RF in RFID: Passive UHF in Practice*, Elsevier, 2013.

[33] R. Abreu, N. Almeida, J. N. Matos, N. B. Carvalho, and J. S. Gomes, "A homodyne low cost uplink receiver for digital short range communication systems," *IEEE 65th Vehicular Technology Conference, VTC2007*, Spring 2007.

[34] H. Nikookar and R. Prasad, *Introduction to Ultra Wideband for Wireless Communications*, 1st Edn., Springer Publishing Company, Inc., 2008.

[35] A. De Angelis, M. Dionigi, R. Giglietti, and P. Carbone, "Experimental comparison of low-cost sub-nanosecond pulse generators," *IEEE Transactions on Instrumentation and Measurement*, **60**, (*1*), 310–318, 2011.

[36] Y. Y. Wei, "Ultra wideband signal generation," *Microwave Journal*, **48**, (*9*), 172–184, 2005.

[37] R. Fernandes, N. Carvalho, and J. Matos, "Low power UWB pulse generator based on a PIN diode," submitted to *IEEE Microwave and Wireless Components Letters*, October 2012.

[38] Klaus Finkenzeller, *RFID Handbook*, 2nd Edn., Wiley, 2010.

6 Integrating tiny RFID- and NFC-based sensors with the Internet

Vasileios Lakafosis, Rushi Vyas, Chiara Mariotti, Taoran Le, and Manos M. Tentzeris

6.1 Introduction

In the upcoming era of ubiquitous cognition, the capability of gathering, processing, and presenting vast amounts of data captured by spatially distributed sensors in an efficient and low-cost way is as essential as ever. Although a plethora of ultra-low cost but yet powerful and physically small sensors have been demonstrated, only a few of these research efforts have managed to integrate these sensing modules with massive wireless sensor network (WSN) platforms: a *sine-qua-non* step.

The necessity of deploying WSN capabilities in sensing systems is highlighted by: (i) the increased physical range and the ability to overcome non-line-of-sight and path loss effects through multi-hopping; (ii) the enhanced reliability due to multipath communication even if some intermediate WSN nodes on the path toward the destination or a gateway fail; and (iii) the decreased power consumption of the radio transmission with multi-hopping, since it is more power efficient to relay packets over lower-strength shorter links than transmitting high strength signals over fewer long single-hop wireless links [1].

Toward that goal, demonstration of the prototypes presented in this chapter that realize real-time and distributed remote sensing within the scope of different applications serves as a proof of concept; it is possible to deploy WSN nodes in-between sensors and internet gateways. As a result of this approach, the primary task of the WSN nodes is not as data generators, as this is almost exclusively carried out by the sensors. Instead, the WSN nodes here serve simply as data routers that relay the sensed information through wireless multi-hop links to one or multiple gateways. It is important to note that the aforementioned prototyped wireless sensors cover the "last mile," or more precisely the last wireless hop the length of which can range from a couple of meters to a few hundreds of meters. Also relevant to this point is that, in this chapter, the terms "RFID" and "NFC" are not constrained and exclusively referring to the standardized EPCglobal Gen2 [2] and NFC ISO 18092 [3], respectively. Rather, the intent is to use the terms RFID and NFC in a way that embraces an array of wireless technologies that provide radio frequency identification and near-field communication for various purposes. Among the networking capabilities of the proposed prototypes are ZigBee [4], i.e. a low-power communication protocol based on the IEEE 802.15 standard, IPv6-enabling 6lowPAN [5], Crossbow's proprietary multi-hop protocol, XMesh [6], and low-power Bluetooth [7].

To name a few, the applications involved with the demonstrated prototypes are location tracking, health monitoring, chemical gas sensing for industrial environments or food quality sensing, secure identity authentification and genuineness verification, and large metallic structure health monitoring.

Conclusively, the ultimate goal in this chapter is to present to the reader a number of design and manufacturing techniques, prototypes, and WSN-incorporating frameworks that can offer a first realization of the Internet of Things [8] on a world scale; the vision under which the tiny devices that sense (and possibly help "actuate" and control) pretty much any aspect of our surrounding world bring the ubiquitous cognition vision closer and closer to reality.

6.2 RFID-based networked prototypes

Radio frequency identification is an emerging and disruptive wireless technology for the remote identification of a vast amount of objects. Among the most prominent challenges that need to be overcome for any large-scale practical implementation are the cost, both for the mass production of the tags and the "reading" infrastructure, regulatory compliance that is country-dependent, power and performance effectiveness and reliability, and development of fabrication and disposal processes that are as environmentally friendly as possible. The prototypes presented in this section have been designed with all the above factors in mind.

The categorization of the prototypes in the following subsections is a direct result of the general classification of RFIDs based on how they are powered, namely active, semi-passive, and passive. In this chapter we are interested in the latter two categories. Passive RFID are completely battery-less. Any power available to them comes exclusively from the power captured from the reader's emission during the illumination phase. This means that passive RFID tags have theoretically "infinite" lifetime, as long as the tag exists in the interrogation zone of the transmitter. Semi-passive RFID tags are a hybrid of the active and passive types. Semi-passive means that the RFID transmission and reception tasks are performed using different power sources. All the communication tasks are performed by backscattering the energy of the reader's transmitted signal, while the computational load is sustained by an embedded power source, such as a supercapacitor that may also be charged from the reader's "illumination." This approach results in a longer lifetime for the device and a lower cost in comparison to the active ones, while featuring a better range than the passive ones.

6.2.1 Semi-passive

6.2.1.1 Solar-powered UHF tag for localization

The physical location information of a device or a user is a vital aspect of an ever-increasing number of applications, such as social networking and crowd sourcing, environmental applications, e-commerce, and others.

Figure 6.1 shows a credit-card-sized, battery-less, solar-powered location-tracking device [9]. The position of this device is tracked in an open-space environment through

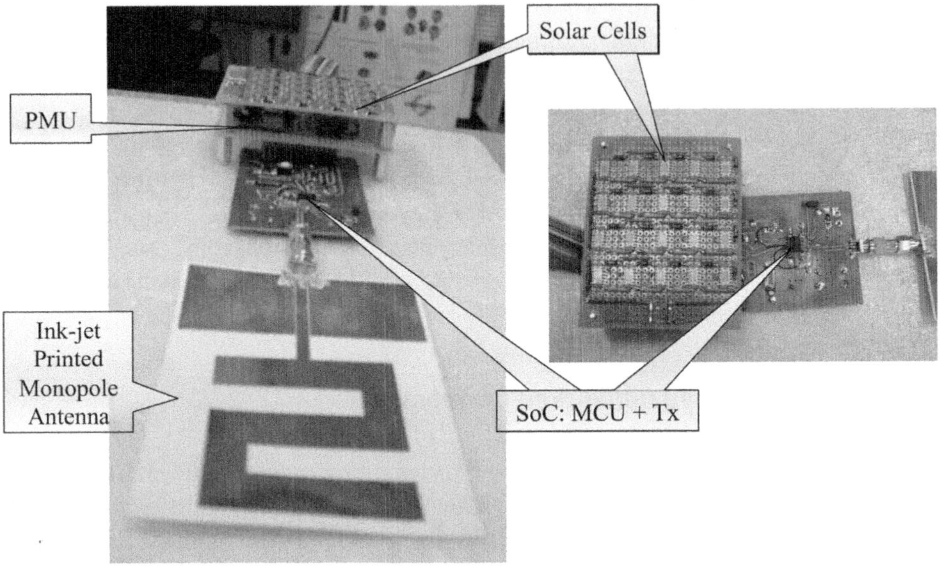

Figure 6.1 Solar-powered battery-less tag prototype.

lateration, a method of estimating the position of the tag by calculating (relying on the received signal strength indicators) its distance from three or more spatially fixed anchor points [10]. In this application, the anchor points, whose exact location coordinates are known beforehand, are the Crossbow MICA2 WSN nodes [11]. Particularly interesting is the fact that the WSN nodes here are not the lowest-level devices in the network infrastructure hierarchy.

An asynchronous communication link between the low-power consuming transmitter of the prototype tag and the MICA2 nodes is required. For the successful link establishment, a set of requirements needs to be met.

From the networking open systems interconnection (OSI) reference layer design [12] perspective, this means that not only should there be physical layer compliance in terms of the RF transmission from the RFID tag, but also the transmitted packet has to implement the proper link layer medium access control encapsulation. More specifically, in regards to the physical layer, the transmission has to be carried out over a supported frequency band from both sides, which in this case is 904.4 MHz, and the transmitted signal has to be frequency shift keying (FSK) modulated meeting specific frequency and modulation requirements. In order to support FSK modulation, passive components are introduced to enable pulling of the phase-locked loop crystal by a controlled amount by the MCU, thereby modulating the RF output going into the antenna with respect to the frequency. As for the link layer, the WSN mote is expecting a specific bit sequence before extracting the payload, which is the actual sensing information. This overhead is not only used for communication protocol purposes but can provide topology-related knowledge and error handling functionality, as described later. The format of the packet transmitted by the tag consists of the following fields: preamble, sync, addr, type, group, length, payload, and

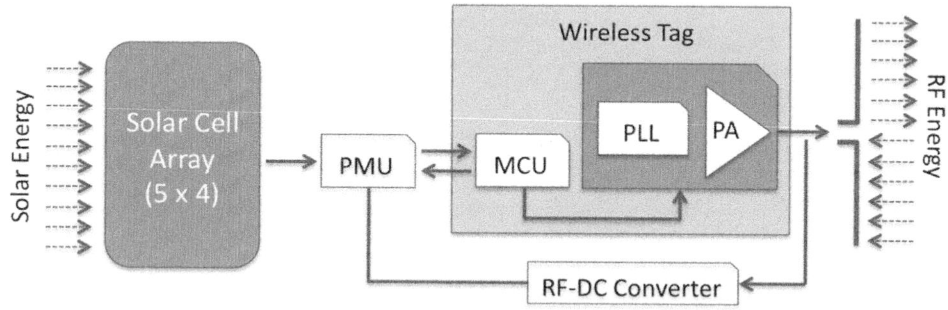

Figure 6.2 The main components of the solar-powered tag prototype.

cyclic redundancy check (CRC). The first two fields are used for the synchronization of the receiver's clock and the last field, CRC, helps eliminate bit errors occurring within the sent bit sequence by successfully recognizing a corrupted packet and discarding it [13]. In addition to cyclic redundancy checks, bit errors on the communication link are noticeably minimized by calibrating the PLL of the module, so that its modulation profile around the center frequency is very closely matched to that required by the receiver MICA2 2 motes. This calibration also has the added advantage of fewer preamble bits required for the receiver to bit synchronize with the tag for Manchester bit encoding [13], which can be very useful given the limited amount of power available per energy duty cycle.

The main functionality of the MCU-enabled wireless connectivity module [14] is the wireless broadcast transmission of identical, unique identification packets in regular intervals triggered and powered using energy present in ambient sunlight without a conventional battery. As expected, the parameters governing the performance of the module are the effective isotropic radiated power (EIRP), the total wireless RF transmit time, and the charging time interval.

This module utilizes supercapacitors, which are much cleaner to dispose of environmentally, have a lower voltage charge threshold, and have much higher recharge lifetimes than batteries. These supercapacitors are charged by the unlimited and clean solar energy harvested by a parallel configuration of GaAs solar cells interfacing to the tag through a power management unit circuit (PMU), which is in turn interfaced to the integrated 8-bit MCU housed in the same chip as the RF transmitter. This is shown in Figure 6.2.

The module firmware ensures that the unique identification packet is transmitted using the same RF power irrespective of the external light conditions, thereby ensuring that the EIRP from the tag remains the same within the area where it is being tracked. The omnidirectional radiation of the power is an equally essential characteristic for the purpose of an accurate localization estimation. Omnidirectionality is achieved by ensuring that the antenna used has a gain as close to 0 dB as possible. The Z-shaped monopole antenna design [15] is fabricated on paper using in-house inkjet-printing technology and is shown as part of the working prototype in Figure 6.1. The antenna design is tweaked to ensure that its impedance is equal to (36.95–j71.77) Ω [16], which is the conjugate of the optimum load impedance of the amplifier in the wireless transmitter of the module prototype. This matching ensures that almost the whole amount

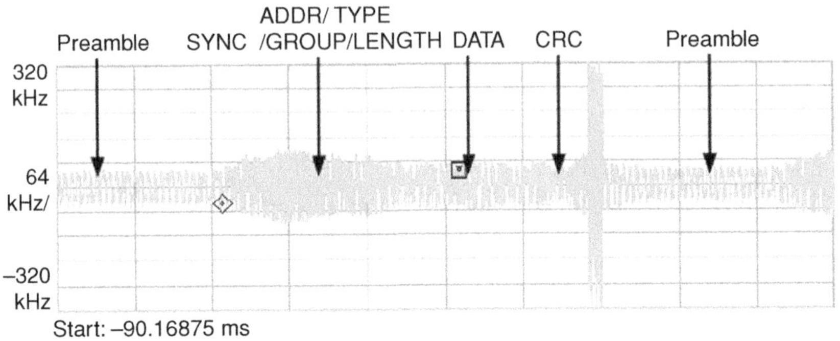

Figure 6.3 The wireless data sequence of a 9 ms packet broadcast transmission by the paper-based wireless module as captured with a real-time spectrum analyzer over time.

of the solar power harnessed is radiated out from the amplifier into the antenna as RF power. Figure 6.3 shows a data sequence transmission.

6.2.1.2 WISP-based

The prototypes presented in this subsection make use of the Intel wireless identification and sensing platform (WISP) tag developed by Intel Research Seattle [17]. The WISP is a tiny computing board consisting primarily of an on-board Texas Instrument microprocessor and a rectifier to harvest the energy needed to carry out sensing and/or computational tasks. It can be powered and read by off-the-shelf UHF RFID readers and, as a result, for an RFID reader the WISP is just a typical RFID tag. Its operation can be considered infinite only as long as the WISP resides within the "illumination" zone of an RFID reader. Its wireless range is confined to a few meters and varies depending on the power consumption incurred by the complexity of the requested tasks, such as analog-to-digital conversion, or some forms of simple sensing, such as temperature or motion sensing.

Concealable and flexible antennas for the WISP module

As a replacement for the conventional non-conformal, long and bulky dipole antenna of the WISP RFID tags [17], and in order to achieve a longer range, paper-based, inkjet-printed antennas have been proposed and fabricated [18]. These antennas have been designed to be as concealable, conformal, miniaturized and, thus, as unobtrusive as possible. In particular, text-based meandered line antenna miniaturization design techniques have been applied to obtain both decreased size and concealment in a way that allows embedding the WISP tags within other everyday objects, such as a product label or a food package.

The above-mentioned antennas have been mounted on the WISP, as shown in Figure 6.4, by first soldering two short conductive wires to the antenna output of the board and afterwards fixing these wires on the paper antenna with conductive epoxy, used instead of traditional soldering on heat sensitive components. A comparison of the fabricated "GATECH" and "LOVE" against the typical dipole WISP antenna in terms

Figure 6.4 The "GATECH" inkjet-printed antenna mounted on the WISP.

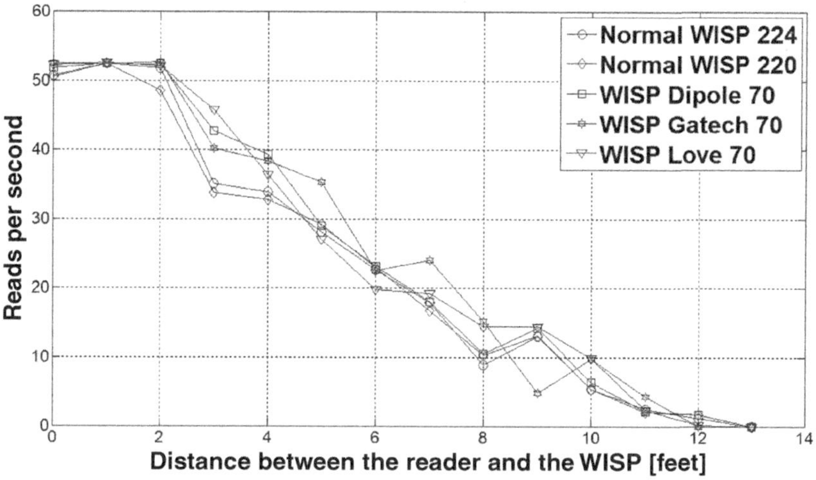

Figure 6.5 Performance comparison of the default WISP dipole antenna versus the optimized paper-based antennas "GATECH" and "LOVE" antennas (1 foot ≡ 30.5 cm).

of rates of successfully repeated read-outs of the tag ID in a single time second, while varying the distance of the device from the reader is included in [18] and presented in Figure 6.5.

Multipacket reception for the RFID EPC Gen2 protocol

A conventional RFID reader possesses no capabilities of extracting useful information from RFID signals from simultaneous tag transmissions that have collided. The reader simply discards this exploitable data and requires a re-transmission of the tag packets, thus incurring identification delays. The dominating UHF EPC Gen2 protocol [2] only specifies collision avoidance algorithms and, although collision recovery techniques have been investigated by researchers in the past [19, 20], no attempt has been made to leverage the application of such techniques with minor modifications to the RFID protocol and, hence, realistically improve its performance.

Toward that end, researchers have put together a flexible test bed [21], made up of an open-source software-defined radio (SDR) reader and programmable Intel WISP tags, to implement techniques for multi-tag acknowledgment under the assumption that collided

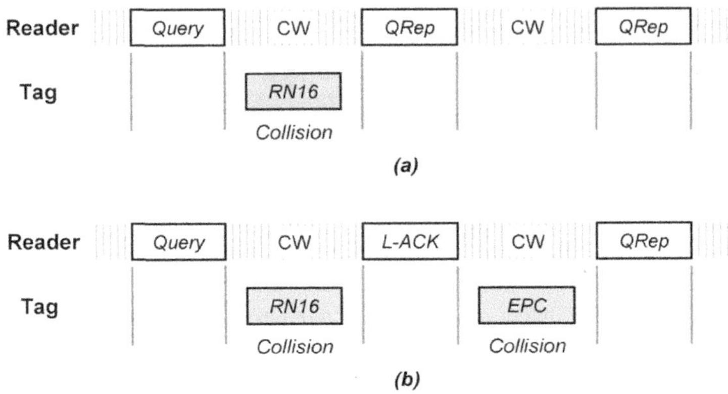

Figure 6.6 RFID handshake when: (a) a conventional; or (b) a collision-recovery capable reader, is considered.

tag packets can be successfully recovered. Specifically, the conventional EPC Gen2 RFID protocol has been modified and a novel collision-aware mechanism to acknowledge multiple tags, i.e. multiple recovered collided packet, has been implemented by sending a new command, referred to as long ACK, as shown in Figure 6.6.

These simple modifications to the EPC Gen2 protocol have been experimentally demonstrated with actual measurements to considerably enhance the performance of current RFID systems: the average throughput per time slot is increased by 72% over the conventional Gen2 MAC scheme and the inventory time is significantly reduced by 26%. Among the numerous benefits from a more reliable RFID reading rate are a 26% speed increase in moving conveyor belt systems while maintaining exactly the same reading reliability and up to 84% of RFID reader power consumption savings.

Chemical gas sensing

Inkjet-printable graphene oxide (GO) prototypes on organic paper and Kapton substrates [22] have been shown to alter their material properties, in particular the load permittivity, conductivity, and impedance when in the presence of a given chemical substance, such as ammonia (NH_3) or nitrogen dioxide (NO_2). By integrating functionalized carbon nanotube materials onto antennas and other passive structures, the changes in material property can be translated into measurable quantities, such as shifts in resonant frequency, backscatter amplitude and voltage, and/or current changes. In addition, the sensitivity of the material can be amplified by appropriate design choices.

To test the material, short microstrip multi-walled carbon nanotube (MWCNT) transmission lines (TL) of 2 mm by 3 mm deposited between silver inkjet-printed traces were used, as shown in Figure 6.7. NH_3 and NO_2, diluted in nitrogen (N_2), were chosen as test gases due to their extensive industrial use, toxicity, and ready availability. The effect of these gases on the MWCNT material properties was examined using a TIN-TEK FlexStream gas standards generator that can provide an extremely stable and accurate mix of gas at very low concentrations using gas permeation tubes of various gas delivery rates.

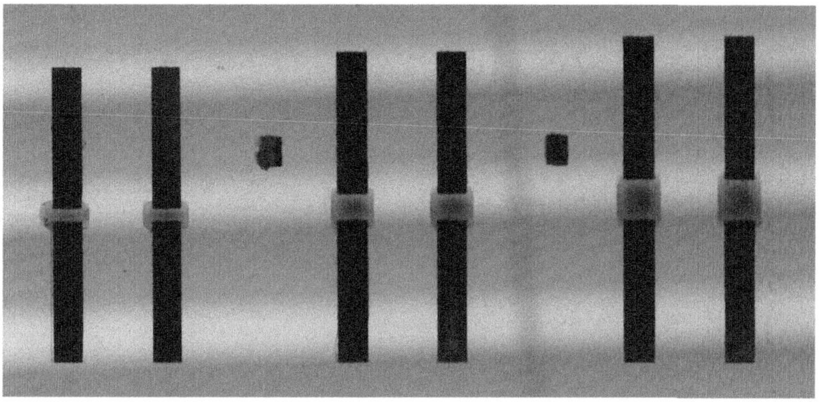

Figure 6.7 Graphene oxide thin films inkjet-printed between and overlapping a silver conductive inkjet-printed trace on a Kapton substrate.

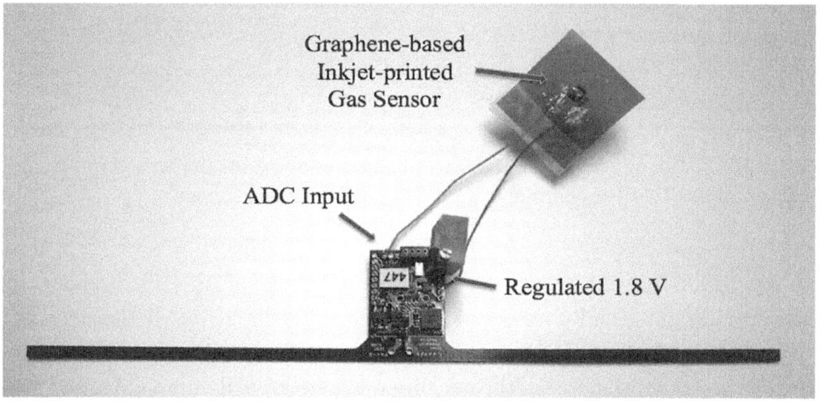

Figure 6.8 The WISP gas sensor prototype platform.

In [22] researchers have modified (hardware- and software-wise) the WISP to create a sensing and communication platform that enables the introduction of the aforementioned gas sensing capabilities to RFID tags. As opposed to the distortion-prone analog-based measurement techniques, such as resonant frequency shift sensing, this flexible WISP platform offers the reliability of wireless digital transmission. This prototype platform is shown in Figure 6.8. The tag is powered by 915 MHz RF energy, which is broadcast by a very popular commercial RFID reader and rectified with a charge pump topology to charge the on-board capacitor. As soon as sufficient energy is accumulated and the voltage across the capacitor monitored by a supervisor exceeds the turn-on voltage of the MCU by an overhead margin, a regulated voltage of 1.805 V powers both the MCU and the organic-based prototype external gas sensor. The analog interface is essentially a voltage divider, in which the voltage drop across the gas sensor module is directly tied to the resistance change of the sensor's graphene film pad and can easily be captured with the WISP's integrated analog-to-digital converter (ADC).

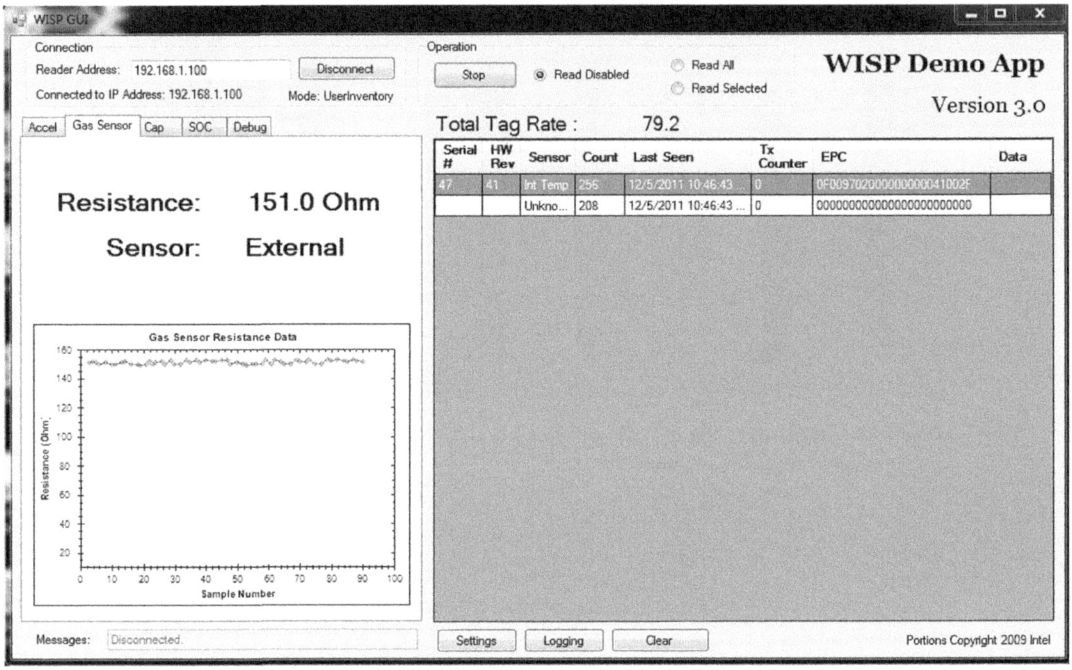

Figure 6.9 The graphical user interface that captures and plots the sensed information from the transmitted RFID EPC communication with the WISP gas sensor.

Whenever located within the interrogation zone of an RFID reader, the WISP gas sensor is automatically detected and begins transmission of the sensed information. Figure 6.9 shows a screenshot of a modified WISP graphical user interface that reports in real time the resistance of the gas film that can be easily mapped to the real gas concentration with the extracted fitting equation. The report rate is only a few milliseconds. For sensing purposes, the sensor yields a 6% normalized resistance change after 15 minutes of exposure to a concentration of 500 ppm of NH3. In addition, excellent recovery time is achieved using the graphene thin films, with over 30% of material recovery observed within 5 minutes.

6.2.2 Passive

6.2.2.1 Paper-based ultra-high frequency sensor

The high demand for and mass production of paper make it the cheapest and most available material ever made. From a manufacturing point of view, paper is well suited for reel-to-reel processing and has a low surface profile that makes it suitable for fast printing processes, such as direct write methodologies instead of the traditional metal etching techniques. After the silver nano-particle droplets are driven through the nozzle, a sintering process follows in order to remove excess solvent and increase the bond of the deposit with the paper substrate. This directly translates to a conductivity of the

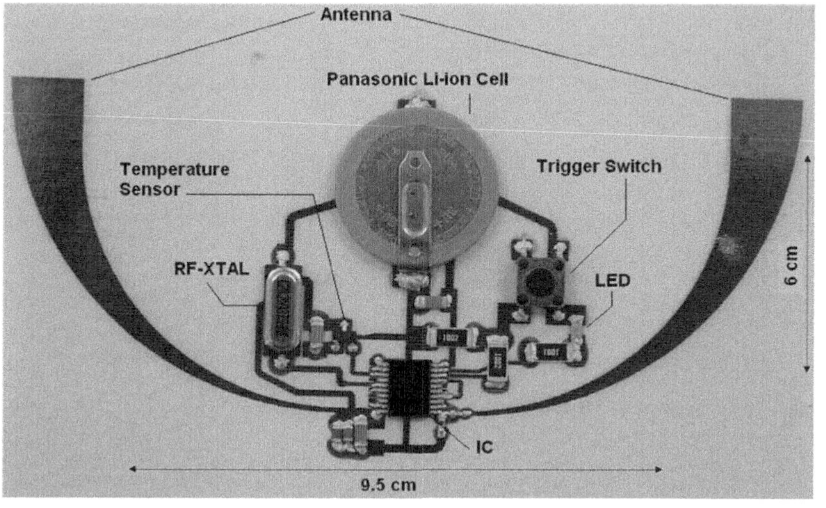

Figure 6.10 RFID-enabled wireless sensor transmitter relying on inkjet-printing on paper.

conductive ink that varies from 0.4×10^7 to 2.5×10^7 siemens/m, depending on the curing temperature and duration time. The paper's high biodegradability, with respect to other ceramic substrates, allows it to turn into organic matter in landfills in only a few months' time and renders it one of the most environmentally friendly materials.

Given the above highly desirable characteristics of paper, it comes as no surprise that paper has been used as a "green" ultra-low-cost organic substrate, on which whole RFID tag circuitry and antenna have been inkjet-printed [16]. This RFID-enabled wireless sensor module prototype uses a U-shaped dipole antenna and was printed and curved on a 2D single photo-paper layer with 12 layers of silver ink. The overall dimensions of the structure, shown in Figure 6.10, are 9.5×6 cm. The wireless range exceeds 300 m and the temperature accuracy is better than 0.2 °C. To ensure maximum conductivity and antenna efficiency, the entire circuit was printed over.

The complete wireless sensor system comprising the TSSOP packaged integrated circuit (IC), including the MCU and the transmitter, its discrete passive components including a crystal oscillator, a temperature sensor, and a Li-ion cell was assembled on the inkjet-printed metallic traces with the use of silver epoxy. Soldering could not be used due the low temperature tolerance of paper, the electronic components used, and the relatively weaker adhesion of printed silver pads on paper.

6.2.2.2 Metallic structural strain sensor

Fatigue-induced fractures and cracks are among the most common concerns for inspectors of large metallic constructions, such as highway bridges. The basic concept behind the operation of the prototype [23, 24] presented in this subsection is to exploit the strain-dependent behavior of the antenna radiation as the sensing mechanism. In particular, when a small piece of 2D antenna is under strain, its resonance frequency changes accordingly, and this can be wirelessly interrogated by an RFID reader.

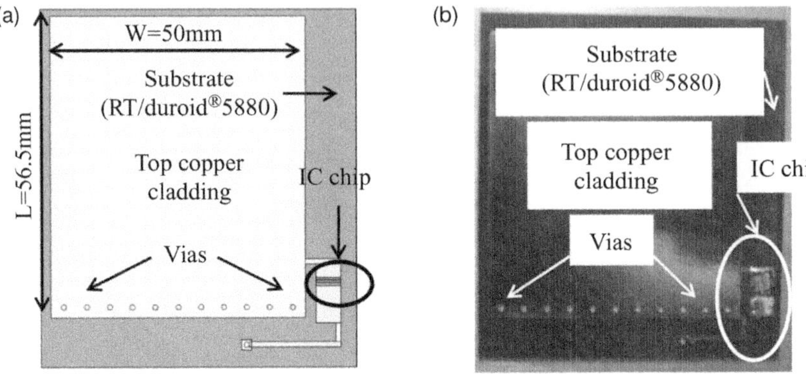

Figure 6.11 Designed (a) and fabricated (b) RFID-enabled prototype smart skin strain sensors.

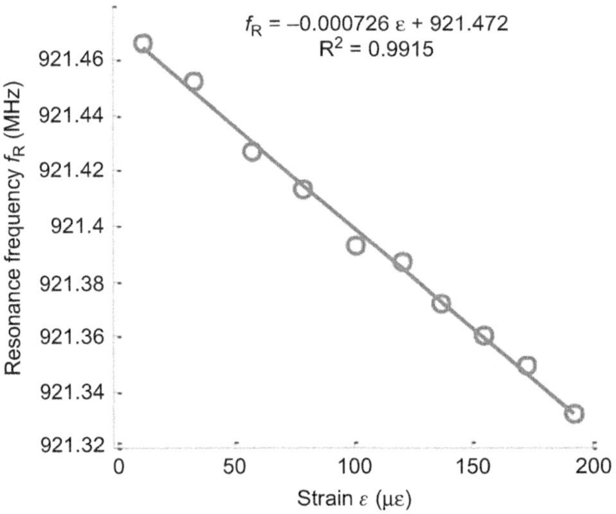

Figure 6.12 Resonance frequency f_R versus strain ε results at ~20 με loading steps at an interrogation distance of 12 inches (305 mm).

The low-profile design, shown in Figure 6.11a, that serves here as a smart skin strain sensor is a quarter wavelength ($\lambda/4$) rectangular folded patch antenna that adopts a specially chosen substrate material with low-dielectric attenuation. The communication signal modulation is performed by an inexpensive off-the-shelf RFID chip for signal modulation. The fabricated smart skin strain sensor is shown in Figure 6.11(b). A 61 by 69 by 0.79 mm^3 glass microfiber reinforced poly-tetra-fluoro-ethylene composite material is chosen as the antenna substrate, on the top and bottom of which are located the 0.0178 mm thickness copper antenna and ground layers.

The distance between the reader antenna and the wireless sensor is set as 12 inches (305 mm). The aluminum specimen together with the smart skin strain sensor is stretched gradually up to 200 με, at ~20 με steps. The mapping of the resonance frequency to the corresponding strain level is depicted in Figure 6.12. The ten data points correspond to ten strain levels. As a result of linear regression, the slope parameter is identified as −726 Hz/με. This parameter corresponds to the strain sensitivity of the smart skin strain sensor, which means that 1 με increment in the aluminum specimen causes 726 Hz decrease in the resonance frequency of the smart skin strain sensor.

6.3 NFC-based networked prototype

Shoe-mounted platform for health-monitoring and localization

The combination of components depicted in Figure 6.13 forms the proposed design of a shoe-mounted platform for health-monitoring and user localization [25]. The 5 cm by 8 cm board mounted on the back of the shoe houses a TI NFC reader chip that is used to detect the position of a person by illuminating a matrix of 13.56 MHz NFC passive tags location-mapped beforehand and embedded into floor tiles [26].

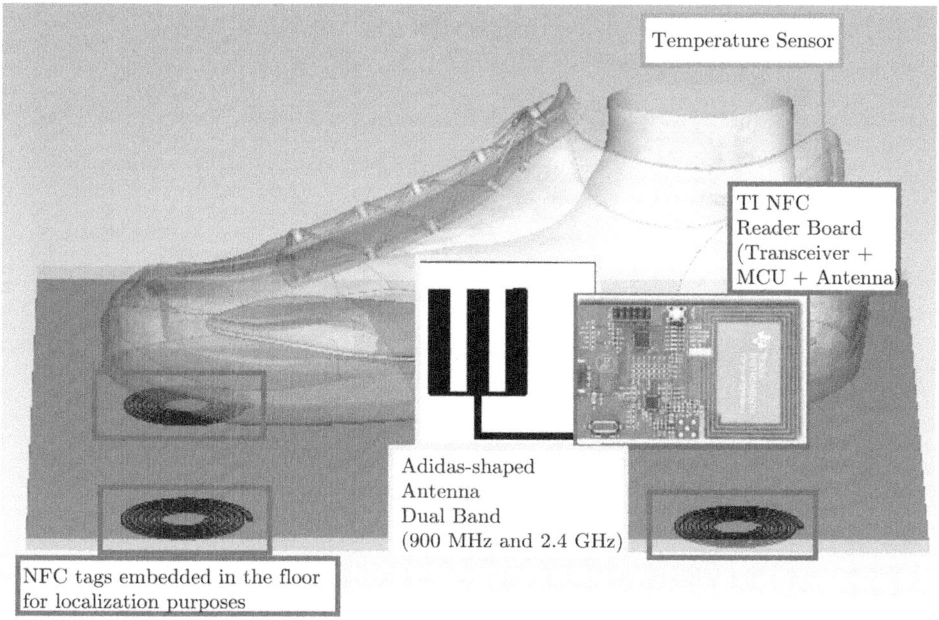

Figure 6.13 Proposed architecture of the shoe-mounted sensor module.

The magnetically coupled lower frequency field not only provides higher isolation from accidentally picking up tags mounted on nearby tiles but also eliminates any interference with the 900 MHz and 2.4 GHz dual-band Adidas^{TM}-logo-shaped antenna. This logo renders the antenna as esthetically unobtrusive as possible and has been fabricated on an organic regular photo-paper substrate.

The NFC reader is used to detect the position of a person moving within a building. The floor is mapped with the help of a matrix of NFC passive tags. On each step the small shoe-mounted NFC reader establishes communication with the NFC tag of the corresponding tile and extracts the unique ID of the tag. The correlation of this ID

Table 6.1 Maximum reading distances of the NFC tag in different positions of a tile.

		Porcelain tile		Ceramic tile	
		Reader orientation			
		Parallel (cm)	Normal (cm)	Parallel (cm)	Normal (cm)
NFC tag positions	1	6.5	8.0	7.5	8.5
	2	6.5	8.0	8.0	8.5
	3	5.5	6.5	8.5	6.5
	4	4.5	5.5	5.0	5.0
	5	8.5	6.0	5.0	6.5
	6	5.5	7.0	5.0	7.0
	7	5.0	8.0	5.0	6.0
	8	6.5	6.0	6.5	5.5

Figure 6.14 NFC sensor mapping for coverage estimation.

number with the real, absolute or relative, position of the human is based on a look-up table where the tag IDs are mapped to the real physical locations of the tiles. This location information, along with the medical-grade temperature of the human object, can be transmitted to an internet-based service, as discussed later in Section 6.4.3, with a transceiver connected to the MCU through an SPI interface.

The results of an analysis of the maximum reading distance of the NFC board are summarized in Table 6.1. Specifically, different positions of the NFC tag under the tile have been tested, for both normal and parallel orientation of the NFC reader antenna, in order to identify the maximum reading distances with the presence of either porcelain or ceramic tiles (Figure 6.14).

The reader consumes no more than 120 mA of peak power. The entire shoe-mounted sensor system can be powered by a hybrid power system that intelligently couples the compatible and very popular technology of rechargeable Li-ion batteries with a flexible piezo transducer (PZT) embedded into the sole [27]. The high, unregulated output voltage of the PZT (100/120 V) is rectified, regulated, and stored on a supercapacitor.

6.4 Using WSNs to interface with the internet

As already discussed in the introduction, the ultimate objective of interconnecting tiny sensors with very low-cost and mature wireless infrastructures is to be able to extract the sensed information not only over long wireless ranges, but even from the other side of the planet. The bridging of the world of sensing devices and RFIDs with that of WSNs enables truly cognitive intelligence over extended physical distances.

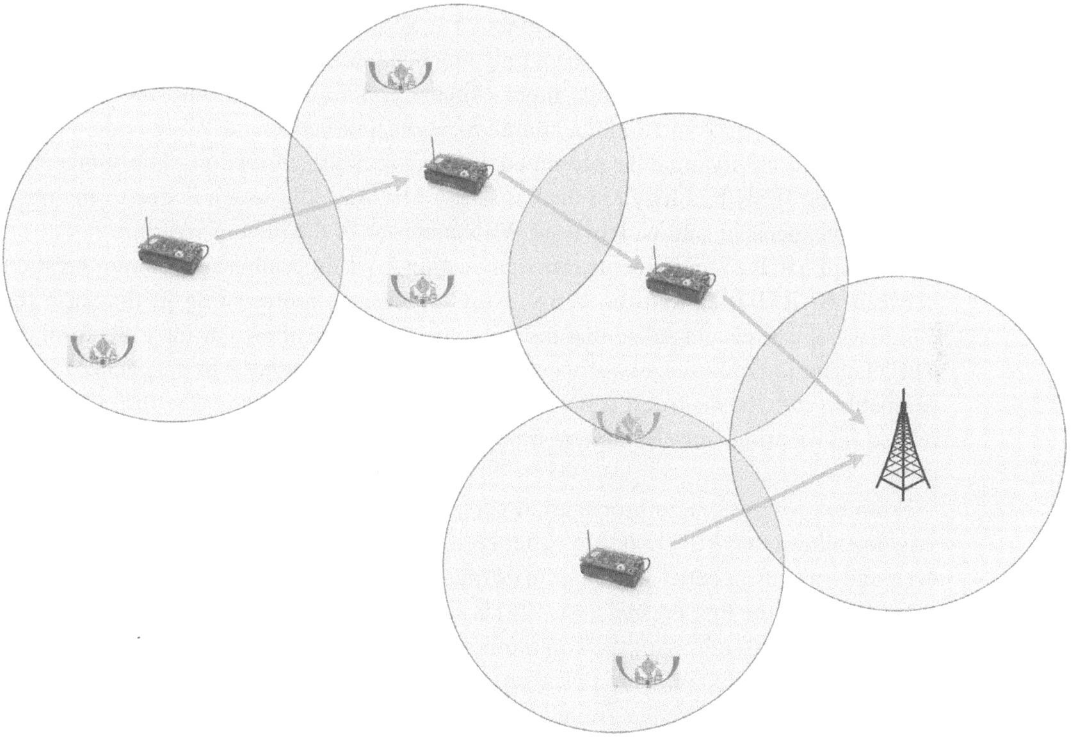

Figure 6.15 An example of a wireless sensor network (WSN) topology, where the WSN motes are no longer the lowest-level network devices in the infrastructure hierarchy.

The motes comprising a WSN are mesh-interconnected with wireless links and are capable of relaying messages to a sink, as graphically depicted in Figure 6.15. Within the multi-hop context of a WSN, instead of transmitting high-strength signals over long single-hop wireless links, it is more power efficient to relay packets a number of times over a higher number of lower-strength shorter links [1]. This decreased power consumption of the radio transmission results in increased power efficiency of the wireless module. At the same time, as long as the WSN nodes are located within direct radio range of others, hopping effectively extends radio communication over higher ranges overcoming non-line-of-sight and path loss effects. Moreover, their multipath topology is inherently self-healing, allowing the network functionalities to be sustained without any interruption because of potential WSN node failures.

Although, for instance, communication between the wireless connectivity solar-powered module of Section 6.2.2.1 and the very popular commercial MICA2 WSN mote [11] has been successfully established using a very simple but power-efficient protocol, the role of the WSN mote can also be assumed alternatively by 802.15.4/ZigBee nodes, DASH7 nodes, Oracle Sun Spot nodes, and others.

6.4.1 Crossbow WSN for location tracking

The key decision that has enabled the cost of the localization solution of Section 6.2.1.1 to be drastically lowered is the deployment of a WSN as the backbone infrastructure, instead of RFID readers which are not only much larger in size than typical WSN motes but also two orders of magnitude more expensive. This is allied with the successful establishment of an asynchronous communication link between a WSN node and the wireless connectivity module presented in the previous subsection. The motes that comprise the WSN topology are the Crossbow MICA2 [11], which is one of the most popular, inexpensive, and widely used WSN motes as of the time of writing.

In addition to the requirements stated in Section 2.1.1 for establishing communication between the RFID tag and the WSN, NesC code [28], representing in this case the middleware, was developed so that the packets transmitted in regular intervals from the RFID tag are not only successfully captured but are also translated to the packet format compatible with the deployed WSN routing protocol and piggybacked with application-related information, such as unique mote ID and received signal strength indicator in dBm.

From the wireless radio propagation perspective, the WSN setup is carried out in a way that allows the WSN nodes to better receive the end device transmitted data, which eventually results in better localization estimation. The requirement for the clearance of at least 80% of the first Fresnel zone [29] is satisfied in most cases for the anchor nodes that are in line-of-sight with the communication module. The MICA2 nodes that form the WSN are mounted on lamp posts; an example is shown in Figure 6.16. The average height of the motes above the ground is 3.5 m and the height of the end device is around 1.5 m. The value of the radius of the cross-section of the ellipsoidal of the first Fresnel zone at the middle of the distance, typically 75 m, is approximately 2.8 m and the same value just 0.5 m away from the mote side is around 0.3 m.

After successful reception of the beacon message broadcast by the end node from a number of WSN anchor nodes, essential information, namely RSSI and unique anchor identity number, is appended to the packet and forwarded to the localization central server, presented in the next subsection, through either the XMesh [6], a fully-featured multi-hop, ad hoc, mesh routing protocol, or over an overlay 2.4 GHz IEEE 802.11 wireless local area network. The latter consists of three Linksys WRT54GL access points, one connected to the central server and the other two connected through UTP cable to each of the eight MICA2 motes, the Ethernet interface of which is provided by an

(a)

(b)

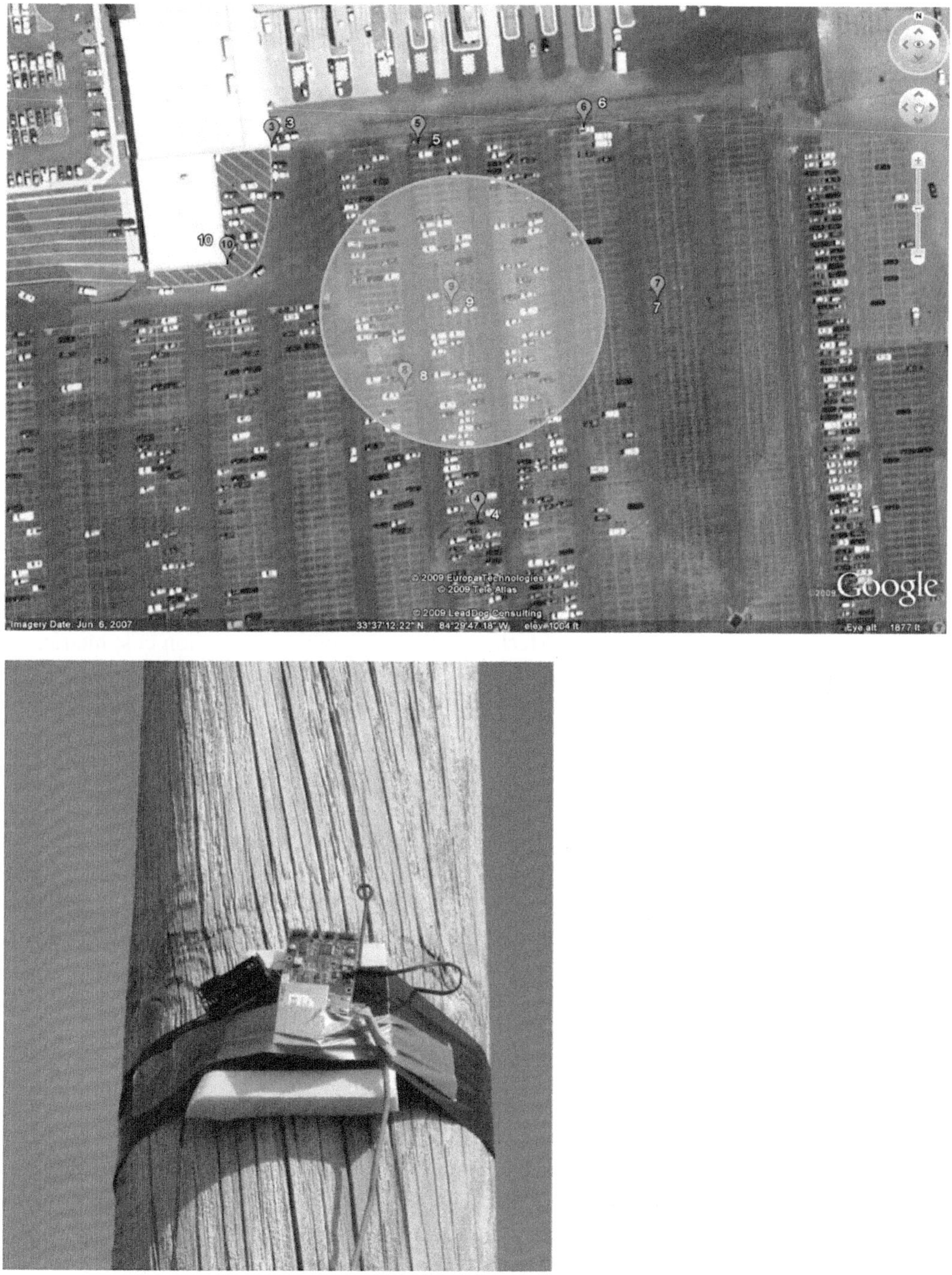

Figure 6.16 (a) The WSN topology and anchor node positions in a large parking lot. (b) A wireless sensor network anchor node mounted on a lamp post.

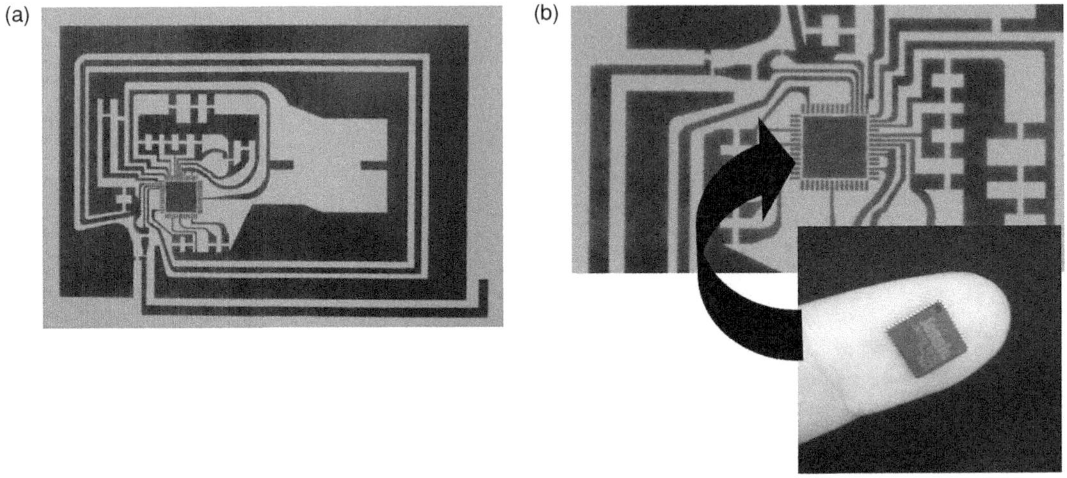

Figure 6.17 (a) Inkjet-printed wireless module on paper substrate. (b) Detail of the SoC module.

additional daughter-board. As is the case with any wireless network, the purpose of the cost metric is to minimize the total cost it takes for any mote to transmit to the base station. It is important to note that, instead of hop count, which is the traditional distance vector routing cost metric, XMesh uses the minimum transmission cost metric, which minimizes the total number of transmissions in delivering a packet over multiple hops to a destination.

6.4.2 ZigBee

IEEE 802.15.4 is a standard defined for WSNs focusing on low-power operation and applications, which do not require high-rate communication. ZigBee is built upon IEEE 802.15.4 and is suited for sensor and control applications focusing on low-cost and low-power design of both star and mesh network topologies [4]. The reliability of ZigBee is largely attributed to the featured packet error detection, hop-level, and end-to-end acknowledgments and abundance of available paths toward a destination due to its mesh networking nature.

6.4.2.1 SWIM – smart wireless integrated module

The design and fabrication of the first IEEE 802.14.4 and ZigBee compliant wireless sensor node on paper substrate using inkjet-printing technology is proposed in [30, 31]. The wireless node [32, 33], shown in Figure 6.17, is a 31 by 60 mm^2 system-on-package solution operating at 2.4 GHz with a printed planar inverted-F antenna (PIFA). The system-on-chip (SoC) chosen to implement the module was the TI CC2431 chip [34] that features an 8051 microcontroller unit with a 128 kB in-system programmable flash and 8 kB RAM, an eight-input 12-bit ADC, a radio based on the CC2420 RF transceiver, and an analog temperature sensor among other features. The on-chip analog temperature sensor output is one of the inputs of the ADC multiplexer, so the MCU can be configured to read and broadcast the sensor readings. The SoC requires a 32 MHz crystal, and an optional 32.768 kHz crystal is added to improve power efficiency by allowing the SoC to switch clocks and enter various power consumption modes.

The entire circuitry tracing has been inkjet-printed on a paper substrate. The smallest dimension that was inkjet-printed is around 125 μm and pertains to the gaps of the traces of the QFN package wireless ZigBee microcontroller.

The differentially fed antenna is interfaced to the chip through one of its 56 pins. IEEE 802.15.4 standards define (in the 2.4 GHz band) 16 ZigBee channels, with each channel requiring 5 MHz of bandwidth, and so an 80 MHz operational bandwidth is required for this antenna. The printed antenna was designed and simulated on a paper substrate in HFSS with a realized gain of 2.8 dBi and return loss at the resonant frequency of −16.4 dB.

6.4.2.2 Near-field certificate of authenticity reader

A complete system [35] that aims to render any typical RFID tag physically unique and hard to closely replicate by just complementing it with a random conductive and dielectric 3D scattering structure, also builds upon WSNs for data transmission.

Refinements of this system include a completely passive operation, interoperability with any wireless tag of any frequency band or embedded storage technology without requiring any reader software modifications, and a cost comparable to that of typical, passive RFID tags. The unique near-field response, or "signature," of the certificate instances is extracted as a set of S_{21} curves by the reader prototype over the 5 to 6 GHz frequency range. Each one of the S_{21} curves of the RF fingerprint corresponds to a single coupling of a pair of antenna elements of the reader's antenna array.

As a means to allow ease of use and portability of the RFID-CoA reader within store premises or industrial facilities, the wireless connectivity capabilities of the incorporated TI MSP F5438 MCU chip [36] are leveraged. The reader design, shown in Figure 6.18, provides two different RF interfaces, where very small ZigBee and Bluetooth modules can be attached. Specifically, two compatible wireless networking module options are the TI CC2530 [37] system-on-chip for 2.4 GHz IEEE 802.15.4/ZigBee and the TI CC2540 [37] for Bluetooth, respectively. As a result, the reader can receive various public keys from different issuers and transmit the extracted identity verification data on top of the standardized and highly reliable ZigBee Pro and Bluetooth wireless networking standards to a central location and a mobile device, respectively, over AES (advanced encryption standard) 128 bit encrypted wireless links.

6.4.3 IPV6-enabling 6lowPAN

The paradigm of the IoT, backed by machine-to-machine (M2M) communication functionality, involves a combination of internet-connected embedded devices with web-based services, so that these devices are universally an integral part of the internet [5].

So far, because of the complexity of the internet communication standards and services, it has not been possible to use internet capabilities natively on low-power devices

(a)

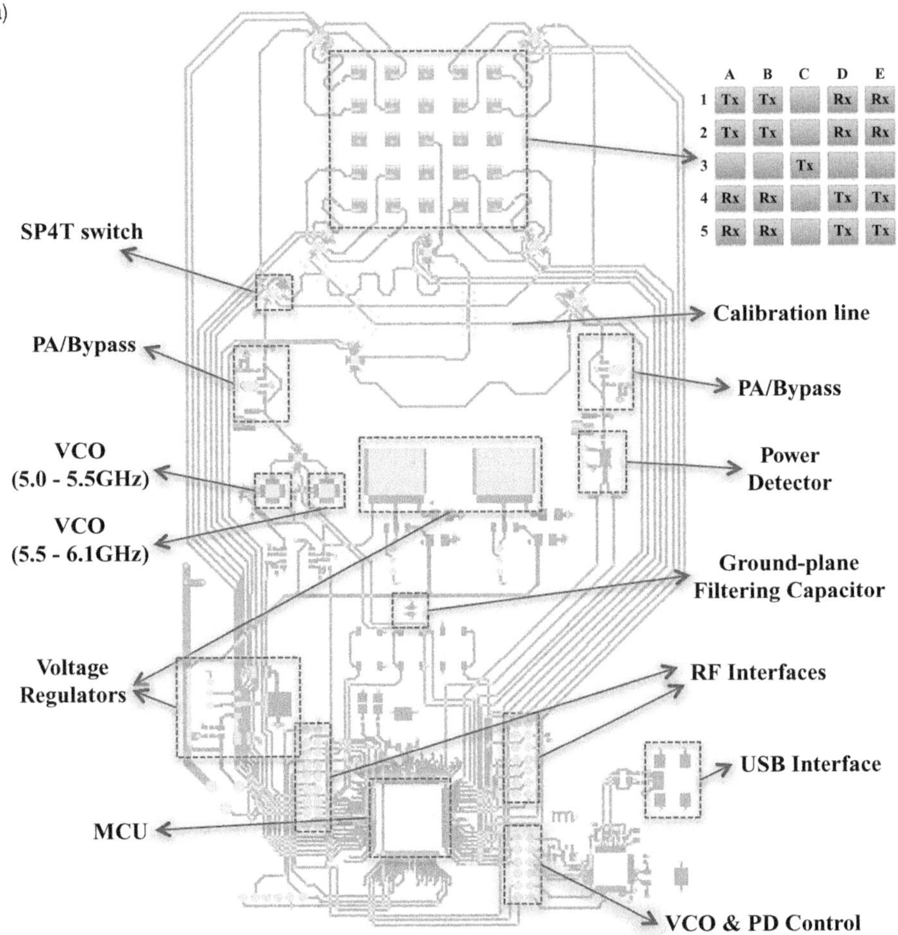

Figure 6.18 (a) The RFID-CoA reader board circuit. (b) The RFID-CoA reader's bottom digital plane, including the MCU.

with limited processing capabilities. ZigBee and a few other proprietary WPAN protocols have not been able to inherently provide internet integration as these protocols translate both address and commands between IP and ZigBee and, thus, the IP stack gets terminated at the gateway points.

The major reason for opting for the 6lowPAN protocol, acronym derived from IPv6 over low-power wireless personal area networks, in applications such as health monitoring performance presented in Section 6.3.1, is the native support of the IPv6 protocol stack on the end device. This has become reality by the encapsulation and header compression mechanisms defined by the 6lowPAN group that allow IPv6 packets to be sent to and received from over IEEE 802.15.4 based networks. In particular, the network, transport, and application layers of the 6lowPAN protocol stack are the same as those of the IPv6 stack and the necessary changes exist in the adaptation layer on top of the IEEE 802.15.4 medium access control and physical layer.

(b)

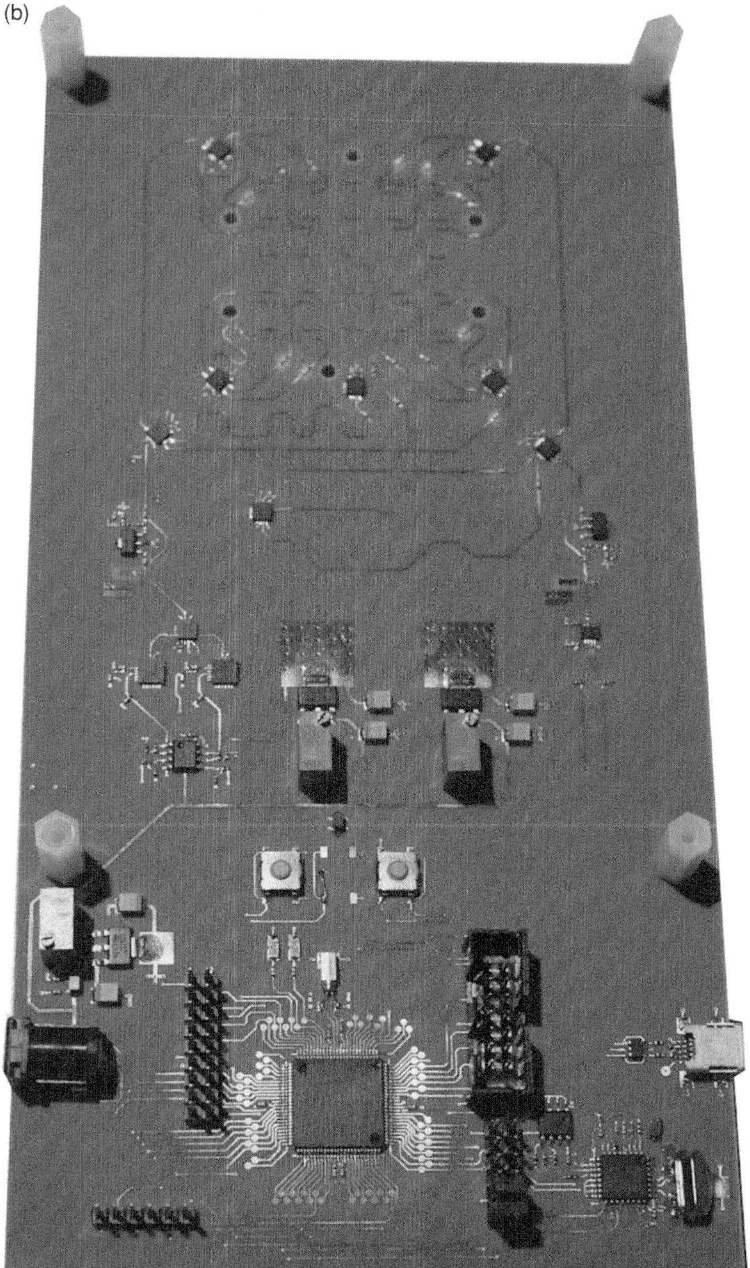

Figure 6.18 *(cont.)*

As a result, the shoe-mounted sensor [25] enjoys all the benefits that stem from IPv6, namely: (i) easy connection to any existing, private or public, IP network infrastructure (e.g. IEEE 802.11 WiFi networks) without translation gateways or proxies; (ii) mesh routing in the WPAN space with simultaneous support of the routability of packets between the IPv6 domain and the PAN domain; (iii) extensive scalability in terms of sensor node density; (iv) remote management and diagnosis using a plethora of long-tested tools; and (v) optionally inherent security support with compressed IPsec [38].

6.4.4 Extensions to mobile and pico-datacenter computing

The WSN scenarios described in the previous section work well if the deployed sensor network is located within a few wireless hops away from a limited-capacity satellite link or a wired internet gateway. What happens, however, when this is not possible, as is the case with sparsely-populated regions of the developing and under-developed world or when the required long-term rental of cloud computing services for processing of the sensed data rental cannot be afforded? This gap can be potentially filled by the more futuristic-sounding infrastructure of intermittently connected mobile devices, i.e. smart phones or pads, that carry out collaboratively all the computations requested [39] or by a framework that consists of long-distance wireless-based, renewable-energy-powered and extremely small datacenter (pico-DC) nodes [40].

Regarding the later proposal of the first pico-DC that promises to bring the "cloud" down to the major populations of the developing world, the reason the datacenters are called pico is that the dimensions and weight of the weather-proof nodes are so small that it is possible to mount them on metal poles or on existing antenna towers. Moreover, in contrast to the containerized micro-DCs [41] of a thousand servers that draw around 500 kW, the pico-DCs operate within a peak power budget of 150 W, which is the typical output of a 1 by 1.5 m^2 solar panel, and thus can rely exclusively on renewable solar power.

A pico-DC network topology can be formed by combining long-distance (10–40 miles), point-to-point, high-bandwidth (5–30 Mbps) backhaul links, at the level of villages and small towns, with medium-range broadcast access links to sensing devices. An example of such a topology is shown in Figure 6.19. Conceptually, this network consists of local cloud clusters of a few hundred or more sensors, the head of which is a pico-DC node. Each such node also serves as a wireless mesh node of the overall pico-DC network.

6.5 Conclusion

The need for integrating the numerous tiny and highly-capable sensors that surround our environment with the internet through wireless sensor networks is nowadays as essential as never before. In this chapter, several WSN networked prototypes that could help achieve this goal are presented in the context of applications that span from localization to health monitoring and security. The aim of the rather strict categorization of these prototypes has been to help the reader identify a big part of the array of wireless technologies that can be applied toward the vision of ubiquitous cognition.

Leveraging the capabilities provided by the "Internet of Everything," which means unsurpassed holistic and efficient exploitation of the enormous volume of sensed data

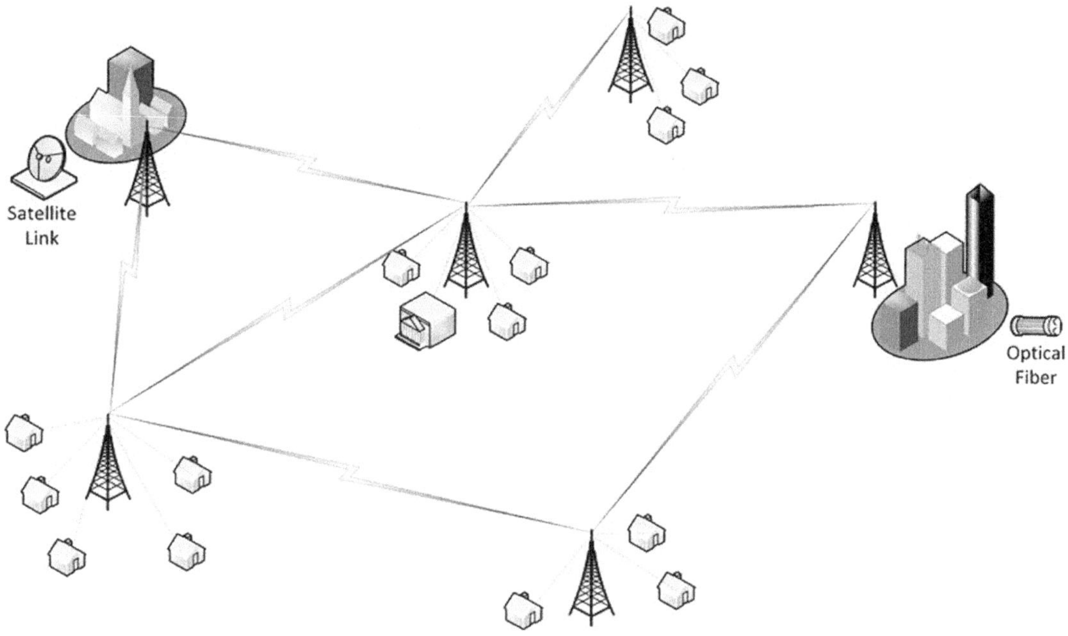

Figure 6.19 A pico-DC wireless network topology.

and corresponding action upon this information, can significantly improve the quality of life.

References

[1] V. Lakafosis and M. M. Tentzeris, "Implementation of multi-hop routing protocols for the dramatic range enhancement of wireless sensor networks," *Antennas and Propagation Society International Symposium, AP-S 2008, IEEE*, pp. 1–4, 2008.

[2] EPCglobal Inc., *EPC Radio-Frequency Identity Protocols Class-1 Generation-2 UHF RFID Protocol for Communications at 860 MHz – 960 MHz Version 1.2.0*, 2008.

[3] ISO 18092 (ECMA-340) *Information Technology – Telecommunications and Information Exchange Between Systems – Near Field Communication – Interface and Protocol (NFCIP-1)*, 2004.

[4] ZigBee Alliance, "ZigBee specification," *ZigBee document 053474r06, version*, vol. 1, p. 378, 2006.

[5] Z. Shelby and C. Bormann, "Introduction," in *6LoWPAN*, pp. 1–25, John Wiley & Sons Ltd, 2009.

[6] Crossbow Technology, *XMesh User's Manual*, 2007.

[7] *Specification of the Bluetooth System, Version 1.1*, http://www.bluetooth.com, 2001.

[8] L. Atzori, A. Iera, and G. Morabito, "The Internet of Things: A survey," *Computer Networks*, **54**, *(15)*, 2787–2805, 2010.

[9] R. Vyas, V. Lakafosis, and M. Tentzeris, "Enabling localization in WSNs with solar-powered end devices," in *Proc-Conf. on Sensor networks, ubiquitous and trustworthy computing*, IEEE, Newport Beach, CA, pp. 155–160, Jun. 2010.

[10] V. Lakafosis, R. Vyas, and M. M. Tentzeris, "A localization and position tracking solution utilizing solar-powered RFID tags," *Proceedings of the Fourth European Conference on Antennas and Propagation (EuCAP)*, pp. 1–4, 2010.

[11] Crossbow Technology, *MICA2 Wireless Measurement System*, 2007.

[12] H. Zimmermann, "OSI reference model–The ISO model of architecture for open systems interconnection," *IEEE Transactions on Communications*, **28**, (*4*), 425–432, 1980.

[13] W. Stallings, *Data and Computer Communications*, 9th Edn., Prentice Hall, 2011.

[14] R. Vyas, V. Lakafosis, and M. Tentzeris, "Wireless remote localization system utilizing ambient RF/solar power scavenging RFID tags," in *Microwave Symposium Digest (MTT), IEEE MTT-S International,* pp. 1764–1767, 2010.

[15] Z. Konstas, A. Rida, R. Vyas, *et al.*, "A novel "Green" inkjet-printed Z-shaped monopole antenna for RFID applications," in *3rd European Conference on Antennas and Propagation, EuCAP 2009*, pp. 2340–2343, 2009.

[16] R. Vyas, V. Lakafosis, A. Rida, *et al.*, "Paper-based RFID-enabled wireless platforms for sensing applications," *IEEE Transactions on Microwave Theory and Techniques*, **57**, (*5*), 1370–1382, 2009.

[17] A. P. Sample, D. J. Yeager, P. S. Powledge, *et al.*, "Design of an RFID-based battery-free programmable sensing platform," *IEEE Transactions on Instrumentation and Measurement*, **57**, (*11*), 2608–2615, 2008.

[18] M. Marroncelli, D. Trinchero, V. Lakafosis, *et al.*, "Concealable, low-cost paper-printed antennas for WISP-based RFIDs," in *IEEE International Conference on RFID*, pp. 6–10, 2011.

[19] J. Kimionis, A. Bletsas, A. G. Dimitriou, *et al.*, "Inventory time reduction in Gen2 with single-antenna separation of FM0 RFID signals," in *IEEE International Conference on RFID-Technologies and Applications (RFID-TA)*, pp. 494–501, 2011.

[20] C. Angerer, R. Langwieser, and M. Rupp, "RFID reader receivers for physical layer collision recovery," *IEEE Transactions on Communications*, **58**, (*12*), 3526–3537, 2010.

[21] D. De Donno, V. Lakafosis, L. Tarricone, *et al.*, "Increasing performance of SDR-based collision-free RFID systems," *Microwave Symposium Digest (MTT), 2012 IEEE MTT-S International*, pp. 1–3, 2012.

[22] L. Taoran, V. Lakafosis, L. Ziyin, *et al.*, "Inkjet-printed graphene-based wireless gas sensor modules," in *62nd Electronic Components and Technology Conference (ECTC), IEEE*, pp. 1003–1008, 2012.

[23] V. Lakafosis, Y. Xiaohua, L. Taoran, *et al.*, "Wireless sensing with smart skins," in *Sensors*, IEEE, pp. 623–626, 2011.

[24] Y. Xiaohua, C. Chunhee, F. Chia-Hung, *et al.*, "Wireless strain and crack sensing using a folded patch antenna," in *6th European Conference on Antennas and Propagation (EUCAP)*, pp. 1678–1681, 2012.

[25] C. Mariotti, V. Lakafosis, M. Tentzeris, *et al.*, "An IPv6-enabled wireless shoe-mounted platform for health-monitoring," in *IEEE Topical Conference on Wireless Sensors and Sensor Networks (WiSNet)*, pp. 46–48, 20–23 Jan. 2013.

[26] C. Mariotti, G. Orecchini, M. Virili, *et al.*, "RFID tag antenna embedded in concrete structures for construction industry," in *6th European Conference on Antennas and Propagation (EUCAP)*, pp. 3469–3472, 26–30 March 2012.

[27] J. Kymissis, C. Kendall, J. Paradiso, *et al.*, "Parasitic power harvesting in shoes," in *Digest of Papers, Second International Symposium on Wearable Computers*, pp. 132–139, 1998.

[28] D. Gay, P. Levis, R. von Behren, *et al.*, "The nesC language: A holistic approach to networked embedded systems," in *Proceedings of the ACM SIGPLAN 2003 Conference on Programming Language Design and Implementation*, New York, NY, USA, pp. 1–11, 2003.

[29] L. W. Barclay and Institution of Electrical Engineers, *Propagation of Radiowaves*, Institution of Electrical Engineers, 2003.

[30] Y. Li, A. Rida, R. Vyas, *et al.*, "RFID tag and RF structures on a paper substrate using inkjet-printing technology," *IEEE Transactions on Microwave Theory and Techniques*, **55**, (*12*), 2894–2901, 2007.

[31] V. Lakafosis, A. Rida, R. Vyas, *et al.*, "Progress towards the first wireless sensor networks consisting of inkjet-printed, paper-based RFID-enabled sensor tags," *Proceedings of the IEEE*, **98**, (*9*), 1601–1609, 2010.

[32] A. Rida, V. Lakafosis, R. Vyas, *et al.*, "Review of technologies for low-cost integrated sensors," in *IEEE International Conference on RFID-Technologies and Applications (RFID-TA)*, pp. 513–520, 2011.

[33] S. Palacios, A. Rida, K. Sangkil, *et al.*, "Towards a smart wireless integrated module (SWIM) on flexible organic substrates using inkjet printing technology for wireless sensor networks," in *IEEE International Workshop on Antenna Technology (iWAT)*, pp. 20–23, 2012.

[34] "System-on-Chip for 2.4 GHz ZigBee / IEEE 802.15.4 with Location Engine," Texas Instruments, 2009.

[35] V. Lakafosis, A. Traille, L. Hoseon, *et al.*, "RF fingerprinting physical objects for anticounterfeiting applications," *IEEE Transactions on Microwave Theory and Techniques*, **59**, (*2*), 504–514, 2011.

[36] "MSP430F543x, MSP430F541x Mixed Signal Microcontroller (Rev. C) – SLAS612C," Texas Instruments, Mar. 2010.

[37] "CC2530 A True System-on-Chip Solution for 2.4-GHz IEEE 802.15.4 and ZigBee Applications – SWRS081B," Texas Instruments, Feb. 2011.

[38] N. Doraswamy and D. Harkins, *IPSec: the New Security Standard for the Internet, Intranets, and Virtual Private Networks*, Prentice Hall, 2003.

[39] C. Shi, V. Lakafosis, M. H. Ammar, *et al.*, *Serendipity: A Distributed Computing Platform for Disruption Tolerant Networks*, Georgia Institute of Technology, 2011.

[40] V. Lakafosis, S. Addagatla, C. Belady, *et al.*, "Prometheus: A wirelessly interconnected, pico-datacenter framework for the developing world," in *10th International Conference on Wired / Wireless Internet Communications (WWIC 2012)*, Island of Santorini, Greece, 2012.

[41] A. Greenberg, J. Hamilton, D. A. Maltz, *et al.*, "The cost of a cloud: research problems in data center networks," *SIGCOMM Comput. Commun. Rev.*, **39**, (*1*), 68–73, 2008.

7 Materials for substrates

Manos M. Tentzeris, Sangkil Kim, Anya Traille, Benjamin S. Cook, and Taoran Le

7.1 Introduction

Selecting proper materials for substrates depending on applications is one of the most important design steps in the field of microwave designs because substrate material determines relative dielectric constant (ε_r), loss ($\tan\delta$), and flexibility. The impedance of transmission lines such as microstrip lines and co-planar waveguides (CPW) is a function of the relative dielectric constant (ε_r) and thickness of the substrate as well as its physical dimensions. Therefore, the dimensions of transmission lines, like width of conductors or gap, are decided by the substrate material. For compact system integration, a substrate which has high relative dielectric constant (ε_r) is preferable since it increases capacitances of the overall microwave circuit components, resulting in small feature sizes and low radiation losses. Otherwise, a material with low relative dielectric constant (ε_r) is a good substrate for structures for radiation like antennas or RFIDs. The thickness of the substrate is also an important design parameter. It is directly related to effective relative dielectric constant (ε_{eff}), which has effects on resonance frequency as well as on feature sizes of the structures on the substrate. It is obvious that the resonance frequencies or poles of RF components like antennas or filters change depending on effective relative dielectric constant (ε_{eff}), which is a function of substrate thickness. In addition, directivity of antennas is affected by the thickness of the substrate because the radiated wave tends to propagate to a material with high dielectric constant (ε_r), which results in an uneven radiation pattern as the substrate thickness increases. The loss of the substrate should be considered when the substrate is chosen too. The loss of the substrate may limit microwave circuit designs since some designs are not compatible with high loss substrates like a cavity filler or an antenna which has high-Q factor. The loss of the substrate can be reduced by utilizing a thin substrate but sometimes it may also be acting as a fabrication limitation for stability of fabrication reasons. For flexibility, the thickness and the natural properties of the substrate material are critical. Flexible substrates are preferred for use in certain applications, such as biomedical. The material and the thickness should be carefully chosen based on application requirements. It is a necessary step to decide how much flexibility is required and what kind of structure is suitable on a flexible substrate since it affects the performance at microwave frequencies.

It is convenient and easy to use well-predefined substrate materials but sometimes it is necessary to characterize a new material. In this chapter, not only mechanical/electrical properties of popular materials (Table 7.1) but also various RF characterization methods

Table 7.1 Properties of materials for substrates.

	BCB[4] [15]	Polyhydroxy-alkanoates[5] (PHA) [19]	Parylene (N, C) [12–14]	Paper [2, 11]	LCP[2] [5]	Polylactic acid[5] (PLA) [18]	PET	Kapton (Polyimide) [9, 10]	SiO_2	FR4[8]	Cellulose acetate[5] [16, 17]	LTCC[3] [18]
Relative permittivity (ε_r)	2.5	2.6 ~ 2.8	2.65 (N) 2.95 ~ 3.15(C)	2.8~3	2.9	3 ~ 3.8	3.4	3.4	3.9	4.0 ~ 5.0	4.8 ~ 7.4[6]	5.7 ~ 9.1
Loss tangent (tan δ)	2.0×10^{-3}	0.02 ~ 0.04	2.0×10^{-4} (N) 0.02 (C)	0.05 ~ 0.06	2.0×10^{-3}	0.02	2.0×10^{-3}	2.0×10^{-3}	2.0×10^{-5}	0.02 ~ 0.03	–	1.2×10^{-3} ~ 6.3×10^{-3}
CTE[1] (ppm/°C)	0.3	130	69 (N) 35 (C)	–	0 ~ 40	–	6	0.5	0.5	14 (x-axis) 13 (y-axis) 175 (z-axis)	180 ~ 270	5.8 ~ 8
Young's modulus (psi)	4.2×10^5	5.1 ~ 5.8 $\times 10^5$	3.5×10^5 (N) 4.0×10^5 (C)	144 ~ 838	1.5 ~ 6×10^6	0.39 ~ 2.3 $\times 10^6$	4.0 ~ 4.5 $\times 10^5$	4.6×10^5	2.6×10^7	3 ~ 3.5×10^6	3.5 ~ 5.9 $\times 10^5$	1.7×10^7
Water absorption (% after 24 hrs)	< 0.2	–	< 0.1	2	0.4 ~ 2.5	–	0.16	< 0.1	negligible	0.25	–	negligible
Compound	B-staged bisbenzo-cyclobutene	polyester from bacterial fermentation of sugar or lipid	para-xylylene	cellulose	polyester	starch	polyethylene terephthalate	polyimide	silicon dioxides	fiber glass, epoxy resin	cellulose	glass, ceramic

1 CTE: Coefficient of thermal expansion
2 LCP: Liquid crystal polymer
3 LTCC: Low-temperature co-fired ceramic
4 BCB: Bisbenzocyclobutene-based monomer
5 Bioplastics
6 Depending on material (L617, L618, L619 [22])

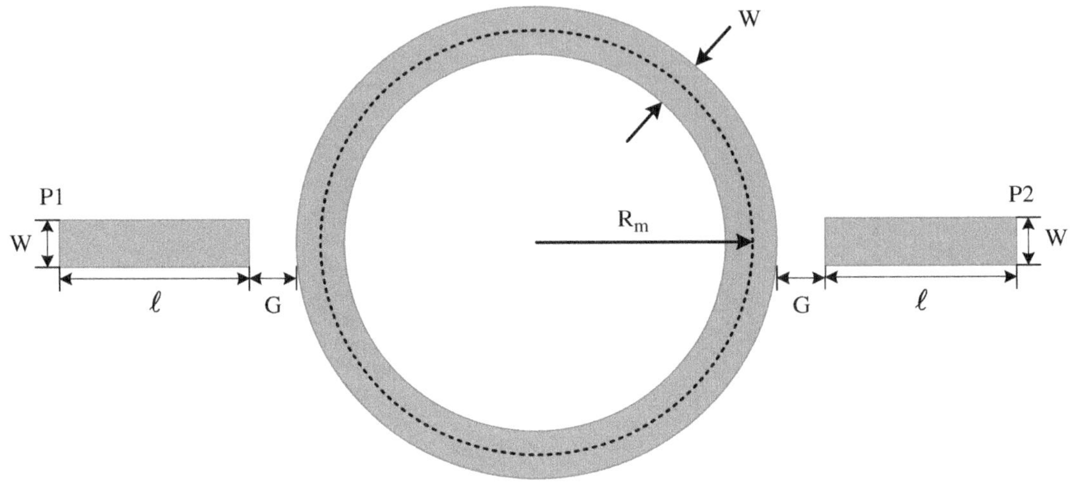

Figure 7.1 Geometry of ring resonator for substrate characterization.

for unknown materials are presented. Ring and T resonator methods for extracting relative permittivity (ε_r) are discussed in detail and the transmission line method for loss tangent (tan δ) extraction is also introduced. On top of that, widely used fabrication processes are introduced and their features are discussed.

7.2 Substrate characterization

Knowing the electrical properties of a substrate is critical to accurate modeling and simulation of microwave devices and antennas, and because of this, substrates must be characterized accurately for permittivity and loss. In this section, several substrate characterization methods utilizing resonators and transmission lines are discussed in detail. High frequency characterization methods utilizing ring resonators, T resonators, and transmission lines are introduced and the theory for each characterization method is presented. A summary of the electrical and mechanical properties of widely used substrate materials for RF applications is presented in Table 7.1.

7.2.1 Ring resonator method

The two-port (P1 and P2) ring resonator is one of the most popular methods for substrate characterization at high frequency [1, 15]. Figure 7.1 shows the structure of a typical microstrip ring resonator. W is the width of the microstrip line, and ℓ and G are the lengths of the feeding lines and the gap between the feeding line and the resonator respectively. R_m is the radius of the ring resonator to the center of the microstrip line, and this is the most critical design parameter. R_m determines the dominant resonance mode frequency of the resonator, and periodic higher order resonance frequencies are decided based on the dominant resonance frequency. The ring resonates when the circumference is approximately a wavelength. The wavelength is determined by the effective permittivity of the microstrip line, which is a function of the substrate thickness, the microstrip line width, and the air above. Because of this, the relative permittivity (ε_r) of a substrate can be extracted from the location of those resonance frequencies using the effective permittivity (ε_{eff}). The loss tangent ($\tan\delta$) of a substrate can be extracted from the quality factor (Q_0) of the resonance peaks, which is the 3 dB bandwidth of the resonance over the resonant frequency.

The dielectric constant (ε_r) of a material can be calculated from:

$$\varepsilon_r = \frac{2\varepsilon_{\text{eff}} + M - 1}{M + 1}, \tag{7.1}$$

$$M = \left(1 + \frac{12h}{W_{\text{eff}}}\right)^{-0.5}. \tag{7.2}$$

ε_{eff} is the effective dielectric constant of the substrate, and W_{eff} is the effective width of the ring resonator. The effective width of the microstrip line is discussed in Appendix 7A. The effective permittivity (ε_{eff}) can be derived from the measured resonant frequencies of the ring resonator, following Equations (7.3) and (7.4):

$$f_n = \frac{n \cdot c}{2\pi \cdot R_m \cdot \sqrt{\varepsilon_{\text{eff}}}}, \tag{7.3}$$

$$\varepsilon_{\text{eff}} = \left(\frac{n \cdot c}{2\pi \cdot R_m \cdot f_n}\right)^2, \tag{7.4}$$

where n is the order of the resonance ($n = 1, 2, 3, \ldots$), c is the speed of light, and f_n is the resonant frequency of the nth resonant mode.

The loss of the substrate can be extracted from the quality factor of the ring resonator at each resonant frequency. The total loss (α_t) includes conductor loss (α_c), dielectric loss (α_d), and radiation loss (α_r) (Equation (7.5)). Q_n is the quality factor of the ring resonator at the nth resonant frequency and λ_{eff} is the effective wavelength (Equation (7.6)). The radiation loss (α_r) and conductor loss (α_c) should be subtracted from the total loss (α_t) to calculate dielectric loss (α_d). The radiation loss and the conductor loss can be theoretically calculated and these are discussed later in the transmission line (TL) method for substrate loss section.

$$\begin{aligned}\alpha_t &= \alpha_c + \alpha_d + \alpha_r \\ &= \frac{\pi}{Q_n \lambda_{\text{eff}}},\end{aligned} \tag{7.5}$$

$$\lambda_{\text{eff}} = \frac{\lambda_0}{\sqrt{\varepsilon_{\text{eff}}}}, \tag{7.6}$$

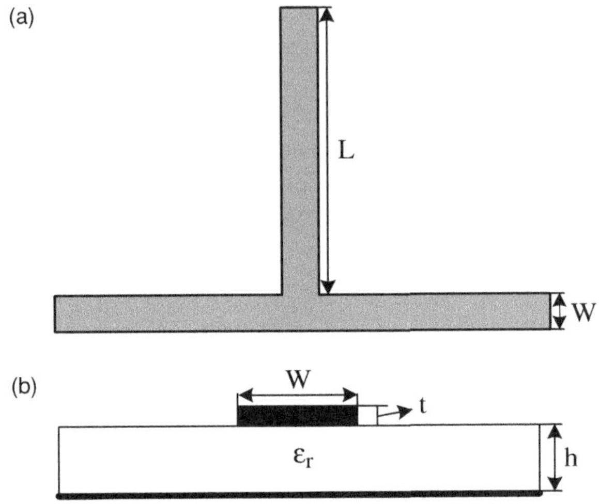

Figure 7.2 (a) Geometry of a microstrip T resonator. (b) Cross-section of the microstrip T resonator for substrate characterization.

$$Q_n = \frac{f_n}{BW_{3dB} \cdot \left(1 - 10^{\frac{-IL}{20}}\right)}, \tag{7.7}$$

$$\tan\delta = \frac{\alpha_d \lambda_0 \sqrt{\varepsilon_{eff}}(\varepsilon_r - 1)}{\pi \varepsilon_r (\varepsilon_{eff} - 1)}. \tag{7.8}$$

The first step to extract the loss tangent is to calculate the quality factor at the measured resonance frequency (Equation (7.7)) where *IL* stands for the insertion loss which corresponds to the magnitude of S_{21} of the ring resonator at the resonance. The total loss of the ring resonator can be calculated according to Equation (7.5), and dielectric loss (α_d) can be derived when the conductor loss (α_c) and radiation loss (α_r) are subtracted from the total loss (α_t). Finally, the loss tangent (tan δ) of the substrate can be extracted from Equation (7.8), the unit of dielectric attenuation is Np/m.

7.2.2 T resonator method

The T resonator, while not as common as the ring resonator in characterization, is much simpler to fabricate as it consists of a microstrip transmission line with an open circuit stub. The structure of the T resonator is shown in Figure 7.2. Due to the open circuit stub with length *d*, a null occurs in the transmission whenever *d* is an odd multiple of a quarter wavelength [3], which is caused by the reflected wave being 180° out of phase once it returns from the open end of the stub.

The equation for resonance is

$$f_{r,n} = \frac{n \cdot c}{4L\sqrt{\varepsilon_{eff}}}, \tag{7.9}$$

where n is the resonance index $(1, 3, 5, \ldots)$, c is the speed of light in vacuum, L is the length of the open circuit stub, and $\varepsilon_{\mathrm{eff}}$ is the effective permittivity of the microstrip line [13]. However, due to the parasitic effects of the T junction and fringing fields at the end of the open circuit stub, a correction factor for the length of the stub needs to be incorporated when calculating $\varepsilon_{\mathrm{eff}}$. A method for doing this is presented in [3]. Shown in Equation (7.10), k is the correction factor which includes the junction and open circuit effects on the resonant frequency. Using the relations of relative to effective permittivity in a microstrip in Appendix 7A, the permittivity of the substrate can be extracted:

$$\varepsilon_{\mathrm{eff},n} = \frac{n \cdot c}{4(L+k) f_{r,n}}. \tag{7.10}$$

The Q-factor of the resonance peaks can be used to extract the loss tangent of the substrate. The loaded Q-factor is calculated in Equation (7.11), and then needs to be converted to the unloaded Q-factor, Equation (7.12), which de-embeds the loading caused by the measurement equipment [3]:

$$Q_{l-n} = \frac{f_{r,n}}{BW_{3\mathrm{dB}-n}}, \tag{7.11}$$

$$Q_{ul-n} = \frac{Q_{l-n}}{\sqrt{1 - 2 \cdot 10^{-IL/10}}}. \tag{7.12}$$

In (7.11), Q_{l-n} is the Q-factor for the nth resonance, $f_{r,n}$ is the resonant frequency, and $BW_{3\mathrm{dB}-n}$ is the 3 dB bandwidth. In (7.12), IL is the insertion loss at resonance which is used to de-embed the loading effects. The total losses in the microstrip resonator, Equation (7.13), can then be extracted [4]:

$$\begin{aligned} \alpha_{\mathrm{tot},n} &= \alpha_{c,n} + \alpha_{d,n} + \alpha_{r,n} \\ &= \frac{8.686\pi f_{r,n}\sqrt{\varepsilon_{\mathrm{eff},n}}}{c \cdot Q_{ul-n}} \left[\frac{\mathrm{dB}}{\text{unit length}}\right]. \end{aligned} \tag{7.13}$$

There is a variety of methods to extract conductor and radiation losses, however, in this work, Agilent's Line-calc utility is used to subtract these two losses from the total loss leaving only the dielectric losses. The dielectric loss can be directly converted to loss tangent through Equation (7.14) using standard microstrip loss equations given in [5]:

$$\tan\delta = \frac{\alpha_{d,n}\lambda_0\sqrt{\varepsilon_{\mathrm{eff},n}}(\varepsilon_{r,n} - 1)}{8.686\pi\,\varepsilon_{r,n}(\varepsilon_{\mathrm{eff},n} + 1)}. \tag{7.14}$$

7.2.3 Transmission line (TL) method for substrate loss

The transmission line (TL) method is a very easy and popular method to extract the loss-tangent of substrates. Unlike other characterization methods using a resonator for the extraction of substrate loss, which only yield data points for permittivity and loss at the resonant frequency, the loss of the substrate can be extracted at every frequency point through the TL method.

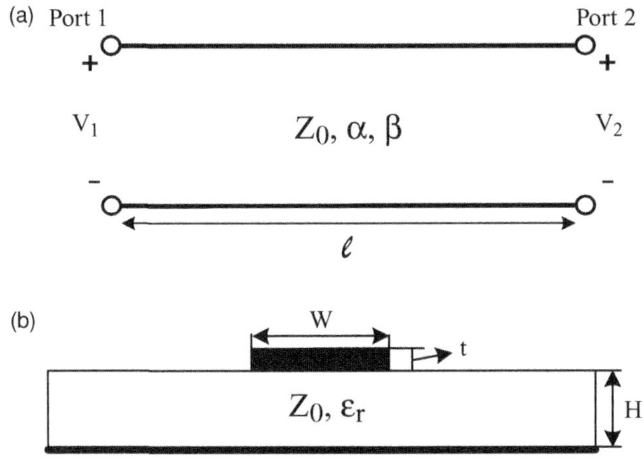

Figure 7.3 (a) Circuit model of transmission lines. (b) Cross-section of a microstrip line.

Figure 7.3 shows a general microstrip line of length ℓ with impedance of Z_0, propagation constant of β, and attenuation constant α. The relationship between voltages at port 1 and port 2 can be expressed as in Equation (7.15). The attenuation constant can be derived from the scattering parameter (S_{21}) of the microstrip line as shown in Equation (7.16) (it should be noted that the S_{21} in (7.16) is the linear scale):

$$|V_2| = |V_1|\,e^{-\alpha\ell}, \tag{7.15}$$

$$\begin{aligned}\alpha &= -\frac{1}{\ell}\ln\left(\frac{|V_2|}{|V_1|}\right)\\ &= -\frac{1}{\ell}\ln(S_{21})\ \text{Np/m}.\end{aligned} \tag{7.16}$$

The attenuation constant (α) in (7.16) is the total attenuation constant (α_t) which includes dielectric attenuation (α_d), radiation attenuation (α_r), and conductor attenuation (α_c) constants (Equation (7.17)). Therefore, dielectric attenuation constant (α_d) can be easily derived from (7.17) as shown in (7.18):

$$\alpha_t = \alpha_d + \alpha_r + \alpha_c, \tag{7.17}$$

$$\alpha_d = \alpha_t - \alpha_r - \alpha_c. \tag{7.18}$$

The radiation (α_r) and conductor (α_c) attenuations of a transmission line are usually theoretically calculated, and subtracted from the total attenuation constant (α_t). The radiation loss of the microstrip line can be theoretically calculated by using simulation and the conductor loss of the microstrip line can be derived from Equations (7.19)–(7.25). The conductor loss of a microstrip line can be calculated in three different ways according to the ratio of the width of the conductor (W) to the substrate thickness (H) (7.19)–(7.21) [6, 7]. W_{eff} is the effective width of the microstrip line's conductor and t is the thickness of the conductor (Figure 7.3). $R_{s,r}$ is the surface roughness resistance of the microstrip line and R_s is the surface resistance of the microstrip line. Z_0 is the characteristic impedance of the microstrip line. $R_{\rm rms}$ is the root-mean-square surface roughness of the conductor. μ and σ are the permeability ($\mu_0 = 4\pi \times 10^{-7}$ H/m) of the substrate and the conductivity of the conductor, respectively. δ_s is the skin depth of the microstrip line's conductor:

i) $\dfrac{W}{H} \le \dfrac{1}{2\pi}$

$$\alpha_c(f) = \frac{R_{s,r}}{2\pi Z_0 H}\left[1-\left(\frac{W_{\rm eff}}{4H}\right)^2\right] \times \left\{1+\frac{H}{W_{\rm eff}}+\frac{H}{\pi W_{\rm eff}}\left[\ln\left(\frac{4\pi W}{t}+1\right)-\frac{1-\frac{t}{W}}{1+\frac{t}{4\pi W}}\right]\right\}, \tag{7.19}$$

ii) $\dfrac{1}{2\pi} < \dfrac{W}{H} \le 2$

$$\alpha_c(f) = \frac{R_{s,r}}{2\pi Z_0 H}\left[1-\left(\frac{W_{\rm eff}}{4H}\right)^2\right] \times \left\{1+\frac{H}{W_{\rm eff}}+\frac{H}{\pi W_{\rm eff}}\left[\ln\left(\frac{2H}{t}+1\right)-\frac{1-\frac{t}{H}}{1+\frac{t}{2H}}\right]\right\}, \tag{7.20}$$

iii) $2 < \dfrac{W}{H}$

$$\alpha_c(f) = \frac{R_{s,r}}{Z_0 H}\frac{\left[\frac{W_{\rm eff}}{H}+\frac{\frac{W_{\rm eff}}{\pi H}}{\frac{W_{\rm eff}}{2H}+0.94}\right]}{\left\{\frac{W_{\rm eff}}{H}+\frac{2}{\pi}\ln\left[2\pi \exp\left(\frac{W_{\rm eff}}{2H}+0.94\right)\right]\right\}^2} \times \left\{1+\frac{H}{W_{\rm eff}}+\frac{H}{\pi W_{\rm eff}}\left[\ln\left(\frac{2H}{t}+1\right)-\frac{1-\frac{t}{H}}{1+\frac{t}{2H}}\right]\right\}, \tag{7.21}$$

$$R_{s,r} = R_s\left\{1+\frac{2}{\pi}\tan^{-1}\left[1.4\left(\frac{R_{\rm rms}}{\delta_s}\right)^2\right]\right\}, \tag{7.22}$$

$$R_s = \sqrt{\frac{\pi\mu f}{\delta_s}}, \tag{7.23}$$

$$\delta_s = \frac{1}{\sqrt{\pi\mu\sigma f}}, \tag{7.24}$$

$$\alpha_c = \frac{R_{s,r}}{Z_0 W}\ \text{Np/m}, \tag{7.25}$$

$$\tan\delta = \frac{\alpha_d\lambda_0\sqrt{\varepsilon_{\rm eff}}(\varepsilon_r-1)}{\pi\varepsilon_r(\varepsilon_{\rm eff}-1)}. \tag{7.26}$$

Equations (7.19)–(7.21) can be reduced to (7.25) when the thickness of the microstrip line's conductor (t) is negligible compared to the thickness of the substrate (H). The loss tangent (tan δ) of the substrate can be derived by plugging the dielectric loss into (7.26). ε_r is the permittivity of the substrate and $\varepsilon_{\mathrm{eff}}$ is the effective permittivity of the substrate. The detailed equations for effective permittivity ($\varepsilon_{\mathrm{eff}}$) are given in Appendix 7A.

7.3 Fabrication method for various substrates

7.3.1 Subtractive processes

Processes such as milling and lithography are typically subtractive in nature, meaning that the material to be patterned is first deposited, and the selected areas are removed to achieve the desired final pattern. These processes are currently the most popular techniques since they offer easy fabrication processing and high resolutions.

7.3.1.1 Milling

The milling process is one of the most common and widely used fabrication methods due to its ability to perform rapid prototyping at low cost. It is a subtractive fabrication method since it removes unwanted material by cutting it away. This process requires a milling machine, a substrate with metallization laminated on one or both sides to remove, and milling tools which have a size equal to or smaller than the smallest feature size of the desired pattern.

Milling can create a variety of features including holes, slots, and 2D/3D structures. This process is compatible with many materials including metals and ceramics, with good tolerance down to 25 μm. However, shape complexity is limited and many adverse features are formed such as chips and rough edges due to the crude cutting of the tool. This method is suitable for hard materials like FR4, but can rarely be used on thin flexible substrates as the milling tool removes part of the substrate along with the metallization.

7.3.1.2 Lithography

Lithography is one of the most popular subtractive fabrication processes in semiconductor and microwave circuit fabrication. Optical projection lithography is generally the utilized process due to its high throughput compared to other lithography techniques like imprint or X-ray proximity lithography. Photolithography works by exposing photoresist (PR) as shown in Figure 7.4 to UV light. Photoresist is a material that either becomes hard/resistant to washing away or soft/becomes easy to dissolve when exposed to UV light. For positive photoresist, the exposed area becomes soft and is easily dissolved in developer. In the case of negative photoresist, the exposed layer becomes polymerized and hard to wash away with developer. To transfer the desired pattern, the substrate is coated with positive/negative photoresist, and the desired mask which blocks UV light in certain areas is placed between the substrate and the UV source. The

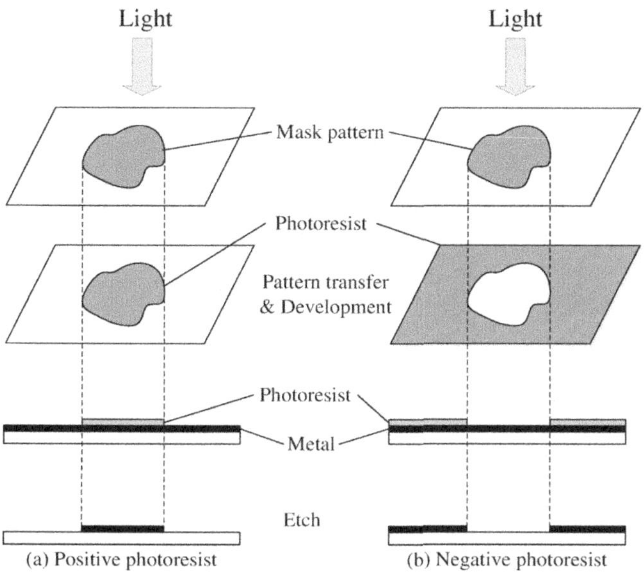

Figure 7.4 Concept of: (a) Positive lithography; and (b) Negative lithography.

substrate with photoresist is then exposed to UV light, and the unwanted photoresist is washed away by developer. Then, the patterned substrate is placed in a metal etching solution and the metal not covered by photoresist is washed away leaving the desired pattern.

Currently, negative photolithography is widely used because of its better resolution. This process can be applied to a wider variety of materials than milling including flexible substrates such as LCP, kapton, and Parylene. In addition, the resolution of this process can go down to several hundred nanometers (193–248 nm).

7.3.2 Additive processes

While the subtractive processes are currently the most popular patterning techniques, subtractive processing produces material waste and in many cases uses harsh chemicals which are difficult to dispose of properly. Additive processing has recently been gaining popularity, as the desired pattern is deposited directly via screen or digital printing methods which decrease material waste and eliminate the requirement for harsh etching chemicals.

7.3.2.1 Inkjet printing

The inkjet printing process has become an area of interest in the fabrication world over the last 10 years because it can be used on a wide array of substrates and can deposit almost any material that can be put into a liquid form including metals, semiconductors, and dielectrics [20]. Inkjet printing is a lithography-free process which removes the need for masks, exposure equipment, and etching, leading to large time cuts and cost reductions in

Table 7.2 Taxonomy of inkjet printing techniques.

Inkjet printing techniques	
Drop on demand (DOD)	thermal inkjet piezo inkjet electrostatic inkjet acoustic inkjet
Continuous	piezo inkjet thermal inkjet

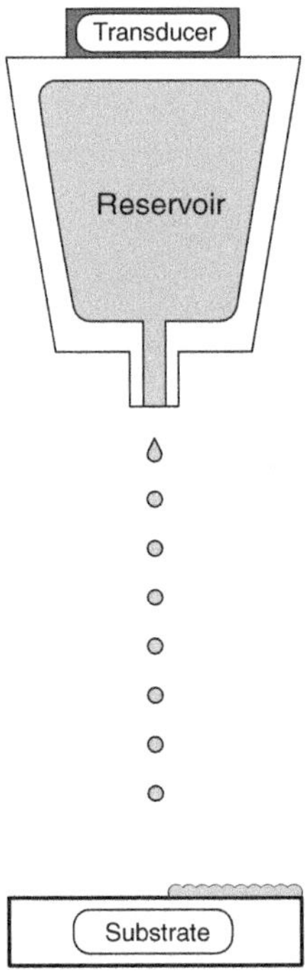

Figure 7.5 Inkjet printing.

the fabrication process. It is mainly used as an additive process, which means it deposits the material that will be part of the final structures, but resist-like material can also be deposited for use as an etching mask and removed after etching. The inkjet printing process has been used to create light emitting diodes, TFTs, conducting lines, and liquid crystal displays [20].

Inkjet printing is a relatively simple concept. A liquid ink is loaded into a reservoir which feeds a nozzle as seen in Figure 7.5. As the nozzle or substrate is moved horizontally, one of several methods is used to eject a drop of ink from the nozzle which then hits and adsorbs to the substrate. By moving the nozzle and ejecting droplets at the correct points, patterned features are created. Inkjet printing comes in several variants. A simple taxonomy of the different processes is shown in Table 7.2.

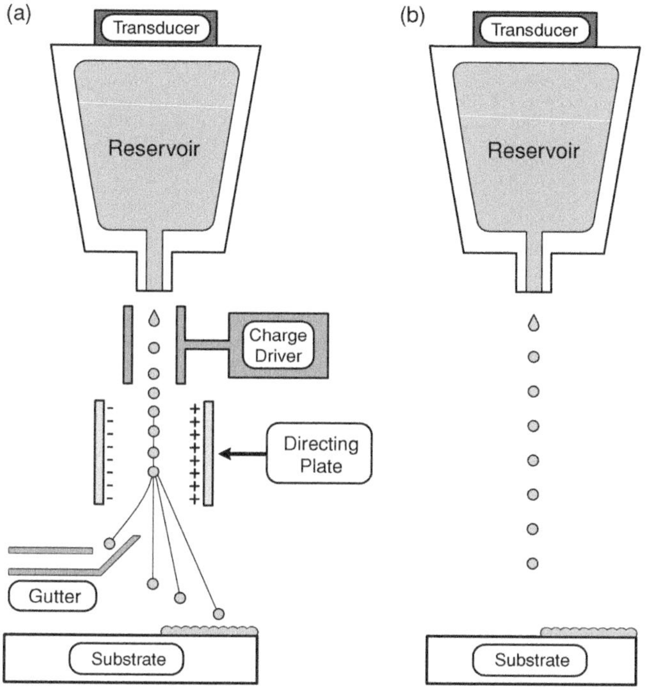

Figure 7.6 Continuous (a) and DOD (b) inkjet printing techniques.

The two main inkjet printing techniques are drop on demand (DOD) and continuous printing. Continuous inkjet printing works by continuously ejecting drops from the reservoir at a constant frequency by setting up an acoustic wave with a transducer. When a drop exits the reservoir it is given an electrostatic charge using charging plates. It is then either directed towards the substrate or into a collection bin with a pair of electrodes. This method allows for patterning without turning off the jet which helps mitigate clogging of the nozzle due to ink drying and inconsistent excitation of the ink, and allows for faster fly velocities [20] (i.e. the velocity at which the substrate moves past the nozzle). This process is best suited for commercial printing as it is usually only compatible with one type of ink and requires inks that can be electrostatically charged [21].

DOD printing is similar to continuous printing in that a transducer generates a pressure wave in the ink reservoir which ejects a drop of ink. However, a signal is only sent to the transducer when a drop is required instead of continuous operation. DOD printing allows for a wider range of inks with varying viscosities and surface tensions, which can be accounted for by changing the signal to the transducer, and can print at higher resolutions [20]. But, clogging is still a prominent issue due to the inconsistent waveform applied to the reservoir and ink drying. Continuous and DOD printing are shown in Figure 7.6.

Within each type of printing, several types of actuators are used including thermal, piezo, electrodynamic, and acoustic. Thermal and piezo actuators work by inducing a pressure wave in the reservoir which expels a drop from the nozzle while electrodynamic

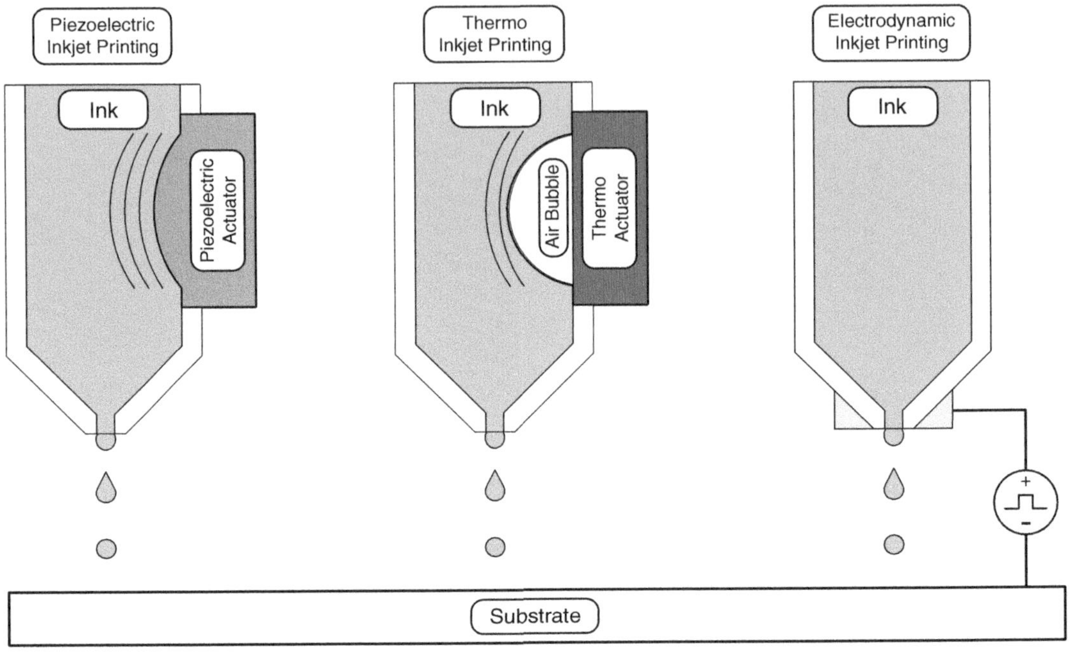

Figure 7.7 Inkjet actuator types.

and acoustic actuators work by disrupting the surface tension of the meniscus formed at the end of the nozzle, causing a drop to jet towards the substrate [21]. Thermal actuators are typically used in commercial and office inkjet printers where the same ink is used. This allows the thermal expansion properties of the ink to be controlled, which results in consistent jetting. For research projects where many different types of inks with different thermal properties may be used, piezo systems are more common. Acoustic systems are virtually exclusive to continuous jet printers found in industrial production and are an extension of the piezo nozzle as the acoustic wave is generated with a piezo crystal [20]. Figure 7.7 demonstrates the different types of actuators available and Figure 7.8 shows inkjet-printed RFID tags on various substrates, such as polyimide and paper.

7.3.2.2 Screen printing

Screen printing is an additive method common in printed circuit board fabrication, where a screen is selectively patterned with holes to allow ink to pass through. The screen printing process is depicted in Figure 7.9. The mask is placed on the substrate and the ink is applied to the mask. A squeegee passes over the mask and pushes the ink through the patterned holes and onto the substrate below. Multiple masks can be used to create multilayer patterned designs with a wide variety of inks. However, unlike inkjet printing, screen printing tends to create more waste in the form of unused ink applied to the mask.

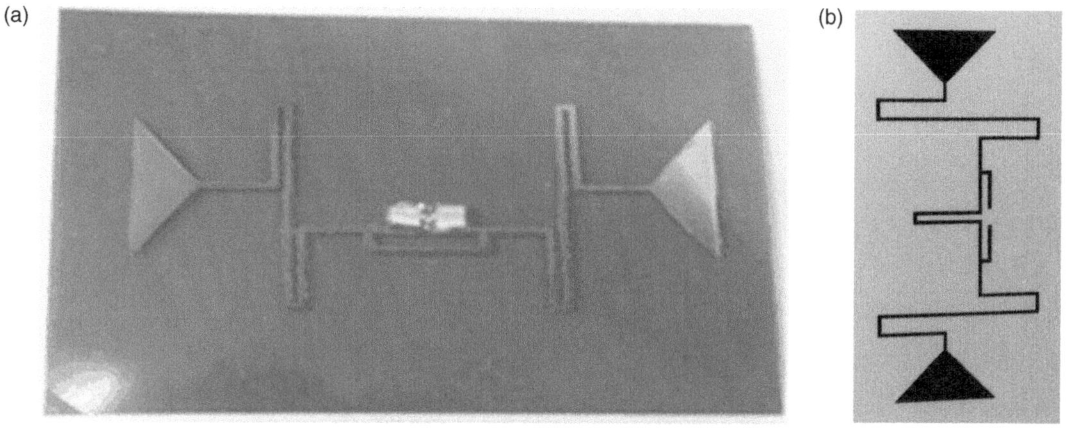

Figure 7.8 Inkjet-printed RFID on: (a) Polyimide; (b) Paper.

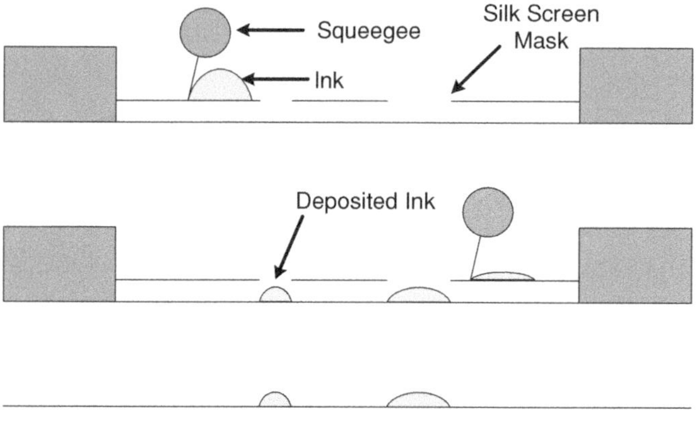

Figure 7.9 Screen printing process.

7.3.3 Mixed processes

Recently a process that can be defined as "mixed" has been introduced for the realization of metal deposition on substrates that, unlike milling, is fully compliant with flexible and soft substrates such as paper [25]. This technique exploits adhesive metal laminates, purposely etched by means of conventional photolithography (to this extent it can be considered a subtractive method), that are then stuck on to the host substrate; since only the metal necessary for the realization of the final circuit is added to the substrate without any etching after deposition, it can be considered additive.

The method is fully described in [25] and consists of five steps:

1. A positive photoresist deposition is applied on top of a multilayer laminate consisting, from the bottom, of a protective sheet, an adhesive layer, and the top metal.
2. An already described photolithography process (see Section 7.3.1.2) is applied directly to this laminate to remove the unwanted metal with respect to the circuit layout.

Figure 7.10 24 GHz single-balanced diode mixer exploiting cellulose-based materials and mixed processes.

3. A sacrificial layer is introduced in the third step and stuck on top of the worked laminate. After this the real metal circuit results sandwiched between the original protective layer at the bottom and the sacrificial layer on top. This phase is crucial to keep the isomorphism of the realized circuit following the sticking on the host substrates.
4. In this step the protective bottom layer is removed and the metal circuit is stuck on the hosting substrate.
5. Finally, the sacrificial layer is lifted off and the circuit is polished.

In order to make this process feasible, the adhesion characteristics of the sacrificial layer and of the protective layer have to be chosen correctly. The adhesion of the protective layer must be weaker than that of the sacrificial layer, in order to allow for the protective layer removal. The adhesion of the metal to the final substrate, in turn, must be stronger than that of the sacrificial layer on the metal in order to allow for the sacrificial layer lift off.

Eventually we end up having a metal circuit realized on a compound substrate consisting of a layer of original substrate material plus a film of adhesive on top. If, instead, we start from a substrate with a plain metal laminate already stuck on the bottom, we will obtain a structure perfectly compliant with microstrip realizations, the real substrate consisting of a sandwich of adhesive, original substrate material, and adhesive. This must be taken into account in the circuit design when the effective propagation constant of the lines has to be evaluated. An example of a 24 GHz single-balanced diode mixer is shown in Figure 7.10.

It is worth emphasizing that this process has both advantages and disadvantages in comparison with conventional photolithography and inkjet printing.

Advantages:

- Excellent for fast prototyping;
- Lower losses in comparison to inkjet printing on paper substrates because of the use of non-cured metal and of the presence of adhesive interposer, the latter being usually a better dielectric than paper;
- No curing needed, meaning good compatibility with very large industrial processes; dimensions are not limited by curing ovens but depend only on the capability to manage, at industrial level, the sticking and removal of laminates; rotary press-like processes are foreseeable;
- Compatibility with conventional soldering process;
- Extremely low cost.

Disadvantages:

- Still photolithography is required;
- Resolution depends on photolithography, but the compliance of the final circuit with the original layout depends on the ability to keep the isomorphism during the sticking and removal phases.

So this method is not to be seen as an alternative to fully additive or subtractive ones, but just as a way to extend the capability of realizing printed circuits on unconventional substrates.

Appendix 7A: The effective width and effective permittivity

The characteristic impedance of a microstrip line in static conditions with given dimensions can be calculated as shown in Equation (A7.1) [22, 23]. The frequency-dependent characteristic impedance can be obtained from Equation (A7.2) based on (A7.1). W is the width of the microstrip line, H is the thickness of the microstrip line, and ε_r is the dielectric constant of the substrate. The error value is less than 1% for the ranges of $0.1 \leq W/H \leq 10$, $1 \leq \varepsilon_r \leq 18$, and $0 \leq H \cdot f \leq 30$ GHz mm:

$$Z_0(0) = \begin{cases} \dfrac{119.9}{\sqrt{2(\varepsilon_r+1)}}\left[\ln\left(\dfrac{4H}{W}+\sqrt{16\left(\dfrac{H}{W}\right)^2+2}\right)-\dfrac{1}{2}\left(\dfrac{\varepsilon_r-1}{\varepsilon_r+1}\right)\left(\ln\dfrac{\pi}{2}+\dfrac{1}{\varepsilon_r}\ln\dfrac{4}{\pi}\right)\right], & (W/H<3.3) \\ \dfrac{119.9\pi}{2\sqrt{\varepsilon_r}}\left[\dfrac{W}{2H}+\dfrac{\ln 4}{\pi}+\dfrac{\ln\left(e\pi^2/16\right)}{2\pi}\left(\dfrac{\varepsilon_r-1}{\varepsilon_r^2}\right)+\dfrac{\varepsilon_r+1}{2\pi\varepsilon_r}\left[\ln\dfrac{\pi e}{2}+\ln\left(\dfrac{W}{2H}+0.94\right)\right]\right]^{-1}, & (W/H>3.3) \end{cases} \tag{A7.1}$$

$$Z_0(f) = Z_0(0)\left(\frac{R_8}{R_9}\right)^{R_{12}}, \tag{A7.2}$$

where $$R_1 = 4.766 \exp\left[-3.228\left(\frac{W}{H}\right)^{0.641}\right],$$

$$R_2 = 0.016 + (0.0514\varepsilon_r)^{4.524},$$

$$R_3 = 1.206 - 0.3144\exp\left(-0.0389\varepsilon_r^{1.4}\right)\left[1-\exp\left(-0.267\left(\frac{W}{H}\right)^7\right)\right],$$

$$R_4 = 1 + 1.275\left[1-\exp\left(-0.00463R_1\varepsilon_r^{1.674}\left(\frac{Hf}{18.37}\right)^{2.745}\right)\right],$$

$$R_5 = \frac{5.086R_2\left(\frac{Hf}{28.84}\right)^{12}}{0.384+0.386R_2}\frac{\exp\left[-22.2\left(\frac{W}{H}\right)^{1.92}\right]}{1+1.3\left(\frac{Hf}{28.84}\right)^{12}}\frac{(\varepsilon_r-1)^6}{1+10(\varepsilon_r-1)^6},$$

$$R_6 = \left[0.0962+\left(\frac{19.47}{Hf}\right)^6\right]^{-1},$$

$$R_7 = \left[1+0.00245\left(\frac{W}{H}\right)^2\right]^{-1},$$

$$R_8 = 0.9408\varepsilon_{\text{eff}}^{R_4}(f) - 0.9603,$$

$$R_9 = (0.9408 - R_5)\varepsilon_{\text{eff}}^{R_4}(0) - 0.9603,$$

$$R_{10} = 0.707(0.00044\varepsilon_r^{2.136}+0.0184)\left(\frac{Hf}{12.3}\right)^{1.097},$$

$$R_{11} = 1 + 0.0503\varepsilon_r^2 R_6\left\{1-\exp\left[-\left(\frac{W}{15H}\right)^6\right]\right\},$$

$$R_{12} = R_3\left\{1-1.1241\left(\frac{R_7}{R_{11}}\right)\exp\left[-0.026(Hf)^{1.1566}-R_{10}\right]\right\}.$$

The effective permittivity (ε_{eff}) at high frequency can be calculated using static effective permittivity ($\varepsilon_{\text{eff}}(0)$) and effective permittivity dispersion [22, 23]:

$$\varepsilon_{\text{eff}}(0) = \frac{\varepsilon_r+1}{2}\left[1-\frac{1}{2H'}\left(\frac{\varepsilon_r-1}{\varepsilon_r+1}\right)\left(\ln\left(\frac{\pi}{2}\right)+\frac{1}{\varepsilon_r}\ln\left(\frac{4}{\pi}\right)\right)\right]^{-2}\left(\text{when } \frac{W}{H}<1.3\right)$$
$$= \frac{\varepsilon_r+1}{2}+\frac{\varepsilon_r-1}{2}\left(1+10\frac{H}{W}\right)^{-0.555},\left(\text{when } \frac{W}{H}>1.3\right) \tag{A7.3}$$

where $$H' = \ln\left(\frac{4H}{W}+\sqrt{16\left(\frac{H}{W}\right)^2+2}\right),$$

$$\varepsilon_{\text{eff}}(f) = \varepsilon_r - \frac{\varepsilon_r-\varepsilon_{\text{eff}}(0)}{1+P(f)}, \tag{A7.4}$$

where $P(f) = P_1 P_2[(0.1844 + P_3 P_4)10fH]^{1.5763}$,

$$P_1 = 0.27488 + [0.6315 + 0.525/(1 + 0.157fH)^{20}]\left(\frac{W}{H}\right) - 0.065683\exp\left(-8.7513\frac{W}{H}\right),$$

$$P_2 = 0.33622\,[1 - \exp(-0.03422\varepsilon_r)],$$

$$P_3 = 0.0363\exp\left(-4.6\frac{W}{H}\right)\left\{1 - \exp\left[-\left(\frac{fH}{3.87}\right)^{4.97}\right]\right\},$$

$$P_4 = 1 + 2.751\left\{1 - \exp\left[-\left(\frac{\varepsilon_r}{15.916}\right)^{8}\right]\right\}.$$

The effective width (W_{eff}) of the microstrip line can be calculated by (A7.5) using (A7.2) and (A7.3) [24]. The effective width (W_{eff}) of the microstrip line can also be written in simpler form as in Equation (A7.6), where t is the thickness of the conductor:

$$W_{\text{eff}}(f) = \frac{H\eta_0}{Z_0(f)\sqrt{\varepsilon_{\text{eff}}(f)}}, \tag{A7.5}$$

$$W_{\text{eff}} = W + \frac{1.25t}{\pi}\left(1 + \ln\left(\frac{2H}{t}\right)\right)^2. \tag{A7.6}$$

References

[1] J.-M. Heinola and K. Tolsa, "Dielectric characterization of printed wiring board materials using ring resonator techniques: a comparison of calculation models," *IEEE Trans. Dielectr. Electr. Insul.*, **13**, (*4*), 717–726, (Aug.) 2006.

[2] L. Yang, A. Rida, R. Vyas, and M. M. Tentzeris, "RFID tag and RF structures on a paper substrate using inkjet-printing technology," *IEEE Trans. Microw. Theory Techn.*, **55**, 2894–2901, (Dec.) 2007.

[3] K. P. Latti, M. Kettunen, J. P. Stoem, and P. Silventoinen, "A review of microstrip T-resonator method in determining the dielectric properties of printed circuit board materials," *IEEE Trans. Ins. Mea.*, **56**, 1845–1850, (Oct.) 2007.

[4] A. R. Fulford and S. M. Wentworth, "Conductor and dielectric-property extraction using microstrip tee resonators," *Microw. Opt. Techn. Let.*, **47**, 14–16, (Oct.) 2005.

[5] D. C. Thompson, O. Tantot, H. Jallageas, *et al.*, "Characterization of liquid crystal polymer (LCP) material and transmission lines on LCP substrates from 30 to 110 GHz," *IEEE Trans. Microwave Theory Tech.*, **52**, (*4*), 1343–1352, 2004.

[6] M. Kirschning and R. H. Jansen, "Accurate model for effective dielectric constant with validity up to millimeter-wave frequencies," *Electron. Lett.*, **18**, 272–273, (Jan.) 1982.

[7] R. A. Pucel, D. J. Masse, and C. P. Hartwig, "Losses in microstrip," *IEEE Trans. Microw. Theory Techn.*, **16**, (*6*), 342–350, 1968.

[8] L. Devlin, G. Pearson, and J. Pittock, "RF and microwave component development in LTCC," in *IMAPS Nordic 38th Annu. Conf.*, pp. 96–110, Sep. 2001.

[9] Z. Wei and A. Pham, "Liquid crystal polymer for microwave/millimeter wave multi-layer packaging," in *IEEE MTT-S Dig.*, **3**, 2273–2276, (Jun.) 2003.

[10] [Online] Available: http://www2.dupont.com/Kapton/en_US/.

[11] B. S. Cook and A. Shamim, "Inkjet printing of novel wideband and high gain antennas on low-cost paper substrate," *IEEE Trans. Antennas Propag.*, **50**, (*9*), 4148–4156, (Aug.) 2012.

[12] C. Karnfelt, C. Tegnander, J. Rudnicki, J. P. Starski, and A. Emrich, "Investigation of arylene-C on the performance of millimeter-wave circuits," *IEEE Trans. Microwave Theory Tech.*, **54**, (*8*), 3417–3425, 2006.

[13] Specialty Coating Systems Co. (SCS). Indianapolis, IN, http://scscoatings.com.

[14] H. Sharifi, R. R. Lahiji, L. Han-Chung, *et al.*, "Characterization of parylene-N as flexible substrate and passivation layer for microwave and millimeter-wave integrated circuits," *IEEE Trans. Adv. Pack.*, **32**, (*1*), 84–92, (Feb.) 2009.

[15] [Online] Available: http://www.dow.com/cyclotene/solution/index.htm.

[16] [Online] Available: http://www.azom.com/properties.aspx?ArticleID=1461.

[17] L. Hartshorn and E. Rushton, "The dielectric properties of cellulose acetate," *Journal of the Institution of Electrical Engineers*, **81**, (*501*), 315–332, (Sep.) 1938.

[18] T. Nakagawa, T. Nakiri, R. Hosoya, and Y. Tajitsu, "Electrical properties of biodegradable polylactic acid film," *IEEE Trans. Ind. Appl.*, **40**, (*4*), 1020–1024, (July/Aug.) 2004.

[19] Z. S. Petrovic, J. Milic, Y. Xu, and I. Cvetkovic, "A chemical route to high molecular weight vegetable oil-based polyhydroxyalkanoate," *Macromolecules*, **43**, (*9*), 4120–4125, (May) 2010.

[20] Microfab Technologies, (2010), available: www.microfab.com.

[21] A. Sridhar, "An inkjet printing-based process chain for conductive structures on printed circuit board materials," PhD, EE, University of Twente, Enshede, 2010.

[22] T. C. Edwards, *Foundations for Microstrip Circuit Design*, Wiley, 1981.

[23] R. K. Hoffmann, *Handbook of Microwave Integrated Circuits*, translation from the German of *Integrierte Mikrowellenschaltungen*, Artech House, 1987.

[24] G. Kompa and R. Mehran, "Planar waveguide model for calculating microstrip components," *Electron. Lett.*, **11**, 459–460, 1975.

[25] F. Alimenti, P. Mezzanotte, M. Dionigi, M. Virili, and L. Roselli, "Microwave circuits in paper substrates exploiting conductive adhesive tapes," *IEEE Microwave Wireless Components Letters*, **22**, (*12*), 660–662, 2012.

8 Organic conductors and semiconductors: recent achievements and modeling

Luca Pierantoni and Paolo Lugli

8.1 Introduction

The idea of using organic materials, namely small molecules and conductive polymers, to realize electronic devices dates back to the beginnings of semiconductor technology. Yet, while silicon has become pervasive at a societal level, allowing the gigantic steps we have witnessed in the last decades, organic electronics has remained a niche field with relatively small impact. In recent years two novelties have generated a lot of interest in organic electronics and revitalized the field. On the one hand, organic light emitting diodes (OLED) displays have been brought to the market and are on their way to a dominant position with respect to liquid crystal. On the other, the discovery of graphene has been demonstrated and has fuelled incredible and unprecedented expectations. Some are already defining the twenty-first century as the "carbon" century, just as the previous century has been the "silicon" century.

This chapter aims to present the properties of conductive polymers and of graphene in the context of RFID development. Thus, we will deal with active components (e.g. diodes and transistors) based on pentacene, the best-known organic semiconductor, and passive components (e.g. antennas) based on graphene. The analyses are based on modeling and simulations, referring to the activities and results obtained by the groups at the Technische Universität München and at the Università Politecnica delle Marche, respectively. Such an approach allows us to discuss some of the physical properties of the underlying materials, to focus on their potential applications and, at the same time, to illustrate the simulation tools that are available for design and optimization. We will refer to the vast available literature for more details on material properties, fabrication techniques, and other applications.

The chapter is organized as follows. We first deal with pentacene active devices, discussing the two basic components of an RFID, namely rectifiers and transistors. Here, commercial drift-diffusion simulators provide the necessary modeling framework. We then move to the description of graphene properties, showing how graphene could provide enhanced performance for a new generation of integrated and multi-functional devices, including wireless applications and RFID antennas. In the second part, we deal with the wonder material, graphene, according to the following paradigm: i) basic electronic properties; ii) dispersion curves, electric tuning; iii) modeling of the electromagnetics-quantum transport phenomena in the ballistic regime: full-wave frequency- and time-domain techniques; iv) modeling and design of graphene antennas for RFID and wireless applications in the non-ballistic regime, at microwave and mm-wave range.

Regarding (iv), it has to be highlighted that the possibility of using graphene-based antennas is a core point in the development of high speed, multi-functional new wireless architecture, leading to a new generation of smart nano-systems for wireless communications, RFID, and wireless power transfer devices. Besides the fact that un-doped graphene is a moderate conductor in the microwave mm-wave range, the possibility of electrical tuning properties leads to unprecedented technological issues, such as: i) tunability of the antenna parameters: radiation pattern, resonant frequency, directivity, efficiency; ii) tunability of the EM coupling with other radiating elements, and of the pattern of an array of graphene patches; iii) possible reduction of the patch size, owing to the "slow wave" behavior (kinetic inductance contribution); iv) integration: in future, a new class of high performance carbon components will hopefully be integrable in the same circuit (amplifiers, switches, mixers, attenuators, etc.).

8.2 Active devices for printed RFIDs

In the following we describe two of the basic building blocks for printed RFIDs, namely inverter circuits (based on organic TFTs) and rectifying diodes. We will refer to conductive polymer, pentacene in our case, as a constituent material, although several other polymers have been demonstrated [1]. Following the philosophy of this chapter, the analysis is based on simulation results. Therefore, we first introduce the simulation framework and then concentrate on the specific applications.

8.2.1 Modeling tools for organic devices

For our simulations, we have chosen the commercial TCAD tool SentaurusTM [2], which we have modified to account for the special nature of organic semiconductors. SentaurusTM is a multidimensional, electrothermal, mixed-mode device and circuit simulator that can handle geometries in up to three dimensions. Generally, the simulator solves the drift-diffusion transport equations self-consistently coupled to the Poisson equation:

$$\nabla \cdot [\varepsilon \nabla V] = q(n - p - N_D^+ + N_A^- + \rho_T), \tag{8.1}$$

$$\frac{\partial n}{\partial t} = \frac{1}{q}\nabla J_n + G_n - R_n, \tag{8.2}$$

$$\frac{\partial p}{\partial t} = -\frac{1}{q}\nabla J_p + G_p - R_p, \tag{8.3}$$

where ε is the electrical permittivity, N_D^+ and N_A^- are the densities of ionized donors and acceptors, respectively, q is the unit charge, n and p are the electron/hole densities, respectively, and ρ_T is the trap density. According to the drift-diffusion model the equations for electron and hole current densities are:

$$J_n = qn\mu_n \nabla V + qD_n \nabla n, \tag{8.4}$$

$$J_p = qp\mu_p \nabla V - qD_p \nabla p. \tag{8.5}$$

The charge carrier transport in the organic film is modeled introducing a field-dependent mobility [3], which has the form of a Poole–Frenkel law for electric fields $E < 2{\cdot}10^6$ V/cm:

$$\mu = \mu_0 \exp\left(\sqrt{\frac{E}{E_0}}\right), \tag{8.6}$$

where μ_0 is the low field mobility and E_0 is a parameter which depends on the material.

The calculations are based on a finite element grid. Main input variables are device geometry, material specifications, and interface properties. Additionally, Sentaurus^{TM} offers a versatile interface which allows the introduction of user-defined physical models for e.g. mobility or recombination. This is of great importance when accurately describing organic semiconductors. A description of the simulation method is given in [4]. Thermionic emission as a boundary condition at the anode and at the cathode has negligible influence on the simulation results since the surface recombination velocity is up to four orders of magnitude smaller than in inorganic devices. The contact barriers are defined as absolute values of the difference between metal work function and the energy of the transport states. For electrons this is the electron affinity χ and for holes the ionization energy $\chi + Eg$.

8.2.2 High frequency rectifiers based on organic Schottky diodes

The rectifier represents the key component in the organic tag since it must be able to follow and rectify an AC signal at the established RFID frequency of 13.56 MHz. This is not an easy task for the low mobility organic materials; in fact this element of the ORFID circuit is the limiting component of the entire system. In the next section we aim to investigate the theoretical limits of organic device-based rectification using advanced drift-diffusion simulations. The results obtained are also compared with available analytical models [5].

A. Analytical model

Recently an analytical model estimating the cut-off frequency has been presented by Steudel *et al.* [5]. The cut-off frequency is defined as the frequency at which the amplitude of the rectified DC signal has decreased by 3 dB compared to the amplitude of the applied input AC signal. For the derivation it is assumed that the organic diodes operate under a space-charge limited regime, and in this case the cut-off frequency has been shown to be:

$$f_{\max} = \frac{9\mu}{12L_D^2 V_{DC}}\left[(V_F - 3V_{DC})\sqrt{V_A^2 - (V_{DC} - V_F)^2} + \left(V_A^2 + 2V_{DC}^2\right)\cos^{-1}\left(\frac{V_{DC} + V_F}{V_F}\right)\right], \tag{8.7}$$

where v_F is the transit voltage, V_A the amplitude of the AC voltage, and V_{DC} the DC voltage.

The analytical model can be extended, taking into account the Poole–Frenkel field-dependent mobility, typical for organic semiconductors, given in Equation (8.6), where E_0 typically has values between 10^4–10^5 V/cm.

Estimation of the cut-off frequency using both the model developed by Steudel *et al.* and the extended analytical model are presented and compared with drift-diffusion results.

B. Drift-diffusion modeling

A more in-detail picture of the operation of organic diodes can be obtained with drift-diffusion simulations by performing a large signal analysis of the actual diode structures accounting for the different barrier heights at the contacts.

The structure of the diode consists of an active layer made out of a p-type semiconductor, sandwiched between two metals: gold (Au) and aluminum (Al). As the organic material we choose the small molecule pentacene, which is extensively used today because of its high mobility for holes – larger than 0.1 cm^2/Vs [6–8]. The results would remain valid even for a different material choice. The pentacene material properties used for our simulations are reported in [4, 9]. A non-zero threshold voltage is accounted for by introducing fixed charges at the semiconductor/insulator interface [4].

The breakdown voltage of the Schottky diode is reported to be greater than −30 V in reverse bias and around 19 V in forward bias [10].

Typical DC supply voltages for modern organic circuits (inverters, ring oscillators) are 10 V [5], or even less (8.5 V [11]). In order to rectify the AC signal ($V_A = 10$ V) at 13.56 MHz we choose a mobility $\mu_0 = 0.2$ cm^2/Vs and a thickness $d = 200$ nm.

Let us now consider a half wave rectifier obtained with a smoothing capacitor in series to the diode. Its schematic is shown in Figure 8.1(a). When a positive voltage is applied on the anode of the Schottky diode, holes are injected in the pentacene film and flow in the upper plate of the capacitor, charging it to a positive potential. This DC positive voltage can be measured between the plates of the capacitor and can be used as supply voltage for the logic circuits in the transponder chip. The value of the series capacitance (C_L) is very important and has to be properly chosen, since it will determine the amplitude of the small AC signal (that will always be present) on top of the rectified DC voltage. For a small modulation this capacitance should be at least 100 times greater [5] than the intrinsic (geometric) capacitance of the diode. This geometric capacitance has a value of around 0.8 pF; therefore the load capacitance should be (at least) equal to 80 pF.

The load capacitance has to be large enough to keep the rectified voltage constant on the half wave when the diode is in reverse bias. On the other hand, if the capacitance is too large it will require a long charging time, making the tag very slow, which is highly undesirable. Figure 8.1(b) shows the rectified voltage for the half wave rectifier with $C_L = 80$ pF, the efficiency (V_{DC}/V_A) lying at around 90%.

A full wave rectifier, which consists of two capacitors and two diodes, as shown in Figure 8.2(a), is similar to the structure presented in [12]. The high frequency AC signal is applied to the pair of horizontal parallel gold electrodes and the rectified DC voltage

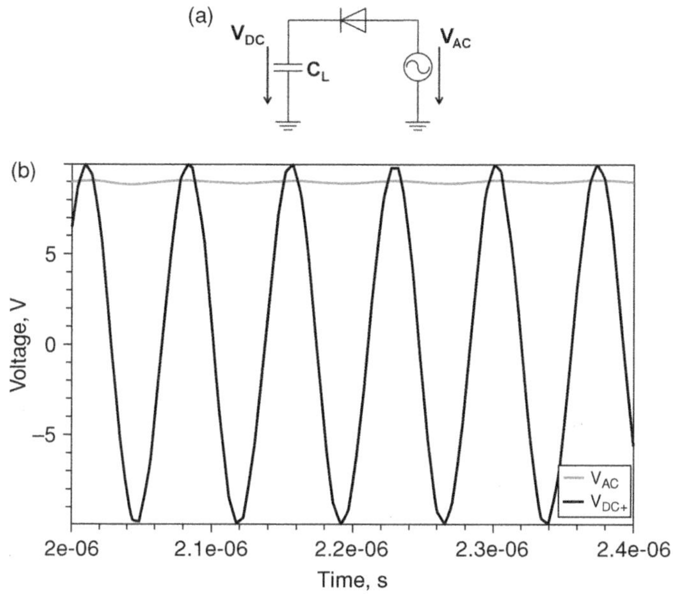

Figure 8.1 (a) Half wave rectifier; (b) Simulated rectified output voltage.

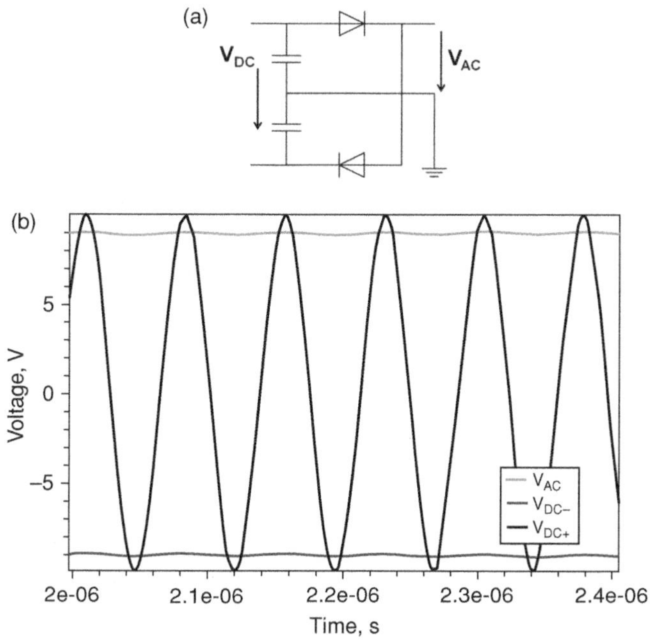

Figure 8.2 (a) Full wave rectifier; (b) Simulated rectification.

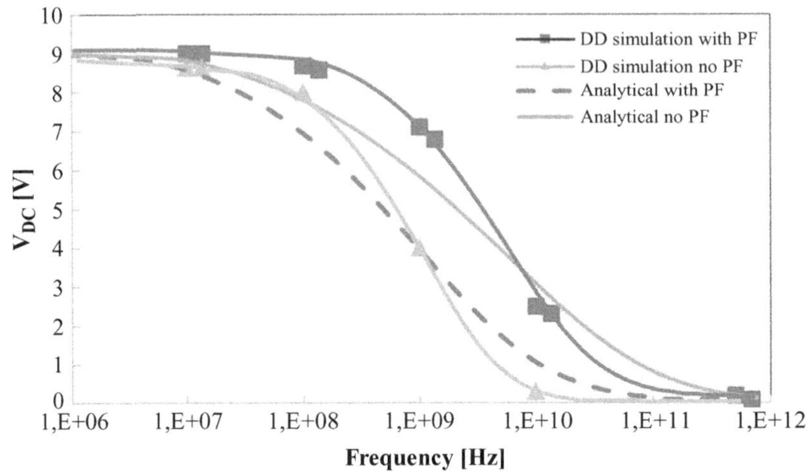

Figure 8.3 Maximum rectifying frequency for SCLC diodes with $L_D = 200$ nm, $\mu_0 = 0.2$ cm^2/V s. The label PF indicates that a Poole–Frenkel mobility, with $E_0 = 10^5$ V/cm, has been used for calculations.

is obtained between the pair of vertical parallel gold electrodes. The two capacitors with the common electrode are sandwiched between gold–parylene–aluminum while the two antiparallel organic Schottky diodes are stacked between the gold and aluminum contacts. The geometry of the gold electrodes is chosen such that the load capacitors have a value of 80 pF, while the diodes have an area of 4500 μm^2.

The DC voltage of the simulated rectifier is around 18 V out of 20 V possible, with an efficiency of 90%. Such a high efficiency is obtained by neglecting any reaction of the materials with the atmosphere. Indeed, in reference [10] very high efficiency values have been obtained with real, fabricated devices but also in vacuum or under a nitrogen atmosphere. It was mentioned there that when the devices were exposed to air and measured, the current densities decreased by almost two orders of magnitude.

Figure 8.3 shows the maximum rectifying frequencies that can be achieved considering constant mobility (triangular markers) and Poole–Frenkel mobility (rectangular markers). The curves show that the simple analytical models described above underestimate the rectified potential up to a certain frequency (700 MHz for constant mobility and around 5.5 GHz for Poole–Frenkel mobility with a maximum spread of 1 V), and overestimate it for higher frequencies (the maximum spread is again around 1V). The cut-off frequency for our configuration is found to be 500 MHz by analytic means, and 600 MHz using the more precise drift-diffusion model in the case of constant mobility and 2 GHz versus 3 GHz in the case of field-dependent mobility.

8.2.3 Basic devices and circuits based on organic TFT

In this section we present developments in hybrid organic/inorganic thin film transistors, as reported in a previous work [4]. Our OTFTs have been fabricated using a bottom

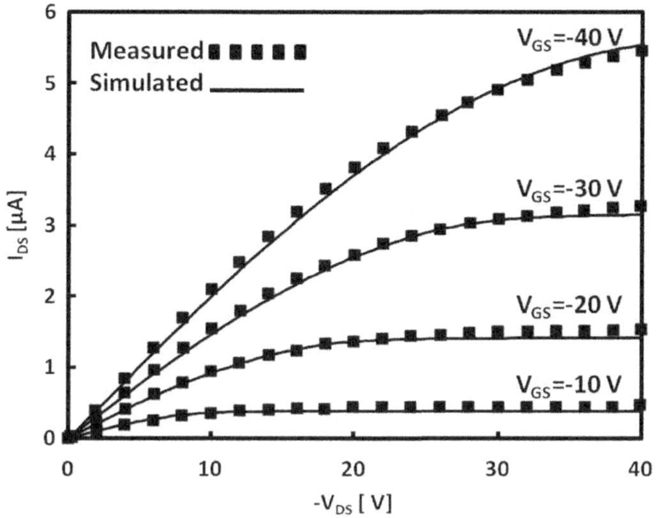

Figure 8.4 Comparison between experimental and simulated output characteristics of pentacene thin film transistor for several gate voltages.

contact configuration and realized on thermally oxidized N-doped Si substrates ($N_D = 10^{21}$ cm^{-3}). The Cr/Au source and drain multifinger contacts were realized by vacuum evaporation using a standard photolithographic process over the silicon dioxide wafer. We realized devices with different dimensions and channel length between 12 μm and 40 μm. The high doped silicon represents the gate of the TFT. After the contact deposition, an about 30 nm thickness layer of pentacene molecules, as obtained from Sigma Aldrich, or porphyrin (octamethylporhyrin (OMP)) has been sublimated (in vacuum, 10^{-6} mbar with a deposition rate of 0.2 Å/s) and patterned in order to be confined in the channel region. The measured output characteristic of an organic thin film transistor with a 30 nm thickness layer of pentacene is shown in Figure 8.4. In order to understand better the possible improvement of OTFT we started a theoretical simulation campaign by setting up a model of charge transport in organic material. It is interesting to compare the experimental results with the simulated (Figure 8.4) and to underline the great agreement. In Figure 8.5(a) one can observe the electrostatic potential for the cross-section of a biased OTFT device, and in Figure 8.5(b) the hole current density J_p. At negative gate bias the holes accumulate at the oxide/pentacene interface where the conductive channel builds up. The simulation is performed using the Sentaurus TCAD software opportunely modified to consider the peculiarity of the organic devices – all simulation details can be found in [13].

For RFID tags we need a complete logic circuitry and so we started to develop the basic electronic units. Active inverters (Figure 8.6) have been realized using OTFT based on pentacene molecules. In this case both the driver and the load transistors are realized using organic material and have the same geometric characteristics (channel length is 20 μm with $W = L = 180$ μm). We have realized two types of configuration following the classical approach used in silicon NMOS inverters.

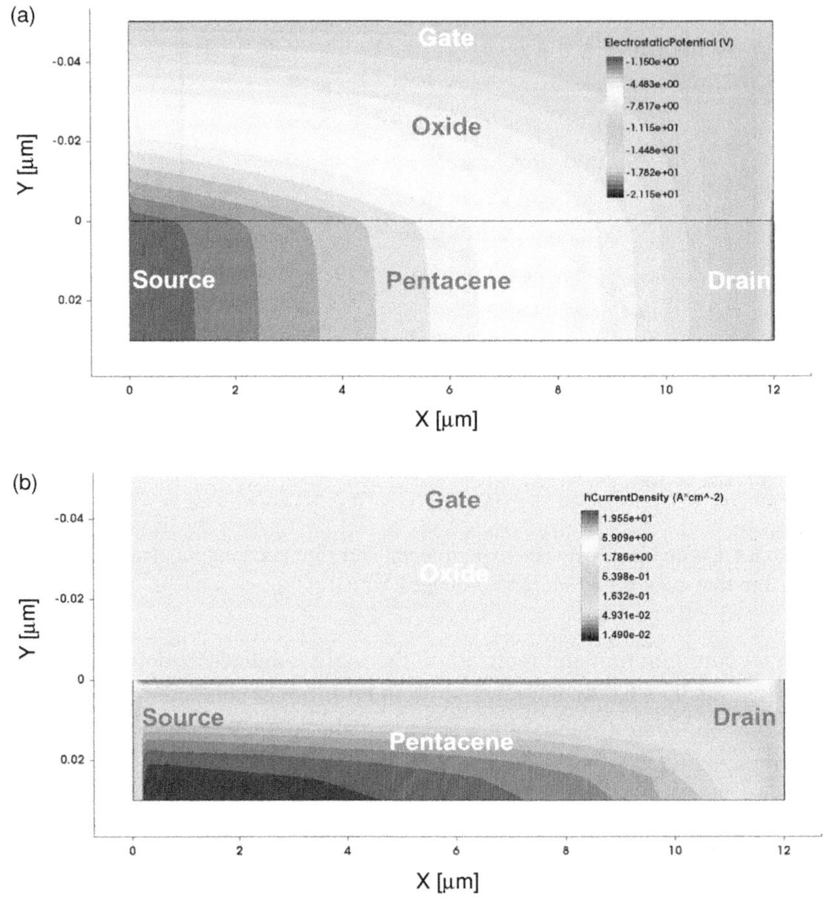

Figure 8.5 Cross-section of a pentacene OTFT with channel length of 12 μm at $V_g = -10$ V and $V_d = -20$ V: (a) Electrostatic potential; (b) Hole current density.

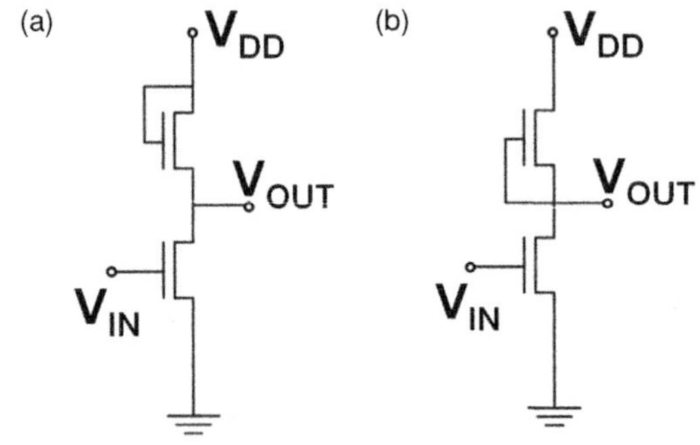

Figure 8.6 Organic active inverter with saturated load (a), and another configuration of active inverter (b).

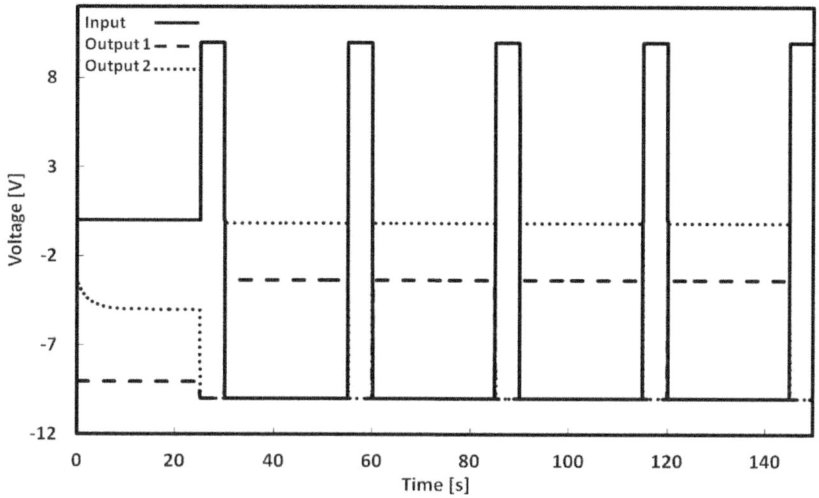

Figure 8.7 Simulated output characteristics of active inverters: saturated load (Output 1) and active load (Output 2).

The supply voltage applied to the drain of load transistor is $V_{DD} = -10$ V. As the input voltage we used a step voltage source (between -10 V and $+10$ V) with a period of $t =$ 30 s. In Figure 8.7 the output voltage of the saturated load configuration as a function of the input voltage is plotted, showing that this circuit acts like an inverter. The output dynamic, however, is lower than the input dynamic and the noise margin is not verified. This is due to the fact that the driver transistor for $V_{in} = 10$ V is not completely closed because of fixed charge and trap states at the interface between oxide and pentacene [14]. In the same figure, the other configuration with load source and gate contact short-circuited is shown (active load). The output voltage, with the same input as the previous configuration, is plotted, as a function of the input voltage, showing that the output dynamic is better [15]. Also in this case the noise margin is not verified. This problem can be resolved using driver and load transistors with different geometric dimensions. Numerical simulation has been a great help in the optimization of the devices.

8.3 Passive RFID components

8.3.1 Graphene: the wonder material

Graphene has recently attracted intense attention in the research community due to its extraordinary mechanical, electronic, and optical properties [16–20]. Graphene is a one-atom-thick two-dimensional carbon crystal retrieved in many materials such as HOPG (highly ordered pyrolytic graphite), which consists of many graphene sheets stacked in a pile, or carbon nanotubes, formed from one or more rolled-up graphene layers. Graphene has many astonishing properties, such as a carrier mobility attaining 200 000 $cm^2V^{-1}s^{-1}$ at room temperature and a Young's modulus of 1.5 TPa, which renders graphene the stiffest material with the highest mobility [16–18]. Graphene is quickly becoming an extremely interesting option for a wide variety of electronic devices, circuits, and systems [21], such as field effect transistors [22], frequency multipliers [23], transparent solar cells [24], meta-materials [25], and graphene plasmonics [26]. It offers the possibility of outstanding performances with much lower power draw, using processing technology compatible with that used in advanced silicon device fabrication (CMOS).

The applications of graphene in passive guided devices and antennas from microwaves to THz have been exploited far less. This is mainly because such structures require electric sizes of the order of the wavelength, while the first available graphene samples were much smaller. However, graphene chemical vapor deposition (CVD) now allows samples up to several centimeters to be obtained, thereby increasing research interest in the realization of passive devices at such frequencies [27–30].

The possibility of using graphene-based antennas is a core point in the development of high speed, multi-functional new wireless architectures, leading to a new generation of smart nano-systems for wireless communications. On the other hand, the growing interest in graphene-based antennas in the microwave, mm-wave range stems also from the real possibility of fabricating low-cost RF sensors and RFID [31].

In fact, most low frequency RFID tags are made from copper or silver, using etching technology. As discussed previously, the etching process is not only environmentally unfriendly, but it also wastes a large proportion of the precious metal deposited or laminated onto a substrate. As an alternative, printed RFID antennas have been pursued by companies as an additive process. Recently, alternative solutions of low cost and high performance conductive coatings based on graphene for printed RFID have been developed. For the aforementioned reasons, the theoretical investigation and modeling of i) graphene antennas and ii) graphene-based (e.g. printed) antennas are key issues in the development and design of wireless communication, RFID systems, and sensors.

8.3.1.1 Basic properties of graphene

Graphene consists of carbon atoms arranged in a 2-dimensional (2D) honeycomb crystal lattice with a bond length of 1.42 Å [16–20]. The carbon atoms are in an sp^2 hybridization state, arranged such that each carbon atom is covalently bonded to three others [20]. A schematic of a single graphene layer is shown in Figure 8.8 (a), including "armchair" and "zig-zag" edges, named after their characteristic appearance on the atomic scale.

In more detail, each carbon atom has four valence electrons:

i) Three of the four valence electrons participate in the bonds to their next neighbors (σ-bonds), as depicted in Figure 8.8(b).
ii) The fourth π-electron orbital is oriented perpendicular to the sheet and delocalized.

The graphene lattice is made up of two equivalent carbon sublattices A and B, which lead to crystal symmetry. As a consequence, the charge carriers can be described by the Dirac equation [8, 16, 18], i.e. the band structure of graphene exhibits a linear dispersion relation for charge carriers, with the linear momentum p (or, the wavenumber k) proportional to the energy E. Finally, the energy bands associated with the sublattices intersect at zero energy E, resulting in a semi-metal with no bandgap ($E_g = 0$ eV). A

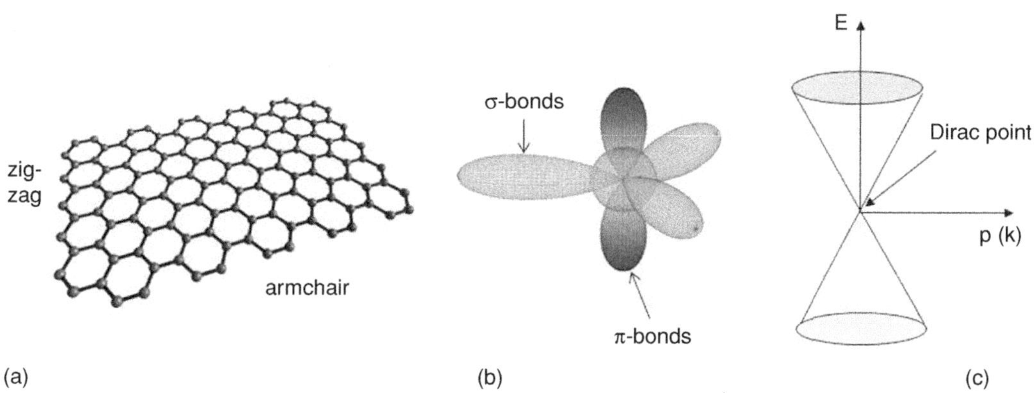

Figure 8.8 (a) A graphene crystallite with characteristic armchair and zig-zag edges. (b) Electron σ- and π-orbitals of one carbon atom in graphene. c) Band diagram of graphene at $k = 0$.

schematic of the band structure in the vicinity of $k = 0$, including the Fermi level E_F, is shown in Figure 8.8(c), [20].

Charge carriers in graphene possess a very small effective mass [16–19], and hence graphene shows extremely attractive material properties relevant to electronic devices. These include carrier mobilities of up to 15 000 cm^2/Vs for graphene on SiO_2 [32], 27 000 cm^2/Vs for epitaxial graphene [33] and 200 000 cm^2/Vs for suspended graphene [19, 34–36]. In addition, a high current carrying capability exceeding 1×10^8 A/cm^2 [37], high thermal conductivity [38, 39], high transparency [40], and mechanical stability [40] have been reported. While similar promising properties have been reported for carbon nanotubes (CNTs), the fact that graphene sheets can be processed with conventional CMOS-technology is potentially a huge advantage over CNTs.

The honeycomb lattice structure of graphene forms the basic structure of other carbon-based materials, in particular, of graphite, which, in turn, results from stacked graphene layers, and of carbon nanotubes, obtained by rolling up one or several graphene sheets.

Graphene is a native 2D crystal, in which the charged particles form a 2D gas. Graphene can be quite easily manipulated and, in particular, can be deposited on an SiO_2 layer grown over a doped silicon substrate. In this configuration, in which the SiO_2 layer has a typical thickness $t = 300$ nm, the Si substrate acts as a gate that induces a surface charge density n [41, 42].

This charge density, related to the applied gate voltage V_g through [41, 42]:

$$n_s = \varepsilon_0 \varepsilon_d \frac{V_g}{de} \simeq \alpha V_g, \quad \alpha = \varepsilon_0 \varepsilon_d \frac{1}{de}, \tag{8.8}$$

shifts the Fermi energy level in graphene, as depicted in Figure 8.9 (a). In Equation (8.8) ε_0 and ε_d are the dielectric permittivities of air and SiO_2, respectively; e is the electron charge, d is the SiO_2 oxide width.

These gate-induced carriers are an example of electrical doping, a process analogous to the chemical doping typical for semiconductor devices. In contrast to semiconducting devices, where the chemical doping cannot be changed once completed, p- or n-doping in

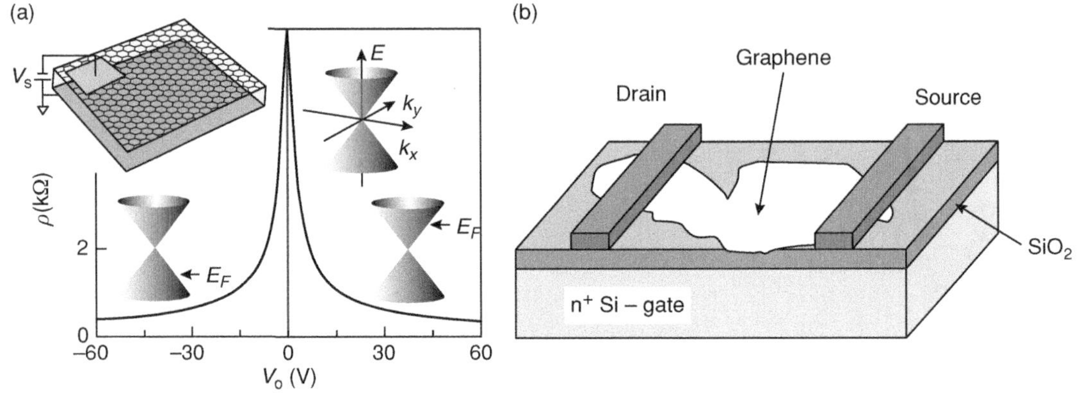

Figure 8.9 (a) Behavior of graphene resistivity (ohm). The electrical doping shifts the Fermi energy level [18]. (b) A typical graphene-based FET-like structure, which represents a generic configuration for many graphene devices [44].

graphene can be electrically tuned by applying negative or positive voltages, respectively. After the graphene sheet is deposited on the Si/SiO_2 structure, electrodes can be patterned over it in order to produce certain devices. Figure 8.9(b) shows a typical graphene-based FET-like structure, which represents a basic configuration for many graphene devices.

8.3.1.2 Analogy of ballistic transport and electromagnetic waves: a rich concept

The mean-free path for ballistic carrier transport is 500–1000 nm at room temperature (this value is being improved continuously by technology), while graphene resistivity depends significantly on the gate voltage, as shown in Figure 8.9(a).

The dispersion relation in graphene is linear for both electrons and holes for small energies E and has the simple form $E = hkv_F$. This relation can be derived starting from the relativistic form of the conservation of energy, in the case of vanishing effective mass m_0 of the carriers [41, 42]:

$$E = \sqrt{p^2 v_F^2 + m_0^2 v_F^4} \;\Rightarrow\; E = pv_F \quad \text{if } m_0 \to 0 \quad p = |\mathbf{p}|, \tag{8.9}$$

where v_F is the Fermi velocity of particles in graphene, and p the magnitude of the linear momentum $\mathbf{p} = p_x\mathbf{x} + p_y\mathbf{y}$. Then, by considering the Planck–Einstein relations

$$\mathbf{p} = \hbar\mathbf{k}, \quad E = \hbar\omega, \quad \hbar = h/2\pi, \tag{8.10}$$

where $h = 6.26 \times 10^{-34}$ [J s] is Planck's constant, and substituting (8.10) in (8.9), we derive:

$$E = \pm\hbar k v_F, \tag{8.11}$$

where k is the magnitude of the wavenumber $\mathbf{k} = k_x\mathbf{x} + k_y\mathbf{y}$ of the charge carriers.

The linear dispersion relationship consists of two lines that cross at a point called the Dirac point. Far from the Dirac point, the carrier transport is unipolar, while in the Dirac

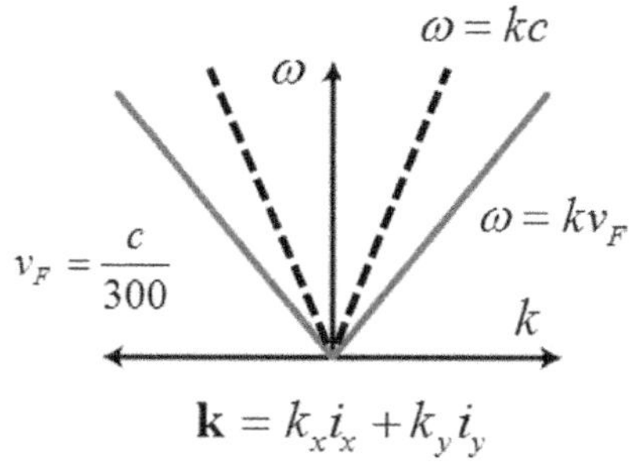

Figure 8.10 Dispersion relation of graphene carriers (solid line) and photons in vacuum (dashed line).

region it is ambipolar and strong recombination processes take place. The situation is depicted in Figure 8.10.

The linear dispersion relation is similar to that of photons propagating in a vacuum, but the significance of this relation is very different. In graphene, the linear relation between the energy and the wavenumber means that the effective mass of charge carriers is practically zero, so that the transport of electrons and holes is ballistic, that is, without any collision in the device if its length is less than the mean-free path, in which case the charge carriers behave as waves instead of particles [16–19, 41, 42]. These waves are slow and propagate with the Fermi velocity v_F:

$$v_F \approx 10^6 \text{ m/s}, \quad v_F \approx c/300, \quad c = 3 \cdot 10^8 \text{ m/s}. \tag{8.12}$$

In contrast, the photons propagate with the speed of light in vacuum, c. The analogy among the ballistic transport of electrons, holes, and electromagnetic waves is a rich concept used in many areas of electronics, optoelectronics, and electromagnetism [16–19, 41, 42].

The behavior of charge carriers in graphene is governed by the Dirac Hamiltonian:

$$H(\mathbf{r}, t) = v_F \begin{pmatrix} \sigma_x k_x + \sigma_y k_y & 0 \\ 0 & -\sigma_x k_x + \sigma_y k_y \end{pmatrix} = v_F \sigma \mathbf{k}, \tag{8.13}$$

where σ is the vector composed of 2D-Pauli matrices. Graphene is the only known solid-state material in which the carriers behave as particles in quantum electrodynamics (QED).

8.3.1.3 Fabrication of graphene

There are currently three main, established fabrication methods for graphene: mechanical exfoliation, epitaxial growth from silicon carbide (SiC) substrates, and chemical vapor deposition of hydrocarbons on reactive nickel or transition-metal-carbide surfaces.

Exfoliation: Novoselov *et al.* have introduced a manual cleaving process of graphite, frequently called "mechanical exfoliation," to obtain single and few layer graphene [16–19]. This process makes use of adhesive tape to pull graphene films off a graphite crystal. These are then thinned down by further strips of tape and finally rubbed against appropriate surfaces like silicon dioxide on silicon, which leaves randomly sized and distributed flakes on the surface. When observed through an optical microscope, single and few layer graphene flakes add to the optical path compared to the bare wafer. If a proper thickness of silicon dioxide is chosen, the resultant visible contrast is sufficient to identify even single graphene layers [15–18].

Epitaxial graphene from silicon carbide: Berger and de Heer have pioneered an epitaxial approach to fabricate graphene from silicon carbide substrates [43–45]. During the process, silicon is thermally desorbed at temperatures between 1250 and 1450 °C. This process is certainly more controllable and industrially relevant than mechanical cleaving. In fact, it has been shown that graphene transistors can be manufactured from epitaxial graphene on a wafer scale [43]. A major disadvantage of epitaxial graphene is the extremely high cost of production-type SiC wafers, their limited size compared to silicon wafers, and the high processing temperatures which are well above current CMOS limits. With these pros and cons, it remains to be seen whether epitaxial graphene will find its way into future nanoelectronics applications.

Chemical vapor deposition: A promising large area deposition method is currently being explored in the form of chemical vapor deposition on metallic surfaces like nickel [27–30], ruthenium, and others. These CVD approaches rely on dissolving carbon into the metal substrates and then forcing it to precipitate out by cooling. Another approach is to grow graphene films directly on iridium or copper [46–48]. Several methods of transferring the CVD graphene films onto relevant substrates have been suggested, including the use of disposable PMMA or PDMS films. After transfer, Raman spectroscopy has been used to verify single layers, and even devices have been fabricated with typical graphene properties. CVD has now almost reached maturity for mass-production. Samples over 50 cm in size have been grown and transferred on target substrates [27–30].

8.3.2 Modeling of the electromagnetics-quantum transport in graphene nanodevices

In the frequency (energy)-domain, the analysis of graphene or graphene nanoribbons (GNR), narrow strips of graphene, can be carried out by using discrete models such as tight binding (TB), and continuous models, such as effective mass and $k \cdot p$ approximations, which stem from the approximation of TB around particular points of the dispersion curves.

The $k \cdot p$ model leads, for the GNR, to a Dirac-like system of equations [49]. Another popular approach for the calculation of quantum transport is given by the use of the Green's function of a GNR region [50–52]. This method is closely related to another approach [53], which breaks up the system into layers perpendicular to the transport direction. All techniques mentioned above are demonstrated to be quite accurate for the analysis of GNR in a variety of problems such as, for example, the effects on charge transport of applied external electric and magnetic fields [54], bending [55], lattice defects and discontinuities [52], edge terminations, and so on.

However, the latter methods require large computational resources, and can hardly include the effect of either i) the self-generated electromagnetic (EM) field, or ii) impinging external EM fields. Recently, we have introduced full-wave techniques both in the frequency (energy)-domain [56–61], and the time-domain [62–71] for the investigation

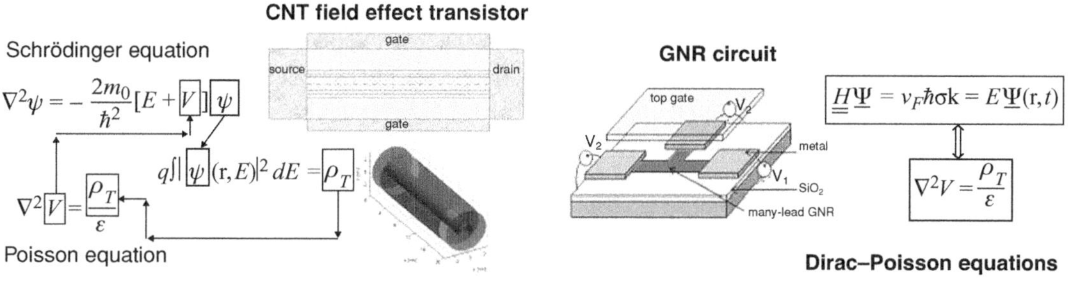

Figure 8.11 Concept of the frequency (energy)-domain technique: the quantum transport is described by the Schrödinger (carbon nanotube-based devices) or Dirac (graphene, GNR-based devices) equation. The EM field provides source terms for the quantum transport equations, that in turn provide charges and currents for the EM field.

of new devices based on carbon materials, namely carbon nanotube (CNT), multi-wall (MW) CNT, graphene, GNR.

In both the approaches, the quantum transport is described by the Schrödinger equation or its Dirac-like counterpart. The electromagnetic field provides source terms for the quantum transport equations, that in turn provide charges and currents for the electromagnetic field.

For the transport phenomena, we derive [56–61] a multimode transmission matrix model, allowing easy simulation of very large structures, despite the possibly high number of electronic channels involved. We make use, however, of an explicit scattering matrix formalism, showing that fundamental physical constraints and consistence relations in quantum transport, such as reciprocity and charge conservation, correspond respectively to reciprocity and power conservation in "microwave engineering." We emphasize that the proposed approach allows handling multi-port graphene systems, where carriers can get into (and out of) many different physical ports, each characterized by its own chirality and possibly by a large number of virtual ports, i.e. electronic channels or sub-bands. The analogy between quantum transport and microwave propagation problems leaves open the possibility of a systematic application of well-known microwave techniques for the synthesis, analysis, and calibration of quantum devices. The Poisson equation is solved by means of a finite-difference in the frequency-domain (FDFD) scheme. The numerical convergence between the transport (Dirac) equation and the Poisson equation is reached in the context of a very stable numerical technique; we have realized a home-made solver, written in "Matlab" code, provided with graphical and visualization tools. The concept of the present scattering matrix–Poisson equation (SM–PE) technique approach is depicted in Figure 8.11.

We have reported realistic examples of GNR-enabled structures: multi-port GNR circuits, scattering by lattice defects, change in width in cascade GNRs, crossing GNRs, T-junctions, metal–carbon contact, field coupling among GNRs [56–61].

8.3.2.1 Frequency domain: the combined Dirac–Poisson problem

In dealing with graphene and graphene nanoribbon, we assume that the EM wavelength is small as compared to the charge spatial variation. This kind of analysis can be applied in order to explore the basic working principles of graphene in its use as interconnect and as a channel for a FET transistor. For instance, a lumped-circuit model with parasitic capacitances, including the quantum capacitance of the graphene channel, can be easily derived from the simulation results. This leads to a rough estimation of the gain and the unit-gain frequency. In practice, non-ballistic effects reduce the conductance of the interconnect, and the unit-gain frequency of the transistor.

The main result of the present work is that it shows, by means of some new examples, the effectiveness of the scattering-matrix approach for the solution of relatively large graphene systems involving thousands of carbon atoms and consequent multichannel transport.

As already mentioned, the examples involve the coupling, through coulomb interaction, of two GNR channels, both semiconducting or both semi-metallic, in a single FET transistor configuration. Differently from simulations performed at the atomic scale, that can only deal with large carbon devices with difficulty, we consider structures made of many thousands of carbon atoms. Other common approaches that, similarly to ours, avoid *ab-initio* calculation, like the non-equilibrium Green's function (NEGF), could be successfully applied.

The NEGF describes the carrier interaction through a spatial charge density that locally perturbs the graphene neutrality. The scattering-matrix approach is equivalent to the NEGF, but, in our opinion, it can better highlight the physical behavior of the system, because: i) it makes use of electronic modes that are easily normalized with respect to the physical current; ii) it considers the physical ports of graphene circuits under the same standard as the virtual ports, i.e. all the accessible electronic wavefunctions; iii) it treats all the ports as adapted ones, i.e. transparent or perfectly absorbing; iv) it allows an easy check of the losslessness and reciprocity constraints involved in charge transport, through the scattering-matrix properties; and v) it allows the use of concepts well known in the microwave and RF community, such as cascade of transmission lines, transformers, lumped-circuit elements, etc., all useful in order to model long graphene structures eventually containing defects, bending, discontinuities, and electrostatic perturbations.

An important parameter for the operation of a FET transistor is the gate differential capacitance: this parameter directly affects the frequency characteristics of the transistor, as well as the unit-gain frequency. The derivative, with respect to the gate bias, of the charge density in the channel, after self-consistent solution of the transport/electrostatic problem, provides the channel contribution to the gate capacitance. Some simple models, based on geometric and dielectric characteristics of the gate and the channel, may provide some partial results, by employing a metallic approximation for the channel. However, it is well known that the electromagnetic screening is much less effective in two dimensions than in three dimensions, and graphene behaves, to some extent, as an ideal two-dimensional electron gas. Hence, a rigorous analysis must include a knowledge of the actual electronic structure of the GNR channel, and its change under external voltage. The result is what is usually referred to as quantum capacitance, that is, the change of the channel energy-levels, and of the corresponding wavefunction distributions, due to an external applied voltage.

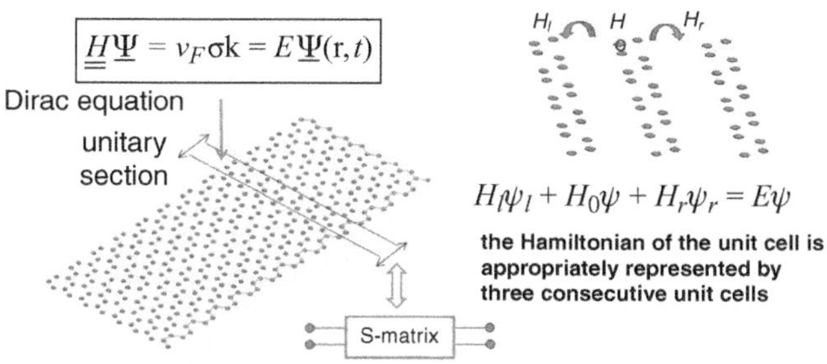

Figure 8.12 Scheme of the calculation of the Hamiltonian of the unit cell, represented by selecting three consecutive unit cells. The scattering matrix of the unit cell is derived.

The Schrödinger equation can be formulated in a discrete version [59–61]:

$$H_l\psi_l + H_0\psi + H_r\psi_r = E\psi, \tag{8.14}$$

where E is the energy, ψ_l, ψ, ψ_r are the wavefunctions of three consecutive GNR unit cells and the Hamiltonian has been decomposed in three matrices which model transfer energies between adjacent cells: matrix H_l (H_r) includes the hopping elements from a unit cell to the previous (following) one, H_0 is the self-energy matrix of the unit cell, as depicted in Figure 8.12. The relations $H_l = H_r^+$ and $H_0 = H_0^+$ hold. The diagonal elements of H_0 contain the electrostatic potential solution of the Poisson equation. Exploiting the lattice periodicity, we have:

$$\begin{aligned} &\left(H_l e^{jkL} + H_0 + H_r e^{-jkL}\right)\psi = E\psi, \\ &\psi = e^{-jky}\sum_i \psi_i \delta(x - x_i, y - y_i), \end{aligned} \tag{8.15}$$

where i spans the atomic sites, k is the wavenumber, and L is the length of the unit cell. The solutions of Equation (8.15), corresponding to different branches of the dispersion curves, can be distinguished as propagating modes for real k, and evanescent modes for complex or imaginary k. In order to provide the correct mode normalization, wavefunctions have to be divided by:

$$A = \sqrt{\frac{f_T(E)}{4\pi \,\mathrm{Im}\left(\psi^+ H_r e^{-jkL}\psi\right)}}, \tag{8.16}$$

where f_T is the Fermi distribution function at T temperature ($T \sim 0$ has been used). The above normalization implies that all propagating modes carry the same current with reference to the Landauer formula.

The scattering matrix **S** of a GNR device, defined as in [33–37], relates incident and reflected mode amplitudes at port sections, which are different terminations of a many-lead GNR circuit. One of the difficulties of the proposed method resides in the fact that the scattering formulation has to be adapted to a periodic waveguide, where the right basis to be used for describing charge transport is given by the complete set of Floquet, or periodic, modes. Within the latter choice the scattering parameters can be shown to obey, for example, the expected reciprocity and losslessness constraints, respectively $\mathbf{S} = \mathbf{S}^{\mathrm{T}}$ and $\mathbf{S}^{+}\mathbf{S} = \mathbf{I}$.

The number of modes at any port depends on the port size, on the Fermi level, and on GNR chirality. Regressive modes, either above or below cut-off, are scattered out of the GNR without reflection: this means perfectly transparent ports.

In the case of a two-port GNR, the charge density is obtained by summing the normalized wavefunctions of carriers injected from the two opposite ports:

$$Q = -e\sum_{i}\int\left(f_1(E)\frac{\left|\psi_1^i(x, y, E)\right|^2}{A^2} + f_2(E)\frac{\left|\psi_2^i(x, y, E)\right|^2}{A^2}\right)\mathrm{d}E, \qquad (8.17)$$

where the upper index "i" of $\psi_{1,2}^i$ indicates the ith band whereas the lower ones indicate the ports, f_1 and f_2 are the Fermi probabilities for the electrons at ports 1 and 2 respectively, $e = 1.6 \cdot 10^{-19}$ C. The charge density (Equation (8.17)) is assumed as the source term in the familiar Poisson equation for the potential V.

For a multi-port device all physical quantities are obtained by including the charge contributions from all ports. For instance, the total current flowing through the pth port includes contributions of all modes transmitted from the port p to all the other ports:

$$I_p = \frac{2e}{h}\sum_{i,j,q}\int (f_p(E) - f_q(E))S_{pq}^{ij}(E)\mathrm{d}E, \qquad (8.18)$$

where S_{pq}^{ij} is the scattering coefficient from mode i to mode j, and from port p to port q.

Once we have derived the multimode basis for the GNR unit cell, the global modeling can be applied. It develops as follows [14–16]:

(i) The quantum devices, CNT and/or GNR are seen as a periodic structures waveguide, isomorphic to photonic crystal slabs.

(ii) A unit cell is chosen, the carbon atoms are identified with potential distribution, the energy bonds among atoms, including edge atoms (in GNRs) are properly described; the Hamiltonian matrix is constructed. The periodic Floquet condition of the unit cell is imposed.

(iii) Device polarization, bias, and any other potential (electric field) distribution are included in the Poisson equation.

(iv) A coupled system involving Poisson–Schrödinger/Dirac equations is constructed. The Poisson equation

$$\nabla^2 V = \frac{\rho_T}{\varepsilon}, \qquad \rho_T = q\int |\psi(\mathbf{r}, E)|^2\mathrm{d}E, \qquad (8.19)$$

is solved by using a proper finite-difference in frequency-domain (FDFD) technique.

(v) The quantum equation, be it Schrödinger or Dirac (Equation (8.13)), provides an updated value for the wavefunction, that is, after an integration over the energy-band, the charge distribution. This charge distribution is the source for the Poisson equation.

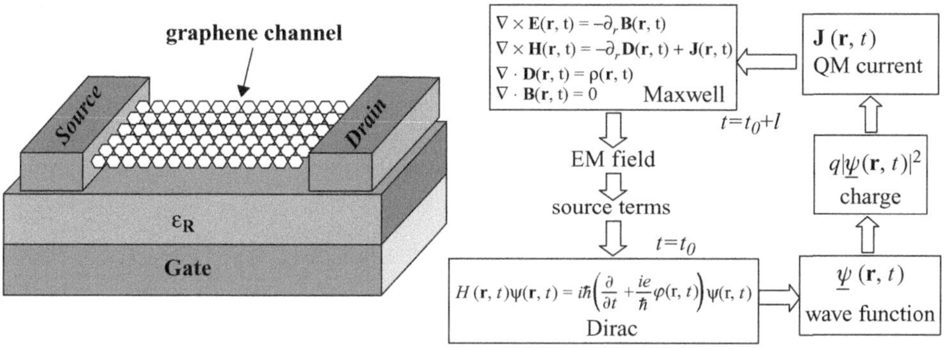

Figure 8.13 (*Left*) Example of graphene-based device: field effect transistor (FET). The graphene region acts as carrier transport channel. (*Right*) The concept of the multi-physics full-wave time-domain technique, dealing with the combined EM/transport problem.

8.3.2.2 Time domain: the combined Dirac–Maxwell problem in the ballistic regime

In the ballistic regime, a time-domain, full-wave approach has been introduced. In this technique, the Maxwell equations, discretized by the transmission line matrix (TLM) method are self-consistently coupled to the Schrödinger/Dirac equations, discretized by a proper finite-difference time-domain or a TLM scheme [62–71], as depicted in Figure 8.13.

For each time step, the Maxwell equations are self-consistently solved together with the Dirac equation, describing the quantum mechanical massless fermion dynamics in a ballistic regime. In the Maxwell equations

$$\begin{aligned}\nabla \times \mathbf{E}(\mathbf{r}, t) &= -\frac{\partial}{\partial t}\mathbf{B}(\mathbf{r}, t) - \mathbf{M}_{eq}(\mathbf{r}, t),\\ \nabla \times \mathbf{H}(\mathbf{r}, t) &= \frac{\partial}{\partial t}\mathbf{D}(\mathbf{r}, t) + \mathbf{J}(\mathbf{r}, t),\end{aligned} \tag{8.20}$$

the electric current source **J** includes the quantum-mechanical contribution, and **M** equivalent magnetic sources account for the external electric field (e.g. plane wave). The Dirac equation, in the presence of an EM field, is defined as

$$\begin{aligned} i\hbar\left(\frac{\partial}{\partial t} + \frac{ie}{\hbar}\varphi\right)\psi_+ &= \sigma \cdot (\hat{\mathbf{p}} - q\mathbf{A})\, c\, \psi_-,\\ i\hbar\left(\frac{\partial}{\partial t} + \frac{ie}{\hbar}\varphi\right)\psi_- &= \sigma \cdot (\hat{\mathbf{p}} - q\mathbf{A})\, c\psi_+.\end{aligned} \tag{8.21}$$

In Equation (8.21), A and φ are vector and scalar potentials, directly related to the electromagnetic (EM) field

$$\mathbf{E} = -\Delta\phi - \frac{\partial \mathbf{A}}{\partial t}, \quad \mathbf{B} = \Delta \times \mathbf{A}, \tag{8.22}$$

through Lorentz gauge:

$$\Delta \cdot \mathbf{A} + \frac{1}{c}\frac{\partial \phi}{\partial t} = 0. \tag{8.23}$$

In Equation (8.21), $\boldsymbol{\sigma}$ are the Pauli matrices and $\mathbf{p}$ is the linear momentum. The solution of the Dirac/graphene Equation (8.21) is the four component spinor complex wavefunction $\psi(\mathbf{r}, t)$:

$$\psi(\mathbf{r}, t) = [\psi_1\ \psi_2\ \psi_3\ \psi_4]^T = [\psi_+\ \psi_-]^T. \tag{8.24}$$

In particular, in the 2D-graphene, the Dirac Equation (8.17) behaves as:

$$H(\mathbf{r}, t)\psi(\mathbf{r}, t) = i\hbar\left(\frac{\partial}{\partial t} + \frac{ie}{\hbar}\varphi(\mathbf{r}, t)\right)\psi(\mathbf{r}, t), \tag{8.25}$$

where the Dirac Hamiltonian is defined as:

$$H(\mathbf{r}, t) = v_F\begin{pmatrix} \sigma_x k_x(\mathbf{r}, t) + \sigma_y k_y(\mathbf{r}, t) & 0 \\ 0 & -\sigma_x k_x(\mathbf{r}, t) + \sigma_y k_y(\mathbf{r}, t) \end{pmatrix} + V_{\mathrm{p}}(\mathbf{r}). \tag{8.26}$$

As in (8.13), in equation (8.26), σ_x and σ_y are two of the three Pauli matrices, V_{p} is the static potential barrier, $i = -j$ is the imaginary unit, v_F is the Fermi velocity, $\mathbf{r} = (r_x, r_y)$ is the in-plane position vector, $k_x(\mathbf{r}, t)$ and $k_y(\mathbf{r}, t)$ are the components of the kinematic momentum $\mathbf{k}(\mathbf{r}, t)$, related to the linear momentum $\mathbf{p}$:

$$\hat{\mathbf{p}} = -i\hbar\nabla, \qquad \hat{\mathbf{k}} = \hat{\mathbf{p}} - q\mathbf{A}(\mathbf{r}, t). \tag{8.27}$$

The basic requirement for the solutions is the following normalization condition, stating that the probability of the particle being somewhere is one, and also corresponding to the charge conservation.

$$\forall t \geq 0 \quad \int_{-\infty}^{+\infty} \|\psi(\mathbf{r}, t)\|^2 \mathrm{d}\mathbf{r} = 1 \Leftrightarrow q\int_{-\infty}^{+\infty} \|\psi(\mathbf{r}, t)\|^2 \mathrm{d}\mathbf{r} = Q_T. \tag{8.28}$$

From the wave solution $\psi(\mathbf{r}, t)$, it is straightforward to derive the corresponding QM current [62–64]. The computational scheme (Figure 8.13) develops as follows:

(i) The Maxwell equations are discretized by using the transmission line matrix (TLM) method [71–73]. The TLM method, like the finite-difference time-domain (FDTD) method, is a well-known technique that allows the EM full-wave modeling of 3D structures with nearly arbitrary geometry for a wide range of applications from EM compatibility to optics. FDTD is a more general technique, suited for discretizing different kinds of equations, e.g. parabolic, hyperbolic, etc. With respect to FDTD, TLM is directly related to the discretization of, mainly, hyperbolic equations (Maxwell, Dirac), but it has the advantage that each portion of the segmented space has an equivalent local electric circuit [71, 72]. The TLM scheme is considered as an implementation of the Huygens principle: propagation and scattering of the wave amplitudes are expressed by operator equations. Wave pulses are scattered by nodes (source points), propagate in the conformal (e.g. rectangular) mesh, and recombine to form a new wave front; at each space-time point of the mesh, the combination of incident/scattered pulses gives the EM field components. In this way, the physical correspondence among EM field components and wave pulses is established.

(ii) Quantum phenomena are introduced in a 2D graphene/nanoribbon region, and described by the Dirac equation.

(iii) At each time step, the Dirac equation is solved by accounting for the quantum device boundary conditions, initial conditions (e.g. injected charge), and additional source terms constituted by the EM field, sampled in the domain of the quantum device(s).

(iv) From the wavefunction (charge) solution of the Dirac equation, we derive the quantum mechanical (QM) current over the device domain. This current is a distribution of local sources for the EM field, that is injected into the TLM nodes, just located on the grid points of the Dirac equation domain.

(v) At the next time step $t+1$, the TLM method provides a new updated distribution of field values that are, again, sampled over the device domain, and so on, iteratively.

(vi) The contribution of external EM field sources, be they impressed and/or dynamic, are accounted for in the Maxwell Equations (8.20).

8.3.3 Graphene antennas for RFID and wireless applications

The possibility of using graphene-based antennas is a core point in the development of high speed, multi-functional new wireless architecture, leading to a new generation of smart nano-systems for wireless communications. On the other hand, the growing interest in graphene-based antennas in the microwave, mm-wave range also stems from the real possibility of fabricating low-cost RF sensors and RFID. Nevertheless, there has been little investigation of the use of graphene in passive guided devices and antennas from microwaves to THz. This is mainly because such structures require electric sizes of the order of the wavelength, while the first available graphene samples were much smaller. However, graphene chemical vapor deposition (CVD) now allows us to obtain samples up to several centimeters, thereby increasing research interest in the realization of passive devices at such frequencies [27–30].

Since graphene is mono-atomic layer, it is perfectly modeled by a surface conductivity tensor. This tensor depends on graphene's unique band structure and on a number of parameters including temperature, scattering rate, Fermi energy, electron velocity, and pre-doping (not all these parameter are independent), as well as electrical and magnetic field bias.

In practice, other parameters such as defects in the potential polycrystalline nature of the graphene can further affect its properties. In general, the conductivity of graphene is very frequency-dependent, and it can have completely different behavior at e.g. microwave and THz.

Though graphene is the best electrical conductor known, it is mono-atomic and thus the surface resistance seems to be high as compared to metals at microwave and mm-wave frequency, even with the possibility of doping and electric field biasing.

Therefore, at microwave and millimeter frequencies, graphene is essentially a moderate conductor: in spite of this, challenging and impacting applications could be addressed, like the design and fabrication of patch antennas, or arrays of antennas [74, 75]. In order to provide a comparison, copper has a surface resistivity as low as 0.03 ohm/square, but this holds for a thickness at least greater than a few times its skin depth, which is a fraction of a micron at 13 GHz. Considering a thin film of copper of 3 Å, in order to compare with graphene (but of course the complex physics of thin films does not allow such a simplification!), the surface impedance should be about 60 ohm/square, which is not too dissimilar to the values that we used for graphene. However, two important points have to be noted:

(i) graphene with such a small thickness can exist, while copper cannot;

(ii) by reducing the in-plane dimensions, the actual copper conductivity would degrade much faster and much more than the actual graphene conductivity.

Moreover, it has also has been demonstrated recently that graphene can be efficiently used as passive substrate with variable resistance, thus permitting tuning of microwave devices in order to obtain the proper matching [42, 76].

In this section, we deal with the design of a first prototype of a graphene-patch antenna, in the 5–20 GHz range, as a building block in the realization of a transmitting/receiving (T/R) nano-scale microwave communication system, as well as the design of advanced, low-cost RFID systems and sensors.

The benefits of this nano-scale system are given by:

(i) Tunability of the antenna parameters: radiation pattern, resonant frequency, directivity, efficiency;
(ii) Tunability of the EM coupling with other radiating elements, and of the array pattern of an array of graphene patches;
(iii) Possible reduction of the in-plane size of the patch owing to the "slow wave" behavior (kinetic inductance contribution);
(iv) integration: in future, a new class of high performance carbon components will be hopefully integrable in the same circuit (amplifiers, switches, mixers, attenuators, etc.).

8.3.3.1 Graphene in the microwave and mm-wave range

As widely analyzed in Section 8.3.1, the graphene surface impedance depends on its intrinsic conductivity, which, in turn, depends on the electric doping of the material, e.g. provided by biasing graphene using an electric field [75–77]. Applying a DC voltage V' affects the graphene surface conductivity by shifting the Fermi level μ_c:

$$\mu_c = e\alpha V' = eV, \tag{8.29}$$

where e is the unit charge, V' is the external applied voltage, α is a parameter depending on the overall geometry and on the dielectric distribution of the device under study. In the latter, as shown later, this voltage is applied across the central and lateral conductors of a co-planar waveguide (CPW).

V is the actual/resulting potential in the graphene layer: in the present analysis it is assumed constant over the patch surface, for simplicity. Considering the large size of the patch under study ($\approx 2.9 \times 9.5\ \text{mm}^2$), the Boltzmann theory seems to be more suited than approaches based on coherent transport. We assume the following hypotheses:

(i) absence of magnetic fields; (ii) pure (undoped and unperturbed) graphene; (iii) room temperature (300 K); (iv) no edge effects; and (v) no interaction of graphene with SiO_2 substrate.

In the above conditions, graphene can be modeled as an infinitesimally thin, local two-sided surface, characterized by a surface impedance with complex tensor conductivity [74, 77]:

$$\underline{\underline{\sigma}}_G = \begin{pmatrix} \sigma_{xx} & \sigma_{xy} \\ \sigma_{yx} & \sigma_{yy} \end{pmatrix} = \begin{pmatrix} \sigma_d & \sigma_o \\ \sigma_{-o} & \sigma_d \end{pmatrix}, \qquad \underline{\underline{\sigma}}_G = \underline{\underline{\sigma}}' + j\underline{\underline{\sigma}}'', \tag{8.30}$$

$$\underline{J} = \underline{\underline{\sigma}}_G \underline{E}, \quad \underline{E} = \underline{\underline{\sigma}}_G^{-1} \underline{J} = \underline{\underline{Z}}_s \underline{J}. \tag{8.31}$$

Then, the graphene surface impedance Z_s can be defined as:

$$\underline{\underline{Z}}_s = \underline{\underline{\sigma}}_G^{-1} \text{ [ohm]}, \qquad \underline{\underline{\sigma}}_G = \underline{\underline{Z}}_S^{-1} \text{ [S]}. \tag{8.32}$$

This conductivity tensor (8.30) depends on graphene's unique band structure and on a number of parameters including temperature (T), scattering rate (Γ), Fermi energy (μ_c), and pre-doping, as well as electrical and magnetic field bias. In practice other parameters such as defects in the potential polycrystalline nature of the graphene can further affect its properties. In general, the conductivity of graphene is very frequency-dependent, and can have completely different behavior at e.g. microwave and THz. Therefore, it is fundamental to describe the conductivity behavior as a function of frequency and all different combinations, when possible, of the above parameters.

In general, the following main feature of graphene conductivity should be noted:

(i) *Isotropic behavior:* an applied *electric field* bias *injects*, or *induces*, more electron or holes carriers in graphene and thereby allows the dynamic control of both real and imaginary parts of the conductivity (Equations (8.30, 8.31)). (In such a case, the graphene conductivity (8.30) is mostly isotropic, in the absence of magnetic bias.)

(ii) *Anisotropic behavior:* if we apply a *magnetic field*, graphene becomes anisotropic. Therefore magnetic bias can be used to dynamically control graphene anisotropy.

(iii) *General isotropic conductivity behavior at microwave and mm-waves.* As outlined before, though graphene is the best electrical conductor known, it is mono-atomic and thus the surface resistance is very high compared to metals at microwave and mm-wave frequency, even with the possibility of doping and electric field biasing. In this frequency range graphene is thus mostly a moderately to badly conductive surface.

On the other hand, graphene conductivity has a totally different behavior at THz frequencies. This is due to its plasma frequency, which is lower than for noble metals. Therefore its conductivity in the THz range, is equivalent to a negative real permittivity for a bulk material. Plasmonic propagation can thus be supported by graphene at THz frequency, leading to interesting properties for potential practical applications.

Let us now focus on case (i), that is of great importance in applications.

We assume that no external magnetic field is present, and so the local conductivity is isotropic, i.e. there is no Hall conductivity. The graphene conductivity (8.30) becomes a scalar complex function, as well as the graphene surface impedance (8.32):

$$\sigma_G = \sigma' + j\sigma'', \quad Z_s = \frac{1}{\sigma_G}\ [\text{ohm}], \quad \sigma_G = \frac{1}{Z_s}\ [\text{S}]. \tag{8.33}$$

In the non-ballistic regime, the graphene conductivity (8.33) can be modeled according to the Kubo–Drude formulation [75]:

$$\begin{aligned}\sigma_G(\omega, \mu_c, \Gamma, T) = &-\frac{\mathrm{j}e^2(\omega - \mathrm{j}2\Gamma)}{\pi^2\hbar}\\ &\times\left\{\frac{1}{\omega(1-\mathrm{j}2\Gamma)}\left[\int_0^{\infty}\varepsilon\left(\frac{\partial f_d(\varepsilon)}{\partial\varepsilon} - \frac{\partial f_d(-\varepsilon)}{\partial\varepsilon}\right)\mathrm{d}\varepsilon\right.\right.\\ &\left.\left.-\int_0^{\infty}\frac{f_d(-\varepsilon) - f_d(\varepsilon)}{(1-\mathrm{j}2\Gamma)^2 - 4(\varepsilon/\hbar)^2}\mathrm{d}\varepsilon\right]\right\},\end{aligned} \tag{8.34}$$

where ω is radian frequency, μ_c is chemical potential, Γ is a phenomenological scattering rate that is assumed to be independent of energy ε, and T is temperature [75]. In Equation (8.34)

$$f_d = \left(\mathrm{e}^{-(\varepsilon-\mu_c)/k_B T} + 1\right)^{-1} \tag{8.35}$$

is the Fermi–Dirac distribution, and k_B is Boltzmann's constant. No external magnetic field is present, and the local conductivity is about isotropic. The first term of Equation (8.30) is due to intraband contributions, and the second term to interband contributions.

For a graphene sheet the carrier density n_s is related to the chemical potential μ_c:

$$n_s = \frac{2}{\pi\hbar^2 v_F^2}\int_0^{\infty}\varepsilon[f_d(\varepsilon) - f_d(\varepsilon + 2\mu_c)]\mathrm{d}\varepsilon. \tag{8.36}$$

The carrier density can be controlled by application of a gate voltage and/or chemical doping. For the undoped, ungated case at $T = 0$ K, carrier concentration n_s is very low.

At microwave, mm-wave frequencies, the intraband term is dominant, while the interband contribution is negligible. More specifically, the imaginary part of $1/\sigma$ is more than one order of magnitude smaller than the real part. Then, in the above frequency range, we can assume that:

$$\sigma_G \approx -\mathrm{j}\frac{e^2 k_B T}{\pi\hbar^2(\omega - \mathrm{j}2T)}\left(\frac{\mu_c}{k_B T} + 2\ln\left(\mathrm{e}^{-\frac{\mu_c}{k_B T}} - 1\right)\right), \tag{8.37}$$

where $\Gamma = 1/\tau$, $\tau_s = 0.5 \div 1$ ps.

In an EM simulation of graphene, the impedance boundary Z_s represents the graphene layer.

As the external voltage V is increased, energy is transferred to graphene thus shifting the Fermi level (electrochemical potential). The charge density, in turns, increases and the surface impedance Z_s decreases. In Figure 8.14(a), we show the dispersive behavior of the graphene conductivity for $\mu_c = 1$, as a function of the frequency. In Figure 8.14(b),

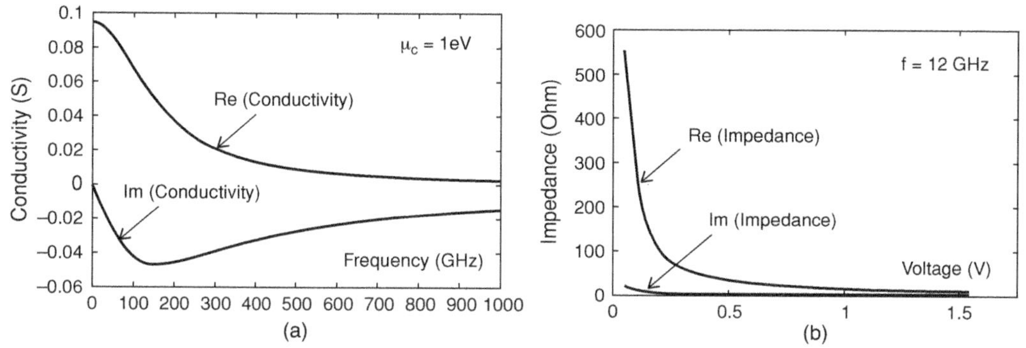

Figure 8.14 (a) Dispersive behavior of graphene conductivity. (b) Real part (gray) and imaginary part (black) of the surface impedance of graphene as a function of the applied voltage V at a fixed frequency of 12 GHz.

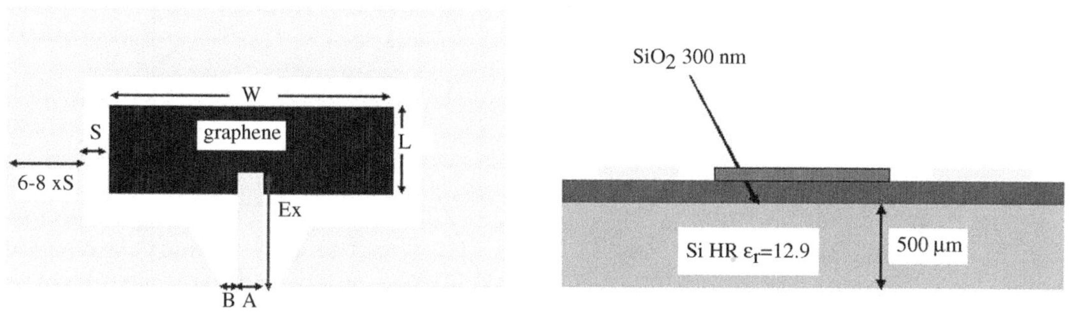

Figure 8.15 Geometry and physical parameters of the analyzed antenna. $W = 0.9$ cm; $L = 0.18$ cm; $S = 0.045$ cm $= 450$ μm; $A = 100$ μm; $B = 50$ μm; $H = 4.2$ mm.

we fix the work frequency ($f = 12$ GHz), and calculate the resulting impedance Z_s, in terms of the applied energy.

8.3.3.2 Antenna design and modeling

A first ptototype of a graphene-based antenna, which is actually a hybrid graphene–metal antenna, is shown in Figure 8.15. As shown, the antenna is excited by the fundamental mode of a co-planar waveguide (CPW). The Si substrate is grounded at the bottom by a perfect metal. Further details are reported in [78]. It is evident that this prototype is quite suited for RFID applications. Simulations have been performed by using HFSS. A preliminary analysis of the graphene-patch antenna is reported in [79].

The radiation mechanism of the proposed antenna is analogous to that of the familiar microstrip-patch antennas: the upper and lower slots, with respect to the patch, radiate in phase; field details are shown in Figure 8.16.

Owing to the physical symmetry of the device, only the half structure is reported: a magnetic wall is imposed in order to correctly describe the even field solution.

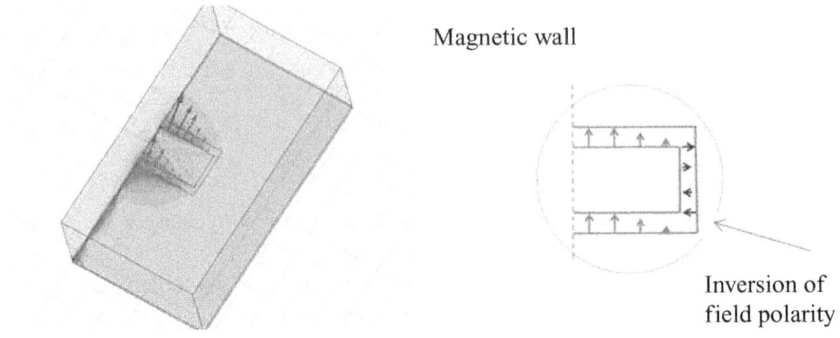

Figure 8.16 The radiation mechanism is similar to that of a standard patch antenna. This is based on the inversion of the field polarity in the lateral slots.

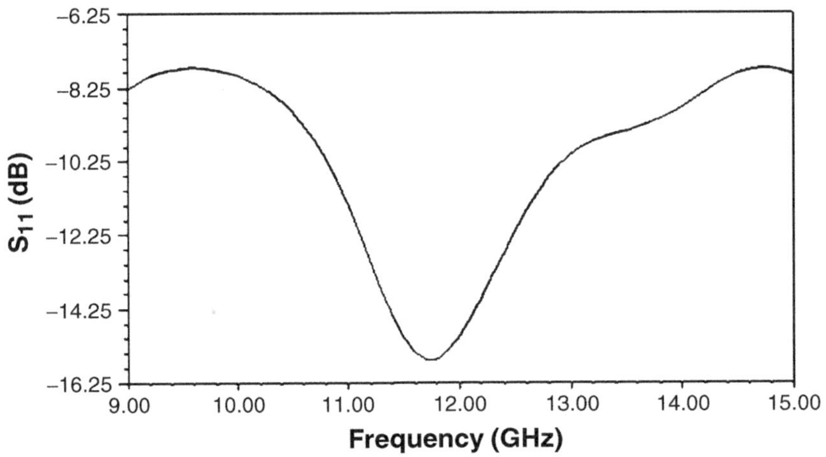

Figure 8.17 Magnitude of the reflection coefficient of the antenna.

With the above geometry, the antenna features a resonating behavior at around 12 GHz, as shown in Figure 8.17. In the present simulation, the external voltage is set high enough to produce a $Z_s \approx 10$ ohm. The possibility of such a small surface impedance has to be experimentally verified.

The radiation pattern of the antenna is shown in Figure 8.18, at 12 GHz: the corresponding curves are the angular E_x distributions in the corresponding planes orthogonal to the patch, depicted in the upper-left part of the figure. The smaller angular plots are the cross-polarized radiated field (E_y): as is evident, the antenna features quite low cross-polarization.

In order to show the effects of changing external voltage on the radiated fields, we analyze the dependence, on the surface impedance, of the electromagnetic coupling between two slightly spaced graphene-patch antennas.

The coupled antennas structure is sketched in Figure 8.19. An electromagnetic simulation of the device shows that the antenna reflection and transmission coefficients

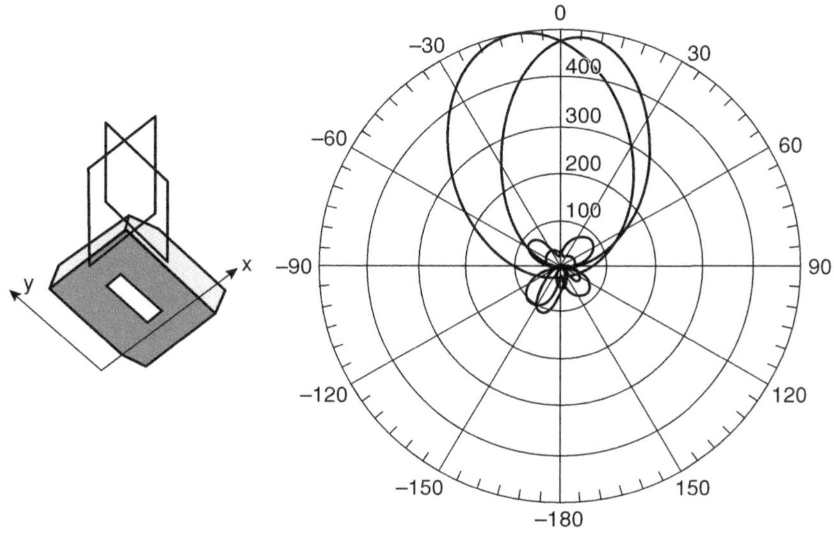

Figure 8.18 Radiation pattern of the antenna: far fields of the main and cross-polarization are shown.

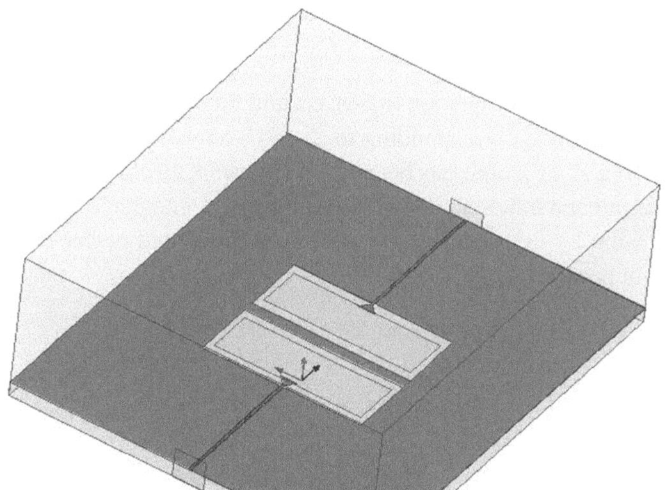

Figure 8.19 Two coupled graphene-patch antennas.

change dramatically with the surface impedance of graphene. This may suggest that modulating the graphene conductivity gives a high degree of freedom in controlling the output of a graphene antenna, or, in general, in controlling the array factor of an array made of closely spaced graphene-patches.

In the simulated example, the surface impedance Z_s is changed from 42 ohm down to 2 ohm, as the external voltage changes from 0.36 V up to 7.5 V. Actually, the Dirac model

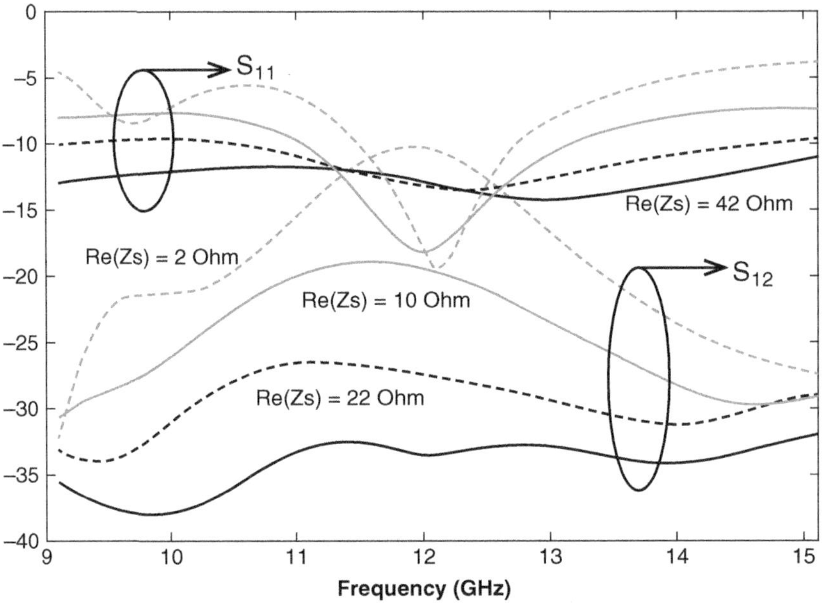

Figure 8.20 Reflection and transmission coefficient of the coupled graphene-patch antennas of Figure 8.19.

of graphene that is implicit in Equation (8.38), is valid for small charge energies (<1 eV), so that only voltages <1 V, corresponding to $Z_s > 10$ ohm, are allowed. However, here an extrapolation to a Z_s of 2 ohm has been made (Figure 8.20).

Regarding the antenna efficiency η_g, it is related to the antenna losses η_L, that in turn can be approximated by calculating the ratio of the dissipated power on the graphene patch vs. the input power:

$$\eta_L \approx \frac{\int\int_{\text{Graphene}} \frac{|E_t|^2}{2Z_s}}{\text{Input power}} \Rightarrow \eta_g \approx 1 - \eta_L, \tag{8.38}$$

where E_t is the tangential electric field sampled on the graphene surface. Typical values are reported in [78].

Then, besides the fact that graphene at microwave frequencies acts as a conductor with moderate losses, by employing a small value of the graphene surface impedance, i.e. equal to or less than 10 ohm, radiation characteristics similar to those of a standard planar metallic-patch antenna can be achieved. This small impedance is likely to be experimentally obtained with very pure and high-quality graphene. We are currently engaged in the fabrication of the device.

The external bias allows tuning of the coupling between slightly spaced graphene-patch antennas, thus suggesting a possible way to control the output of the array factor of an array made of closely spaced graphene-patches.

Acknowledgment

The work reported in Section 8.3.3 has been supported by the European Project "Carbon based smart systems for wireless applications," (NANO-RF, n. 318352).

References

[1] A. Dodabalapur, "Organic and polymer transistors for electronics," *Materials Today*, **9**, (*4*), 24–30, 2006.

[2] Synopsis, Sentaurus TCAD, *Sentaurus Device User's Guide* G-2012.06, 2012.

[3] G. Horowitz, "Organic field-effect transistors," *Advanced Materials*, **10**, (*5*), 365–377, 1998.

[4] A. Bolognesi, M. Berliocchi, M. Manenti, *et al.*, "Effects of grain boundaries, field dependent mobility, and interface trap states on the electrical characteristics of pentacene TFT," *IEEE Trans. Electron Devices*, **51**, (*12*), 1997–2003, 2004.

[5] S. Steudel, S. De Vusser, K. Myny, *et al.*, "Comparison of organic diode structures regarding high-frequency rectification behavior in radio-frequency identification tags," *Journal Of Applied Physics*, **99**, (*11*), 4519, 2006.

[6] T. W. Kelley, D. V. Muyres, F. Baude, T. P. Smith, and T. D. Jones, *High Performance Organic Thin Film Transistors*, Materials Research Society, 2003.

[7] O. D. Jurchescu, J. Baas, and T. M. Palstra, "Effect of impurities on the mobility of single crystal pentacene," *Appl. Phys. Lett.*, **84**, 3061, 2004.

[8] http://www.sigmaaldrich.com/materials-science/organic-electronics.html.

[9] M. Nakamura, H. Ohguri, H. Yanagisawa, *et al.*, "Electrical characterization of pentacene thin films using four-point-probe field-effect measurement and atomic force microscope potentiometry," *Proc. Int. Symp. Super-Functionality Organic Devices*, IPAP Conf. Series 6, pp. 130–135, 2005.

[10] S. Steudel, K. Myny, V. Arkhipov, *et al.*, "50 MHz rectifier based on an organic diode," *Nature Materials*, **4**, (August) 2005.

[11] D. S. Vusser, S. Steudel, K. Myny, J. Genoe, and P. Heremans, "High-performance low voltage organic thin-film transistors," *Mater. Res. Soc. Symp. Proc.*, vol. 870E, 2005.

[12] K. Myny, S. Steudel, P. Vicca, J. Genoe, and P. Heremans, "An integrated double half wave organic Schottky diode rectifier on foil operating at 13.56 MHz," *Appl. Phys. Lett.*, **93**, 093305, 2008.

[13] A. Bolognesi, A. Di Carlo, P. Lugli, and G. Conte, "Large drift-diffusion and Monte Carlo modeling of organic semiconductor devices," *Synthetic Metals*, **138**, (*1–2*), 95–100, 2003.

[14] A. Bolognesi and A. Di Carlo, "Influence of carrier mobility and interface trap states on the transfer and output characteristic of organic thin film transistors," *Journ. Comp. Elect.*, **2**, (*2–4*), 297–300, (Dec.) 2003.

[15] J. Millman and A. Grabel, *Microelectronics*, McGraw-Hill, 1987.

[16] K. S. Novoselov, A. K. Geim, S. V. Morozov, *et al.*, "Two-dimensional gas of massless Dirac fermions in graphene," *Nature*, **438**, 197–200, 2005.

[17] K. S. Novoselov, A. K. Geim, S. V. Morozov, *et al.*, "Electric field effect in atomically thin carbon films," *Science*, **306**, 666–669, 2004.

[18] G. K. Geim and K. S. Novoselov, "The rise of graphene," *Nature Mater.*, **6**, 183–191, 2007.

[19] S. V. Morozov, K. S. Novoselov, M. I. Katsnelson, *et al.*, "Giant intrinsic carrier mobilities in graphene and its bilayer," *Phys. Rev. Lett.*, **100**, (*1*), 16602–16605, 2008.

[20] M. Lemme, "Current status in graphene transistors," *Solid State Phenom.*, **499**, 156, 2010.
[21] P. Blake, E. W. Hill, A. H. Castro Neto, *et al.*, "Making graphene visible," *Appl. Phys. Lett.*, **91**, 063124, 2007.
[22] L. Pierantoni, F. Coccetti, and P. Russer, "Nanoelectronics: the paradigm shift," *IEEE Microwave Magazine*, **11**, (*17*), 8–10, (Dec.) 2010.
[23] Y. M. Lin, K. A. Jenkins, A. Valdes-Garcia, *et al.*, "Operation of graphene transistors at gigahertz frequencies," *Nano Lett.*, **9**, 422–426, 2009.
[24] H. Wang, D. Nezich, J. Kong, and T. Palacios, "Graphene frequency multipliers," *IEEE Electron. Device Lett.*, **30**, 547–549, 2009.
[25] X. Wang, L. Zhi, and K. Mullen, "Transparent conductive graphene electrodes for dye-sensitized solar cells," *Nano Lett.*, **8**, 323–327, 2008.
[26] A. Vakil and N. Engheta, "Transformation optics using graphene," *Science*, **332**, 1291–1294, 2011.
[27] M. Jablan, H. Buljan, and M. Soljacic, "Plasmonics in graphene infrared frequencies," *Physical Review B*, **80**, 1–7, 2009.
[28] A. N. Obraztsov, E. A. Obraztsova, A. V. Tyurnina, *et al.*, "Chemical vapor deposition of thin graphite films of nanometer thickness," *Carbon*, **45**, (10), 2017–2021, 2007.
[29] Q. Yu, J. Lian, S. Siriponglert, *et al.*, Graphene segregated on Ni surfaces and transferred to insulators," *Appl. Phys. Lett.*, **93**, (*11*), 113103, 2008.
[30] X. Li, W. Cai, J. An, *et al.*, "Large-area synthesis of high-quality and uniform graphene films on copper foils," *Science*, **324**, (*5932*), 1312–1314, 2009.
[31] I. Latzer, C. Kremers, A. Carbellos-Aparicio, *et al.*, "Graphene-based nano-patch antenna for terahertz radiation," *Photonics and Nanostructures*, **10**, (*4*), 353–358, 2012.
[32] L. Pierantoni, D. Mencarelli, T. Rozzi, *et al.*, "Multiphysics analysis of harmonic RFID tag on paper with embedded nanoscale material," *Proceedings of the 5th European Conference on Antennas and Propagation*, Rome, Italy, 11–15 April 2011.
[33] C. Berger, Z. Song, X. Li, *et al.*, "Electronic confinement and coherence in patterned epitaxial graphene," *Science*, **312**, (*5777*), 1191–1196, 2006.
[34] K. I. Bolotin, K. J. Sikes, Z. Jiang, *et al.*, "Ultrahigh electron mobility in suspended graphene," *Solid State Communications*, **146**, (*9–10*), 351–355, 2008.
[35] X. Du, I. Skachko, A. Barker, *et al.*, "Approaching ballistic transport in suspended graphene," *Nat. Nano*, **3**, (*8*), 491–495, 2008.
[36] J. Moser, A. Barreiro, and A. Bachtold, "Current-induced cleaning of graphene," *Appl. Phys. Lett.*, **91**, (*16*), 163513, 2007.
[37] A. A. Balandin, S. Ghosh, W. Bao, *et al.*, "Superior thermal conductivity of single-layer graphene," *Nano Letters*, **8**, (*3*), 902–907, 2008.
[38] S. Ghosh, I. Calizo, D. Teweldebrhan, *et al.*, "Extremely high thermal conductivity of graphene: Prospects for thermal management applications in nanoelectronic circuits," *Appl. Phys. Lett.*, **92**, (*15*), 151911, 2008.
[39] P. Blake, P. D. Brimicombe, R. R. Nair, *et al.*, "Graphene-based liquid crystal device," *Nano Letters*, **8**, (*6*), 1704–1708, 2008.
[40] T. J. Booth, P. Blake, R. R. Nair, *et al.*, "Macroscopic graphene membranes and their extraordinary stiffness," *Nano Letters*, **8**, (*8*), 2442–2446, 2008.
[41] Z. H. Ni, H. M. Wang, J. Kasim, *et al.*, "Graphene thickness determination using reflection and contrast spectroscopy," *Nano Letters*, **7**, (*9*), 2758–2763, 2007.
[42] M. Dragoman, D. Neculoiu, D. Dragoman, *et al.*, "Graphene for microwave," *IEEE Microwave Magazine*, 81–86, (Dec.) 2010.

[43] M. Dragoman and D. Dragoman, "Graphene-based quantum electronics," *Progress in Quantum Electronics*, **33**, 165–214, 2009.

[44] W. A. de Heer, C. Berger, X. Wu, *et al.*, "Epitaxial graphene," *Solid State Communications*, **143**, (*1–2*), 92–100, 2007.

[45] J. Kedzierski, J. P.-L. Hsu, P. Healey, *et al.*, "Epitaxial graphene transistors on SiC substrates," *IEEE Transactions on Electron Devices*, **55**, (*8*), 2078–2085, 2008.

[46] J. Rohrl, M. Hundhausen, K. V. Emtsev, *et al.*, "Raman spectra of epitaxial graphene on SiC(0001)," *Appl. Phys. Lett.*, **92**, (*20*), 201918, 2008.

[47] A. Reina, X. Jia, J. Ho, *et al.*, "Large area, few-layer graphene films on arbitrary substrates by chemical vapor deposition," *Nano Letters*, **9**, (*1*), 30–35, 2009.

[48] K. S. Kim, Y. Zhao, H. Jang, *et al.*, "Large-scale pattern growth of graphene films for stretchable transparent electrodes," *Nature*, **457**, (*7230*), 706–710, 2009.

[49] P. W. Sutter, J.-I. Flege, and E. A. Sutter, "Epitaxial graphene on ruthenium," *Nat. Mater.*, **7**, (*5*), 406–411, 2008.

[50] L. Brey and H. A. "Electronic states of graphene nanoribbons," *Cond-mat.*, 0603107, 5 Mar. 2006.

[51] S. Raulot, "Green functions for the Dirac operator under local boundary conditions and applications," *Cond-mat.*, 0703197, 2007.

[52] Y. Yoon, G. Fiori, S. Hong, G. Iannacone, and J. Guo, "Performance comparison of graphene nanoribbon FETs with Schottky contacts and doped reservoirs," *IEEE Transaction on Electron Devices*, **55**, (*9*), 2314, 2008.

[53] Z. F. Wang, R. Xiang, Q. W. Shi, *et al.*, "Modeling STM images in graphene using the effective-mass approximation," *Physical Review B*, **74**, 125417, 2006.

[54] T. Ando, "Quantum point contacts in magnetic fields," *Physical Review B*, **44**, 8017, 1991.

[55] E. V. Castro, N. M. R. Peres, and J. M. B. Lopes dos Santos, "Gapped graphene bilayer: disorder and magnetic field effects," *Phys. Stat. Sol.* (b), **244**, (*7*), 2311–2316, 2007.

[56] R. Akhmerov and C. W. J. Beenakker, "Boundary conditions for Dirac fermions on a terminated honeycomb lattice," *Cond-mat.*, 0710.2723v1, 2008.

[57] D. Mencarelli, L. Pierantoni, and T. Rozzi, "Optical absorption of carbon nanotube diodes: Strength of the electronic transitions and sensitivity to the electric field polarization," *Journal of Applied Physics*, **103**, (*6*), 0631–03, March 2008.

[58] D. Mencarelli, T. Rozzi, C. Camilloni, *et al.*, "Modeling of multi-wall CNT devices by self-consistent analysis of multi-channel transport," *IOP Science Nanotechnology*, **19**, (*16*), (April) 2008.

[59] D. Mencarelli, T. Rozzi, and L. Pierantoni, "Coherent carrier transport and scattering by lattice defects in single- and multi-branch carbon nanoribbons," *Physical Review B*, **77**, 1954351, (May) 2008.

[60] D. Mencarelli, T. Rozzi, and L. Pierantoni, "Scattering matrix approach to multichannel transport in many lead graphene nanoribbons," *IOP Science Nanotechnology*, **21**, (*15*), 5570–15580, (March) 2010.

[61] D. Mencarelli, L. Pierantoni, A. Di Donato, and T. Rozzi, "Self-consistent simulation of multi-walled CNT nanotransistors," *Int. Journal of Micr. and Wireless Technologies*, **2**, (*5*), 453–456, (Dec.) 2010.

[62] D. Mencarelli, L. Pierantoni, M. Farina, A. Di Donato, and T. Rozzi, "A multi-channel model for the self-consistent analysis of coherent transport in graphene nanoribbons," *ACS Nano*, **5**, (*8*), 6109–6128, 2011.

[63] L. Pierantoni, D. Mencarelli, and T. Rozzi, "A new 3D-transmission line matrix scheme for the combined Schrödinger–Maxwell problem in the electronic/electromagnetic characterization of nanodevices," *IEEE Trans. on Microwave Theory and Techniques*, **56**, (*3*), 654–662, 2008.

[64] L. Pierantoni, D. Mencarelli, and T. Rozzi, "Boundary emittance operators for the Schrödinger–Maxwell problem of carrier dynamics in nanodevices," *IEEE Trans. Microw. Theory Tech.*, **57**, (*5*), 1147–1155, 2009.

[65] L. Pierantoni, D. Mencarelli, and T. Rozzi, "Modeling of the electromagnetic/coherent transport problem in nano-structured materials, devices and systems using combined TLM-FDTD techniques," *Microwave Symposium Digest, 2011 Int. Microwave Symposium*, Baltimore, MA, USA, pp. 1–4, 5–10 June 2011.

[66] T. Rozzi, D. Mencarelli, and L. Pierantoni, "Deriving electromagnetic fields from the spinor solution of the massless Dirac equation," *IEEE Trans. Microw. Theory Tech.*, **57**, (*12*), 2907–2913, 2009.

[67] T. Rozzi, D. Mencarelli, and L. Pierantoni, "Towards a unified approach to electromagnetic fields and quantum currents from Dirac spinors," *IEEE Transactions on Microwave Theory and Techniques*, special issue on "RF nanoelectronics," **59**, (*10*), 2587–2594, 2011.

[68] D. Mencarelli, L. Pierantoni, and T. Rozzi, "Graphene modeling by TLM approach," *Microwave Symposium Digest, 2012 International Microwave Symposium*, Montreal, QC, Canada, pp. 1–3, 17–22 June 2012.

[69] L. Pierantoni, D. Mencarelli, A. Di Donato, and T. Rozzi, "Multiphysics techniques for the electromagnetic/coherent-transport problem in carbon nanodevices: Analysis of the metal-carbon transition," *Proceedings of the IEEE International Symposium on Antennas and Propagation (AP-S) and USNC-URSI National Radio Science Meeting*, Chicago, IL, USA, 8–14 July 2012.

[70] L. Pierantoni, D. Mencarelli, and T. Rozzi, "Advanced techniques for the investigation of the combined electromagnetic-quantum transport phenomena in carbon nanodevices," *Proceedings of the International Conference on Electromagnetics in Advanced Applications (ICEAA) 2012-IEEE APWC 2012-EEIS 2012*, Cape Town, South Africa, pp. 873–876, 2–7 Sept. 2012.

[71] G. Vincenzi, G. Deligeorgis, F. Coccetti, *et al.*, "Extending ballistic graphene FET lumped element models to diffusive devices," *Solid-State Electronics*, **76**, 8–12, (Oct.) 2012.

[72] L. Pierantoni and T. Rozzi, "A general multigrid-subgridding formulation for the transmission line matrix method," *IEEE Transactions on Microwave Theory and Techniques*, **55**, (*8*), 1709–1716, 2007.

[73] L. Pierantoni, A. Massaro, and T. Rozzi, "Accurate modeling of TE/TM propagation and losses of integrated optical waveguide," *IEEE Transactions on Microwave Theory and Techniques*, **53**, (*6*), 1856–1862, 2005.

[74] A. Naeemi and J. D. Meindl, "Conductance modeling for graphene nanoribbon (GNR) interconnects," *IEEE Electron Device Letters*, **28**, (*5*), 428–431, 2007.

[75] P. Sharma, J. S. Gomez-Diaz, A. M. Ionescu, and J. Perruisseau-Carrier, "Determination of minimum conductivity of graphene from contactless microwaves measurements," *IEEE NANO 2012, 12th Intern. Conf.on Nanotech.*, Birmingham, UK, 2012.

[76] J. S. Gomez-Diaz and J. Perruisseau-Carrier, "Microwave to THz properties of graphene and potential antenna applications," *2012 International Symp. on Antennas and Propagation* (ISAP2012), Nagoya, Japan, 2012.

[77] M. Dragoman, A. A. Muller, D. Dragoman, *et al.*, "Terahertz antenna based on graphene," *J. Appl. Phys.*, **107**, 104313, 2010.

[78] G. W. Hanson, "Dyadic Green's functions and guided surface waves for a surface conductivity of graphene," *J. Appl. Phys.* **103**, 064302, 2008.

[79] D. Mencarelli, M. Dragoman, L. Pierantoni, T. Rozzi, and F. Coccetti, "Design of a coplanar graphene-based nano-patch antenna for microwave applications," *2013 International Microwave Symposium (IMS), Microwave Symposium Digest (MTT)*, Seattle, WA, USA, 4–6 June 2013.

9 RFID enabling new solutions

Federico Alimenti, Chiara Mariotti, and Luca Roselli

9.1 Introduction

The short-range wireless transmission of sensor information finds application in several fields ranging from the monitoring of biological parameters in medicine [1–6], to the measurements of mechanical quantities in industrial applications [7–10] and robot guidance [11]. In the last years several technologies have been developed to this purpose, and the emerging one is based on the Radio Frequency IDentification (RFID) concept [12]. This is due to the convergence of several new ideas and approaches like RF energy harvesting [13, 14], RF carrier re-use, the load modulation method, and organic [15–23] and inkjet-printed [24–29] electronics. RF energy harvesting and RF carrier re-use, for example, make possible battery-less (i.e. passive) RFID sensors that can operate for years without any maintenance.

A detailed description of a common RFID system is given in Chapter 2; however, it is worth recalling that it is usually composed of: a reader (or interrogator), that sends an interrogation signal to an RFID tag; a tag (or transponder) which identifies the object with a code, basically implemented by an IC and an antenna; and a host computer that codes and encodes the data from the reader. Moreover, a wireless connection to a global network can be provided [30]. Figure 9.1 shows a block diagram of a typical RFID system, including the tag, the reader, and the host computer connected to a global network.

A classification of the available tag technologies is sketched in Figure 9.2. The passive systems can be divided in two families, namely: chip-based and chipless tags. The first family exploits usually a low-power CMOS chip to implement the main tag functions (RF carrier rectification, DC voltage regulation, ASK demodulation, ID decoding, load modulation, etc.) and thus can be adapted to sensor applications by a few additional circuit blocks (signal conditioning, ADC, etc.). The main advantage of such an approach is the digital modulation of the transmitted signal, and thus the reliability of the data treatment, as shown in [31, 32]. The production costs of chip-based tags are mostly associated with the heterogeneous integration of the silicon chip with the antenna [33, 34], the latter being typically fabricated on a flexible substrate or textile substrate [35–40].

The RFID technology is indeed well consolidated with its main advantages of high flexibility and versatility in terms of the application [41]. Nowadays, however, new functionalities and properties are being pursued following the new paradigm of the

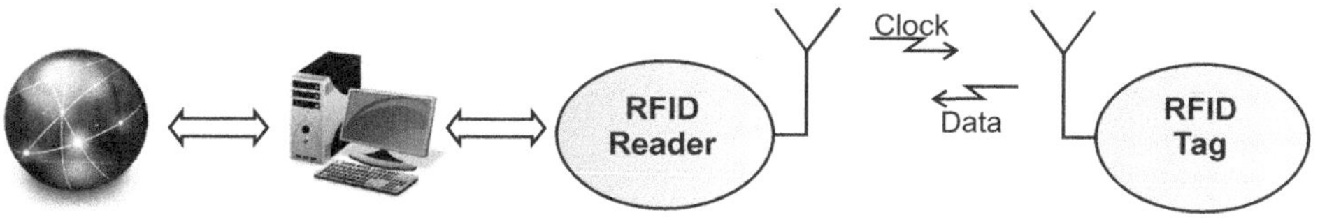

Figure 9.1 A typical RFID system.

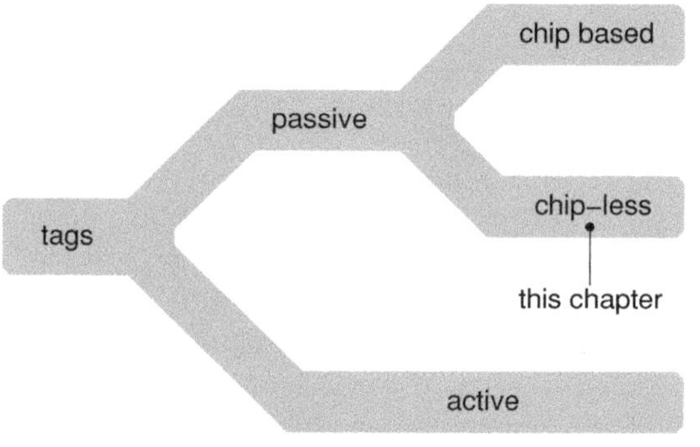

Figure 9.2 Classification of RFID tag technologies.

Internet of Things (IoT), for which the recyclability of the materials, the lower cost of production, and the energy autonomy of the tags are needed. In this context the main weaknesses of the traditional chip-based RFID technologies are the presence of the IC in the tag and the necessity of a connection between the antenna and the IC itself. For these reasons, a technology that exhibits very low cost, high robustness, and the ability to be used in severe environments is very attractive and highly desirable for many applications in the IoT and ubiquitous sensing scenarios [41].

An alternative to the common IC-based RFID technology is the traditional barcode, very cheap because it is simply printed in marks and spaces, but with some disadvantages from the point of view of the IoT scenario: static behavior, no sensing capability nor any kind of "intelligence," LoS operation, short-range readability and so on [42]. Ideally, for the purposes of the IoT applications, one of the best solutions might be mixing the low fabrication costs of barcode-like systems with the long readability range and the information capabilities of RFIDs.

Significant investment and research have been focused on lowering the price of RFID tags by reducing the cost of the RFID IC, which represents the expensive part of the tag itself. However, the price of an RFID tag is still much higher than a barcode, and this is the main reason why RFID technology has not completely replaced the barcode yet and new technologies are being proposed.

Recently, the interest of the scientific community in this field has moved towards the development of the already cited *chipless* RFID tags, thus reducing the fabrication cost by eliminating the IC and keeping at the same time the long-range and non-line-of-sight (N-LoS) readability of the common RFID tags. This is made possible also thanks to some growing technologies such as, for example, nano-particle silver inkjet printing

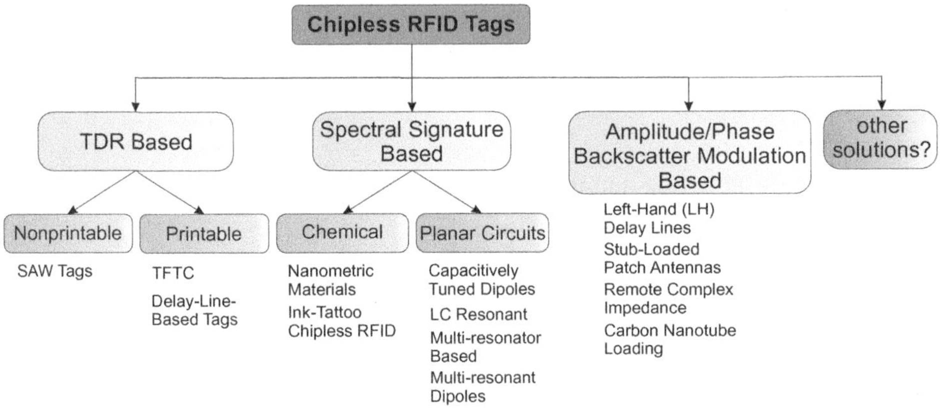

Figure 9.3 Classification of chipless RFID tags.

[24] or copper adhesive laminate techniques [43] on photographic paper or other flexible substrates.

The basic chipless technology simply applies radar principles: the information is embedded into the electromagnetic signature (EMS) of the structure based on the assumption that "every object inherently has a specific EMS that can be associated with an ID" [42], being to this extent an RF evolution of the barcode but without electronics involved.

As of now, one of the most challenging design steps for an RFID chipless tag is to find an optimal method to encode data, possibly in a non-static way, when no IC is present. In order to address this issue, researchers are proposing some different approaches. It is possible to categorize chipless RFID tags as follows (see Figure 9.3):

- time-domain reflectometry (TDR)-based chipless tags;
- spectral signature-based chipless tags;
- amplitude/phase backscatter modulation-based chipless tags;
- other solutions?

9.2 Time-domain reflectometry (TDR)-based chipless tags

In time-domain reflectometry (TDR)-based chipless RFIDs, the reader interrogates the tags by sending a pulse and listening to the echoes reflected by them. The time of arrival of the echoes is used to encode the data. TDR-based chipless tags are low cost and have higher reading ranges compared to other chipless tags, however, the number of encodable bits is low and the reader is complex. TDR-based chipless tags can be nonprintable, such as surface acoustic wave (SAW) tags, or printable, such as for instance, thin film transistor circuit (TFTC)-based tags, or delay-line-based tags. In the SAW-based circuit architecture tag, illustrated with an example in Figure 9.4(a), the interrogation pulse is received by the antenna and then converted to an SAW by means of an interdigital

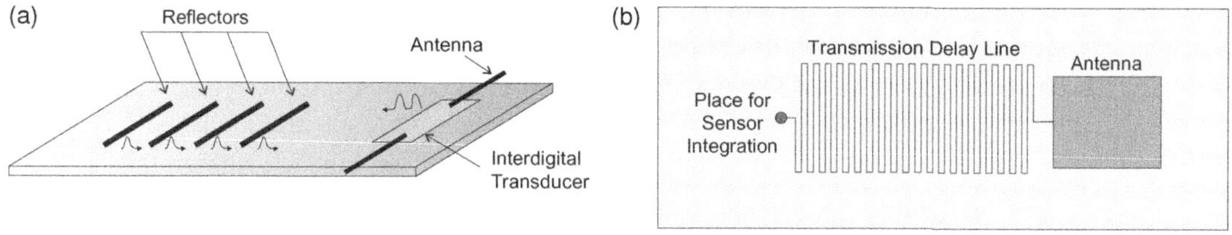

Figure 9.4 Examples of TDR-based chipless tags architecture: (a) The SAW-based tag (nonprintable); (b) The delay-line-based tag (printable).

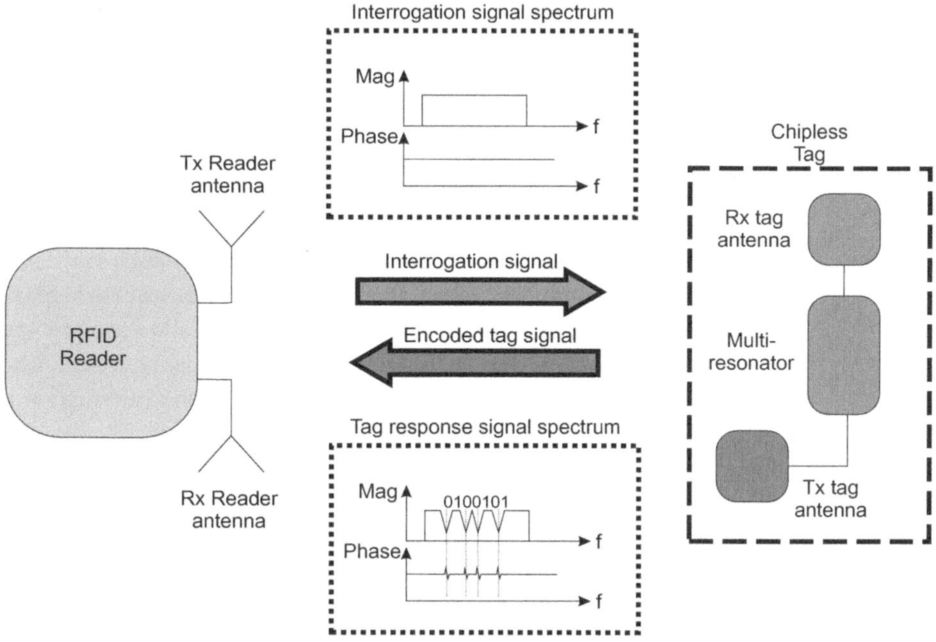

Figure 9.5 Spectral signature-based chipless tag operating principle.

transducer. The SAW propagates to a number of reflectors, which create a number of echoes with phase shift [42].

In contrast, Figure 9.4(b) shows an example of delay-line-based chipless tag architecture, in which there is a receiving antenna and a transmission delay line that reflects at various points (discontinuities) the received pulse, creating its multiple echoes and thus encoding the information. This type of tag replicates the operating principle of SAW-based chipless tags using microstrip technology, making it fully printable.

9.3 Spectral signature-based chipless tags

Spectral signature-based chipless tags work by encoding the data into the spectrum by means of resonant structures. Figure 9.5 shows the operating principle of this type of chipless tag: the reader sends an UWB signal, received by the Rx antenna of the tag; the tag encodes the information using a multi-resonating structure that substantially filters some frequencies of the original spectrum. The encoded spectrum is then sent back to the interrogator through the tag transmitting antenna. This spectrum is usually encoded assuming that the filtered frequencies represent the "0" and the not filtered the "1."

The advantages of this approach are that the tags are fully printable and the data storage capability is higher compared to other chipless tags, the number of bits being, in principle, equal to the number of resonators. However, a large spectrum is needed for data encoding and, on the reader side, wideband dedicated RF components are thus required.

Signal-based chipless tags can also be classified as follows: chemical, such as ink-tattoo chipless RFID [44] or nanometric materials-based [45]; and planar circuits (for example: capacitively tuned dipoles [46], LC resonant circuits [47], multi-resonator-based [48], multi-resonant dipoles-based [49]).

9.4 Amplitude/phase backscatter modulation-based chipless tags

In amplitude/phase backscatter modulation-based chipless tags, the data encoding is achieved by changing the amplitude or the phase of the backscattered signal through the variation of the tag antenna loading. This variation is usually due to the presence, as a load on the antenna, of a material which modifies its impedance according to the change of the information to transmit. The reader then encodes the information based on the variation of the amplitude/phase of the backscattered signal.

An example of this type of chipless tag is the one reported in [50], consisting of an on-photographic-paper-inkjet-printed antenna working at 868 MHz and loaded by a single wall carbon nanotube (SWCNT), sensitive to the gas variation in the neighboring atmosphere. Some measurements have proved that the impedance of the SWCNT is modified when a certain quantity of ammonia is introduced in the air, thus varying the antenna load and as a consequence the backscattered signal. The interrogator is then able to detect the presence of the ammonia.

There are other types of amplitude/phase backscatter modulation-based chipless tags, such as left-hand delay-lines-based [51], stub-loaded patch antenna-based [52], and remote complex impedance-based [53].

In general, the advantages of these chipless tags are that they operate over narrow bandwidths and the architecture is simple; the disadvantages are that they can detect a lower number of bits compared to other chipless tags and that the encoding might be performed by means of lumped/chipped elements, thus increasing the tags' fabrication costs.

9.5 Other solutions?

In the already described scenario for chipless tags, there is still space for novel solutions. In the following sections an innovative approach, recently proposed, to solve the issue of the absolute accuracy required by most of the passive chipless RFID sensors is illustrated. To this purpose the sensor information is encoded as the phase difference between two signals, acting as the reference signal to each other. First the tag receives a carrier at frequency f_0. Then two equal signals at frequency $2f_0$ are generated by means of a diode-based frequency doubler and a power divider. At this point one of the two signals is phase-shifted using a passive sensing element. Finally the $2f_0$ signals are re-radiated by exploiting two orthogonally polarized antennas. With this approach the sensor information can be extracted by a suitable reader equipped with two complex (I/Q) receivers. The above tag concept is compatible with paper substrate and inkjet printing technology since antennas [54], diodes [55], and passive sensing elements [56–59], i.e. all the main tag components, will soon be available on paper material. In addition, substrate integrated waveguide (SIW) structures [60], a way to implement standard waveguide geometries [61] exploiting planar processes and dielectric substrates, will be soon available on paper. As a consequence, a further cost reduction of standard PCB technologies [62] could be directly achieved.

The remaining part of this chapter is organized as follows. The sensor architecture is described in Section 9.6, where both tag and reader subsystems are considered. Then, Section 9.7 illustrates the theory of operation and the basic equations needed to recover the information. Finally, possible applications of the developed theory are proposed in Section 9.8 for four significant cases.

9.6 Novel RFID sensor

The proposed RFID sensor [63] is based on the harmonic radar concept [64], i.e. on a tag that, being illuminated by a carrier at frequency f_0, is capable of generating the second harmonic $2f_0$. Such a tag, however, is also responsible for the sensor information encoding, as shown in Figure 9.6.

To describe the tag operation, let us consider the signal flow shown in the same figure. The incoming electromagnetic wave at frequency f_0 is received by a spiral or a helical antenna [65]. In this way the power at the antenna output is maximized regardless of the polarization or the relative reader–tag orientation. The received power is then fed into a varactor or a Schottky diode frequency doubler [66] and the second harmonic is generated. At this point the $2f_0$ signal is split by a power divider. The first part is directly re-radiated in vertical polarization (e.g. E_η in Figure 9.6) in such a way as to form a reference signal component. The second part, instead, is phase-shifted by the angle $\Delta\varphi$ and then re-radiated in horizontal polarization (e.g. E_ξ in Figure 9.6). Phase-shifters based on the concept proposed in [67] could, for example, be used to this purpose.

The phase angle $\Delta\varphi$ encodes the sensor information and must be recovered by the reader. To this purpose the reader is composed of four sub-systems, as depicted in Figure 9.7. A phase-locked loop (PLL) oscillator is used to generate both the f_0 and the $2f_0$ signals in a synchronous way.

The f_0 carrier is managed by the transmitter (TX) and serves to remotely illuminate the tag. The transmitter operates in continuous wave (CW) mode or by exploiting a

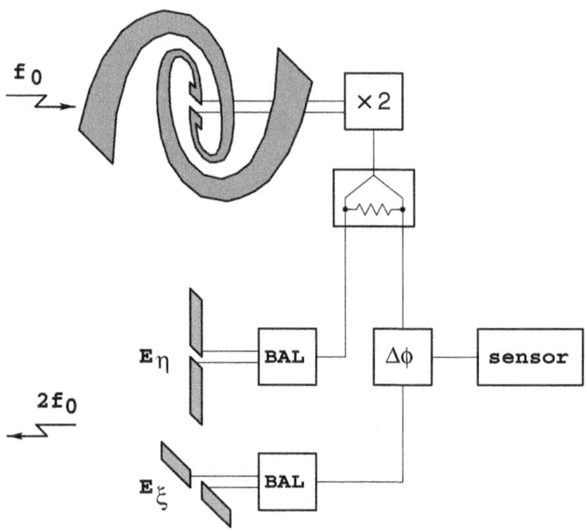

Figure 9.6 Harmonic tag block diagram (after [63]).

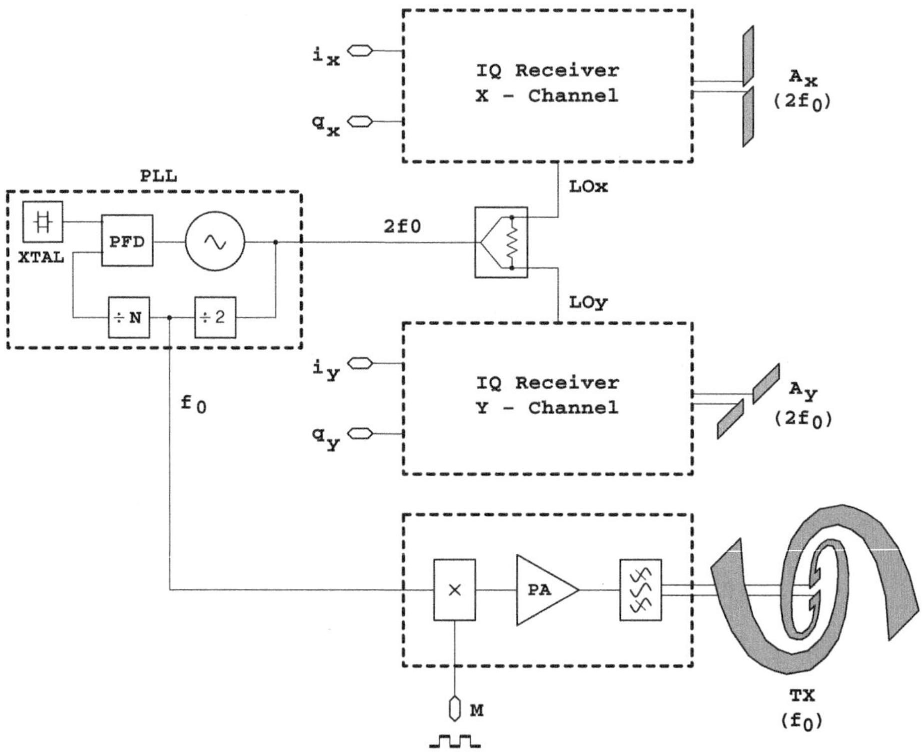

Figure 9.7 Reader block diagram (after [63]).

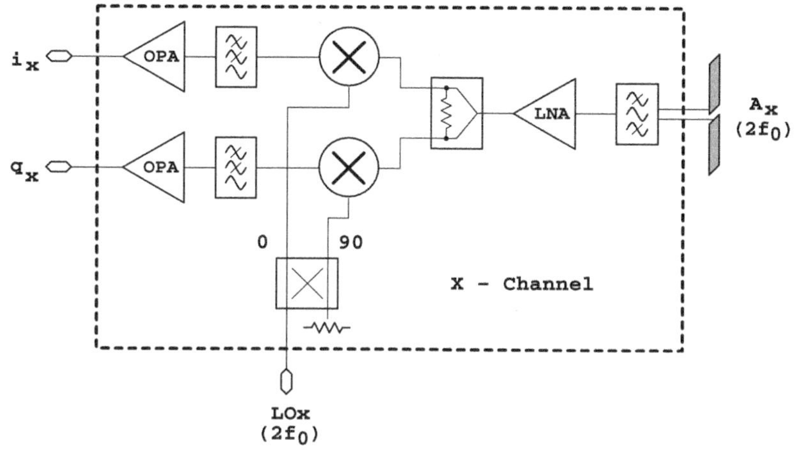

Figure 9.8 IQ receiver (x-channel) block diagram (after [63]).

modulated carrier (see M input in Figure 9.7). It is composed of a modulator and a power amplifier, the latter followed by a band-pass filter at f_0 (or a stop-band filter at $2f_0$). The filter is necessary to significantly reduce the second harmonic generated by the power amplifier, thus eliminating the risk of blinding the receiver. The second harmonic of the TX, in fact, should be kept below the noise floor of the receiver.

The transmitting antenna is again of spiral or helical type in order to maximize the illumination power regardless of the orientation.

The $2\,f_0$ component, instead, is used as the local oscillator of the receiver. Such a signal is thus divided into two parts and applied to both the x- and y-receiver channels. Each channel is sensitive to a particular wave polarization, achieved by exploiting dipole antennas. In particular, the x-channel is adapted to receive the vertical polarization signal while the y-channel is exploited for the horizontally polarized one. Because the sensor information is encoded as phase difference $\Delta\varphi$ between the two polarizations, a vector receiver is needed for each channel. This vector receiver is based on a zero-IF in-phase and quadrature (IQ) architecture, as illustrated in Figure 9.8.

9.7 Basic theory

This section is devoted to the operation theory of the novel RFID sensor system. First, the detection of the two orthogonal $2\,f_0$ plane waves is considered. These waves are generated and radiated by the tag and thus the relevant phase and amplitude information associated with them is recovered by the reader. The illumination of the tag with a carrier at f_0, the second harmonic generation, and the phase information coding are also treated in this section.

The theory will be developed under the basic assumption of a tag antenna plane parallel to the reader antenna plane. In addition, without lack of generality, the two antenna systems will be considered axially aligned along the propagation direction.

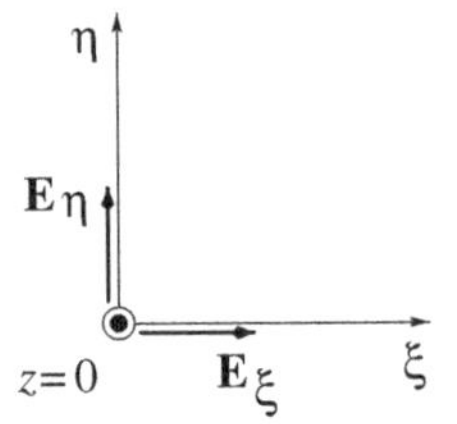

Figure 9.9 Coordinate system adopted in the tag plane (i.e. $\xi-\eta$ plane) and $2f_0$ electric field components. These electric fields are radiated by the two equal orthogonal tag antennas after [63].

This hypothesis is satisfied in many industrial applications, such as the contact-less measurement of mechanical quantities in rotating systems like engines and turbines [68] or monitoring of objects above a conveyor belt in a distribution chain. In the latter case the tag is assumed to be interrogated when it is exactly perpendicular to the reader.

9.7.1 Tag information encoding

The proposed RFID sensor relies on the phase and amplitude information written on two orthogonally polarized plane waves at frequency $2f_0$. These waves are fully determined by a knowledge of the associated electric field vectors as in Figure 9.9. In particular, the $\xi-\eta$ coordinate system is used in the tag plane while the two electric fields E_ξ and E_η are radiated by two equal antennas, linearly polarized along ξ and η respectively.

Adopting the complex vector formalism one can write:

$$\begin{aligned}\mathrm{E}_\xi &= E_\xi e^{j\varphi_\xi}\mathrm{u}_\xi,\\ \mathrm{E}_\eta &= E_\eta e^{j\varphi_\eta}\mathrm{u}_\eta,\end{aligned} \tag{9.1}$$

where E_k, φ_k, and u_k, with $k=\xi,\eta$, are the electric field amplitudes, the electric field phases and the unit vectors respectively. In particular the amplitudes E_k will be assumed real and positive having concentrated all the phase relationships between the two fields in the φ_k values.

As a consequence of Equation (9.1) the total electric field E_T, radiated by the tag at $2f_0$, is given by:

$$\begin{aligned}\mathrm{E}_T &= \mathrm{E}_\xi + \mathrm{E}_\eta\\ &= E_\xi e^{j\varphi_\xi}\left[\mathrm{u}_\xi + \frac{E_\eta}{E_\xi}e^{j\varphi_\eta}\mathrm{u}_\eta\right],\end{aligned} \tag{9.2}$$

with $\mathrm{E}_\xi = 0$ and $\Delta\varphi$ being the phase difference between the two fields:

$$\Delta\varphi = \varphi_\eta - \varphi_\xi. \tag{9.3}$$

It is worth noting that (9.2) is the general expression for an elliptical polarized plane wave. As particular cases both purely linear ($\Delta\varphi = 0,\ \pi$) and purely circular ($\mathrm{E}_\eta/\mathrm{E}_\xi = 1$ and $\Delta\varphi = \pm\,\pi/2$) polarizations are allowed. This means that the relevant tag information can be encoded either in the phase difference $\Delta\varphi$ or in the amplitude ratio $\mathrm{E}_\eta/\mathrm{E}_\xi$.

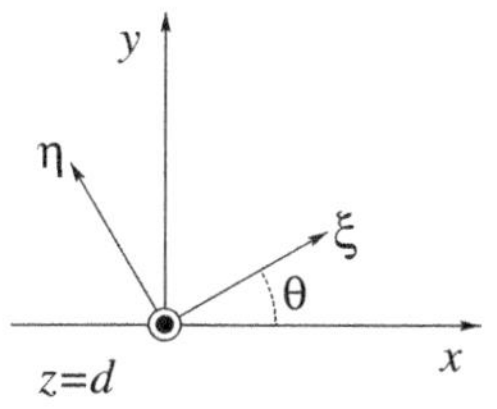

Figure 9.10 Coordinate system adopted in the reader plane (i.e. $x-y$ plane). The reader plane is parallel to the tag plane. The tag $\xi-\eta$ coordinate system is rotated by an angle ϑ with respect to the reader $x-y$ coordinate system. The distance between reader and tag is equal to d (after [63]).

9.7.2 Received reader voltages

Let us now consider a reader placed at a distance d from the tag. The reader exploits two equal linear polarized antennas in orthogonal directions to detect the elliptical polarized wave generated by the tag. This situation is summarized in Figure 9.10, where the reader antenna plane has been assumed parallel to the tag antenna plane. In this scheme the two reader antennas are placed along the x- and the y-axis. Moreover, an angle ϑ is formed between ξ- and x-axis. Such an angle will be, in general, not known since it depends on the way the tag is attached to the object to be monitored.

As a final assumption, let us consider the case when the reader and the tag antenna systems are perfectly aligned along a common z-axis. In this case the electric field at the reader location E_R is simply given by the tag field E_T multiplied for the transfer function of the channel:

$$\mathrm{E}_R = \alpha \mathrm{e}^{-\mathrm{j}\beta d}\mathrm{E}_T, \tag{9.4}$$

where α is the path loss, β is the propagation constant, and d is the reader–tag distance. If a free-space radiation model can be applied to the considered problem, α and β are:

$$\begin{aligned} \alpha &= \frac{\lambda}{4\pi d}\sqrt{D_R D_T}, \\ \beta &= \frac{2\pi}{\lambda}. \end{aligned} \tag{9.5}$$

In (9.5) the wavelength λ is evaluated at $2f_0$, and D_R and D_T are the reader and the tag antenna directivity, respectively. The electric field at the reader location E_R can then be projected along the x-oriented receiver antenna:

$$\begin{aligned} E_x &= \mathrm{E}_R \cdot \mathrm{u}_x \\ &= E_0\left[\mathrm{u}_\xi \cdot \mathrm{u}_x + \frac{E_\eta}{E_\xi}\mathrm{e}^{\mathrm{j}\Delta\varphi}\mathrm{u}_\eta \cdot \mathrm{u}_x\right]. \end{aligned} \tag{9.6}$$

Similarly, for the y-oriented receiver antenna one gets:

$$\begin{aligned} E_y &= \mathrm{E}_R \cdot \mathrm{u}_y \\ &= E_0\left[\mathrm{u}_\xi \cdot \mathrm{u}_y + \frac{E_\eta}{E_\xi}\mathrm{e}^{\mathrm{j}\Delta\varphi}\mathrm{u}_\eta \cdot \mathrm{u}_y\right]. \end{aligned} \tag{9.7}$$

In (9.6) and (9.7) the phasor E_0 is defined as:

$$E_0 = \alpha e^{j(\varphi_\xi - \beta d)} E_\xi. \tag{9.8}$$

Considering that the projections of the ξ and η unity vectors along the x and y directions are given by:

$$\begin{aligned} \mathrm{u}_\xi \cdot \mathrm{u}_x &= \cos\vartheta, \\ \mathrm{u}_\eta \cdot \mathrm{u}_x &= -\sin\vartheta, \\ \mathrm{u}_\xi \cdot \mathrm{u}_y &= \sin\vartheta, \\ \mathrm{u}_\eta \cdot \mathrm{u}_y &= \cos\vartheta, \end{aligned} \tag{9.9}$$

one obtains:

$$E_x = E_0 \left[\cos\vartheta - \frac{E_\eta}{E_\xi} e^{j\Delta\varphi} \sin\vartheta\right], \tag{9.10}$$

$$E_y = E_0 \left[\sin\vartheta + \frac{E_\eta}{E_\xi} e^{j\Delta\varphi} \cos\vartheta\right]. \tag{9.11}$$

The E_x and E_y fields can be finally related to the received voltages processed by the reader's two channels. To this purpose one can exploit the effective antenna length l_e as defined in [69, p. 305]. The input received voltages are:

$$V_k^i = l_e E_k, \tag{9.12}$$

where $k = x, y$ and the index i stands for input. If, for example, a half-wave dipole antenna is considered, the maximum l_e is:

$$l_e = \frac{\lambda}{\pi}, \tag{9.13}$$

which is smaller than $\lambda/2$ due to the sinusoidal distribution of the current along the dipole itself. Inserting Equations (9.10) and (9.11) into (9.12) and using the Euler's formula to express $e^{j\Delta\varphi}$ one gets:

$$V_x^i = l_e E_0 (A_x + jB_x), \tag{9.14}$$

$$V_y^i = l_e E_0 (A_y + jB_y), \tag{9.15}$$

where the A and B quantities contain all the relevant tag information, i.e. $\Delta\varphi$, E_η/E_ξ, and ϑ:

$$\begin{aligned} A_x &= \cos\vartheta - \frac{E_\eta}{E_\xi} \cos\Delta\varphi \sin\vartheta, \\ A_y &= \sin\vartheta + \frac{E_\eta}{E_\xi} \cos\Delta\varphi \cos\vartheta, \end{aligned} \tag{9.16}$$

$$\begin{aligned} B_x &= -\frac{E_\eta}{E_\xi} \sin\Delta\varphi \sin\vartheta, \\ B_y &= \frac{E_\eta}{E_\xi} \sin\Delta\varphi \cos\vartheta. \end{aligned} \tag{9.17}$$

The time-domain signals are easily obtained from the phasors in (9.14), (9.15), and considering the expression for E_0 quoted in (9.8). The x-channel input voltage is given by:

$$\begin{aligned} v_x^i(t) &= \Re\{V_x^i \mathrm{e}^{\mathrm{j}2\omega_0 t}\} \\ &= V_0 \Re\{(A_x + \mathrm{j}B_x)\mathrm{e}^{\mathrm{j}(2\omega_0 t + \varphi_\xi - \beta d)}\} \\ &= V_0[A_x \cos(2\omega_0 t + \varphi_\xi - \beta d) - B_x \sin(2\omega_0 t + \varphi_\xi - \beta d)], \end{aligned} \tag{9.18}$$

where V_0 is the amplitude of the input referred voltage:

$$V_0 = l_e \left|E_0\right| = \alpha l_e E_\xi . \tag{9.19}$$

Similarly, the y-channel input voltage can be written as:

$$v_y^i(t) = V_0[A_y \cos(2\omega_0 t + \varphi_\xi - \beta d) - B_y \sin(2\omega_0 t + \varphi_\xi - \beta d)]. \tag{9.20}$$

9.7.3 Conversion products

Once the time-domain expressions have been obtained, it is quite straightforward to derive the conversion products and, from them, the in-phase $i(t)$ and quadrature $q(t)$ output signal components. For the x-channel one obtains:

$$\begin{aligned} i_x(t) &= \mathrm{LPF}\{2G_v^{rx} v_x^i(t)\cos(2\omega_0 t - \psi)\} \\ &= G_v^{rx} V_0(A_x \cos\psi_R - B_x \sin\psi_R), \end{aligned} \tag{9.21}$$

$$\begin{aligned} q_x(t) &= \mathrm{LPF}\{-2G_v^{rx} v_x^i(t)\sin(2\omega_0 t - \psi)\} \\ &= G_v^{rx} V_0(A_x \sin\psi_R + B_x \cos\psi_R), \end{aligned} \tag{9.22}$$

where G_v^{rx} is the overall receiver gain, ψ is the local oscillator phase, and LPF {} is the low-pass operator that rejects the $4\omega_0$ mixing product. The quantity ψ_R is defined as:

$$\psi_R = \psi + \varphi_\xi - \beta d. \tag{9.23}$$

Similarly, the y-channel outputs are:

$$i_y(t) = G_v^{rx} V_0(A_y \cos\psi_R - B_y \sin\psi_R), \tag{9.24}$$

$$q_y(t) = G_v^{rx} V_0(A_y \sin\psi_R + B_y \cos\psi_R). \tag{9.25}$$

In the case of a modulated carrier (signal injected at the M input in Figure 9.7), Equations (9.21)–(9.25) represent the amplitude of the base-band signal at the receiver output. As a final observation it is worth noting that the same receiver gain G_v^{rx} and local oscillator phase ψ have been assumed for both the x and the y readers' channels. In practice this condition can be met by calibrating the two receivers with a set of preliminary measurements.

9.7.4 Information recovery

The relationships (9.21), (9.22) and (9.24), (9.25) constitute a set of four equations that can be written at each time instant. The known terms are given by the in-phase and quadrature signals at the output of the two readers' channels, i.e. by measurable values. The unknowns are the received amplitude V_0, the overall received phase ψ_R, and the tag quantities, namely ϑ, $\Delta\varphi$, and E_η/E_ξ. As a result five unknowns are obtained, this means that additional knowledge of the system is needed. For example, if the tag is designed to have a specific E_η/E_ξ ratio, all the information is encoded within $\Delta\varphi$.

In order to eliminate V_0 and ψ_R from the above system of equations the following ratio is computed at each time instant:

$$U = U_R + \mathrm{j}U_I = \frac{i_x + \mathrm{j}q_x}{i_y + \mathrm{j}q_y}. \tag{9.26}$$

Developing the above expression one obtains:

$$\begin{aligned} U_R &= \frac{i_x i_y + q_x q_y}{i_y^2 + q_y^2} \\ &= \frac{A_x A_y + B_x B_y}{A_y^2 + B_y^2}, \end{aligned} \tag{9.27}$$

$$\begin{aligned} U_I &= \frac{q_x i_y - i_x q_y}{i_y^2 + q_y^2} \\ &= \frac{B_x A_y - A_x B_y}{A_y^2 + B_y^2}. \end{aligned} \tag{9.28}$$

The U_R and U_I values can be directly evaluated from the measured output voltages (known terms), whereas the tag unknowns are within the A and B quantities. The relationship between U_R, U_I, and these unknowns can be found by inserting Equations (9.16), (9.17) into (9.27), (9.28). After some manipulation one obtains:

$$U_R = \frac{1}{D_u}\left\{\left[1 - \left(\frac{E_\eta}{E_\xi}\right)^2\right]\cos\vartheta\,\sin\vartheta + \frac{E_\eta}{E_\xi}\cos\Delta\varphi(\cos^2\vartheta - \sin^2\vartheta)\right\}, \tag{9.29}$$

$$U_I = -\frac{1}{D_u}\frac{E_\eta}{E_\xi}\sin\Delta\varphi, \tag{9.30}$$

where the denominator D_u is given by:

$$D_u = \sin^2\vartheta + \left(\frac{E_\eta}{E_\xi}\right)^2\cos^2\vartheta + 2\frac{E_\eta}{E_\xi}\cos\vartheta\,\sin\vartheta\,\cos\Delta\varphi. \tag{9.31}$$

9.8 Applications

Possible applications of the above general theory are analyzed below in a number of particular cases of practical interest. The first case is that of a tag with fixed (an *a-priori* known) orientation angle ϑ in which the information is encoded by the amplitude only. This assumption means that $\Delta\varphi$ is constant, whereas $\rho = E_\eta/E_\xi$ is varied according to the physical quantity to be measured. If $\vartheta = 0$ and $\Delta\varphi = 0$, Equations (9.27) and (9.28) reduce to:

$$U_R = \frac{E_\xi}{E_\eta}, \tag{9.32}$$

$$U_I = 0. \tag{9.33}$$

As a consequence the information can immediately be derived using U_R, thus identifying a very simple method of zero-power wireless transmission of sensor data.

In the second case the tag again has a fixed orientation angle $\theta = 0$, but the information is encoded as a variation of the phase angle $\Delta\varphi$ only. Here an unit amplitude ratio between the two tag channels is assumed: $\rho = E_\eta E_\xi = 1$. As a result Equations (9.27) and (9.28) reduce to:

$$U_R = \cos\Delta\varphi, \tag{9.34}$$

$$U_I = -\sin\Delta\varphi. \tag{9.35}$$

Again the information can immediately be recovered from U_R and U_I. Note that digital information could be transmitted by simply switching $\Delta\varphi$ between 0 ($U_R = 1$, $U_I = 0$) and $-\pi/2$ ($U_R = 0$, $U_I = 1$).

Lets now consider a situation where θ is not an *a priori* known and, possibly, variable with time. As discussed above, one can decide to encode the sensor information either in the amplitude ratio $\rho = E_\eta/E_\xi$ or in the phase difference $\Delta\varphi$ of the tag. The third case refers to the amplitude encoding mechanism. To this purpose $\Delta\varphi \neq 0$ because two quantities (i.e. ρ and θ) now have to be extracted from the two measured quantities (i.e. U_R and U_I). Thus, in the third application case, $\Delta\varphi = -\pi/2$ is considered. As a results one obtains:

$$U_R = \frac{(1-\rho^2)\sin(2\theta)}{(1+\rho^2)-(1-\rho^2)\cos(2\theta)}, \tag{9.36}$$

$$U_I = \frac{2\rho}{(1+\rho^2)-(1-\rho^2)\cos(2\theta)}. \tag{9.37}$$

The previous equations are reported in Figure 9.11 as a function of the rotation angle θ and for ρ equal to 0.3, 0.5, and 0.7. At this point it is interesting to note that the proposed reader–tag system can also be used as a wireless rotation sensor. In such a situation the only information is the angle θ, so it is sufficient to set ρ at a fixed value, for example $\rho = 0.5$.

Finally, in the fourth application case, the information is encoded in the phase difference $\Delta\varphi$, from where it is extracted from the measured quantities along with the orientation angle θ. If $\rho = E_\eta/E_\xi = 1$, Equations (9.27) and (9.28) become:

$$U_R = \frac{\cos\Delta\varphi\cos(2\theta)}{1+\cos\Delta\varphi\sin(2\theta)}, \tag{9.38}$$

$$U_I = \frac{-\sin\Delta\varphi}{1+\cos\Delta\varphi\sin(2\theta)}. \tag{9.39}$$

The function U_R is singular whenever its denominator is equal to zero. This occurs for particular values of the phase $\Delta\varphi$ (i.e. of the phase encoding the sensor information) and of the rotation angle θ. The critical angles can easily be computed and are reported

Table 9.1 Critical angles.

$\Delta\varphi$	θ
$0, 2\pi \ldots$	$\frac{3}{4}\pi, \frac{7}{4}\pi \ldots$
$\pi, 3\pi \ldots$	$\frac{1}{4}\pi, \frac{5}{4}\pi \ldots$

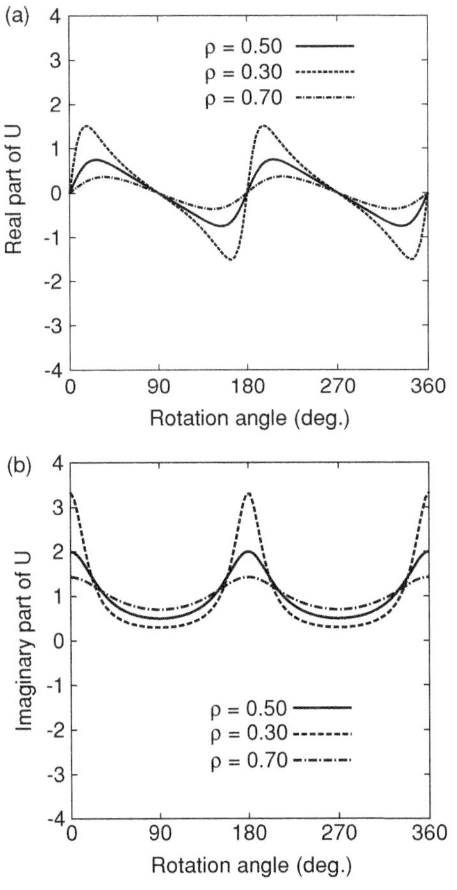

Figure 9.11 (a) Real U_R, and (b) imaginary U_I parts of the U function versus the rotation angle θ. The function is drawn in the particular case $\Delta\phi = -\pi/2$ and for three values of $\rho = E_\eta/E_\xi$. Such a parameter has been assumed to be in the range between 0.3 to 0.7 to avoid singularities when ρ approaches zero (after [63]).

in Table 9.1. The function U_I, instead, is never singular since, for $\Delta\varphi = n\pi$ with n integer, its numerator is always zero. Nevertheless, for $\Delta\varphi$ close to the critical angles of Table 9.1, the U_I values can also be very high.

The singularity problem occurs for those values of $\Delta\varphi$ producing a purely linear wave polarization. Such a problem can always be solved in the following way. When a high value of U_R is detected, the ratio between $(i_y + jq_y)$ and $(i_x + jq_x)$ is evaluated instead of Equation (9.26). Alternatively, the critical values of $\Delta\varphi$ can be avoided, by conditioning

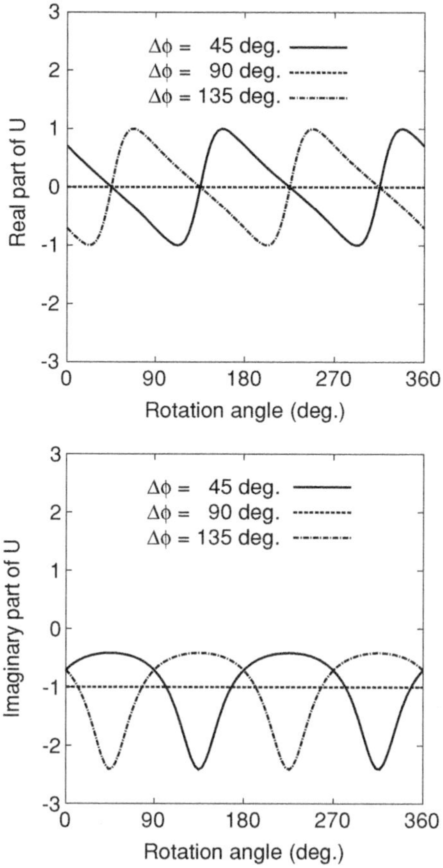

Figure 9.12 Real U_R and imaginary U_I parts of the U function versus the rotation angle θ. The function is drawn for the particular case $\rho = E_\eta / E_\xi = 1$ and for three values of $\Delta\phi$. Such a parameter has been assumed to be in the range between 45 to 135 degrees to avoid singularities.

the modulating signal. The latter approach is illustrated in Figure 9.12, where U_R and U_I are drawn versus the rotation angle θ and for $\Delta\varphi$ in the 45 to 135 degrees range. It is interesting to note that $\Delta\varphi = 90$ degrees implies a purely circular polarization and thus both U_R and U_I are independent of the rotation angle.

9.9 Conclusions

In this chapter the basic technologies for chipless RFID tags were reviewed. Then a novel zero-power wireless sensor was conceived. The sensor is based on harmonic generation and orthogonally polarized waves. The information is encoded in the phase difference or in the amplitude imbalance between two signals that are transmitted in orthogonal polarization. The mathematical development shows that the rotation angle between the sensor and the reader can also be retrieved on the basis of measurable data. Although the developed theory is absolutely general, practical applications could appropriately be targeted in the UHF frequency range ($f_0 = 960$ MHz and $2 f_0 = 1920$ MHz) or in the

WLAN band ($f_0 = 2.45$ GHz and $2f_0 = 4.90$ GHz). The results obtained demonstrate the feasibility of the proposed zero-power chipless sensor tag and of the corresponding reader electronics.

References

[1] C. Sauer, M. Stanacevic, G. Cauwenberghs, and N. Thakor, "Power harvesting and telemetry in CMOS for implanted devices," *IEEE Trans. on Circuits and Systems I*, **52**, (*12*), 2605–2613, 2005.

[2] K. Opasjumruskit, T. Thanthipwan, O. Sathusen, *et al.*, "Self-powered wireless temperature sensors exploit RFID technology," *IEEE Pervasive Computing*, **5**, (*1*), 54–61, 2006.

[3] H. M. Lu, C. Goldsmith, L. Cauller, and J.-B. Lee, "MEMS-based inductively coupled RFID transponders for implantable wireless sensor applications," *IEEE Trans. on Magnetics*, **43**, (*6*), 2412–2414, 2007.

[4] C. Occhiuzzi and G. Marrocco, "The RFID technology for neurosciences: Feasibility of limbs' monitoring in sleep diseases," *IEEE Trans. on Information Technology in Biomedicine*, **14**, (*1*), 37–43, 2010.

[5] A. Vaz, A. Ubarretxena, I. Zalbide, *et al.*, "Full passive UHF tag with a temperature sensor suitable for human body temperature monitoring," *IEEE Trans. on Circuits and Systems II: Express Briefs*, **57**, (*2*), 95–99, 2010.

[6] S. R. Munnangi, G. Haobijam, M. Kothamasu, R. Paily, and R. S. Kshetrimayum, "CMOS capacitive pressure sensor design and integration with RFID tag for biomedical applications," in *TENCON 2008*, Hyderabad, pp. 1–6, Nov. 2008.

[7] B. Todd, M. Phillips, S. M. Schultz, *et al.*, "Low-cost RFID threshold shock sensors," *IEEE, Sensors Journal*, **9**, (*4*), 464–469, 2009.

[8] B. Todd, M. Phillips, S. M. Schultz, *et al.*, "RFID threshold accelerometer," *AUTOTESTCON, IEEE*, pp. 55–59, 811, Sept. 2008.

[9] A. Sample, D. Yeager, P. Powledge, A. Mamishev, and J. Smith, "Design of an RFID-based battery-free programmable sensing platform," *IEEE Trans. on Instrumentation and Measurement*, **57**, (*11*), 2608–2615, 2008.

[10] R. Bhattacharyya, C. Floerkemeier, and S. Sarma, "Low-cost, ubiquitous RFID-tag-antenna-based sensing," *Proceedings of the IEEE*, **98**, (*9*), 1593–1600, 2010.

[11] M. Kim, K. Kim, and N. Chong, "RFID based collision-free robot docking in cluttered environment," *Progress In Electromagnetics Research*, **110**, 199–218, 2010.

[12] K. Finkenzeller, *RFID Handbook: Fundamentals and Applications in Contactless Smart Cards and Identifcation*, John Wiley & Sons, Inc., 2003.

[13] J. Hagerty, F. Helmbrecht, W. McCalpin, R. Zane, and Z. Popovic, "Recycling ambient microwave energy with broad-band rectenna arrays," *IEEE Trans. on Microwave Theory and Techniques*, **52**, (*3*), 1014–1024, 2004.

[14] A. Costanzo, A. Romani, D. Masotti, N. Abizzani, and V. Rizzoli, "RF/baseband co-design of switching receivers for multiband microwave energy harvesting," *Sensors and Actuators A: Physical*, **179**, 158–168, (Jun.) 2012.

[15] E. Cantatore, T. Geuns, G. Gelinck, *et al.*, "A 13.56 MHz RFID system based on organic transponders," *IEEE Journal of Solid State Circuits*, **42**, (*1*), 84–92, 2007.

[16] V. Subramanian, J. Frechet, P. Chang, *et al.*, "Progress toward development of all-printed RFID tags: Materials, processes, and devices," *Proceedings of the IEEE*, **93**, (*7*), 1330–1338, 2005.

[17] S. Steudel, K. Myny, V. Arkhipov, *et al.*, "50 MHz rectifier based on an organic diode," *Nature Materials*, **4**, 597–600, (Aug.) 2005.

[18] E. Fortunato, N. Correia, P. Barquinha, *et al.*, "High-performance flexible hybrid field-effect transistors based on cellulose fiber paper," *IEEE Electron Device Letters*, **29**, (*9*), 988–990, 2008.

[19] T. Sekitani, Y. Noguchi, U. Zschieschang, H. Klauk, and T. Someya, "Organic transistors manufactured using inkjet technology with sub-femtoliter accuracy," *Proceedings of the National Academy of Sciences of the USA*, **105**, (*13*), 4976–4980, (Apr.) 2008.

[20] M. Cardinali, L. Valentini, J. Kenny, and I. Mutlay, "Graphene based composites prepared through exfoliation of graphite platelets in methyl methacrylate/poly(methyl methacrylate)," *Polymer International*, **61**, (*7*), 1079–1083, 2012.

[21] A. Loi, L. Basirico, P. Cosseddu, *et al.*, "Organic bendable and stretchable field effect devices for sensing applications," *IEEE Sensors Journal*, **13**, (*12*), 4764–4772, 2013.

[22] F. Arca, M. Sramek, S. F. Tedde, P. Lugli, and O. Hayeden, "Near-infrared organic photodiodes," *IEEE Journal of Quantum Electronics*, **49**, (*12*), 1016–1025, 2013.

[23] P. Cosseddu, S. Lai, M. Barbaro, and A. Bonfiglio, "Ultra-low voltage, organic thin film transistors fabricated on plastic substrates by a highly reproducible process," *Appl. Phys. Lett.*, **100**, (*9*), 093 305–093 305–5, (Feb.) 2012.

[24] L. Yang, A. Rida, R. Vyas, and M. Tentzeris, "RFID tag and RF structures on a paper substrate using inkjet-printing technology," *IEEE Trans. on Microwave Theory and Techniques*, **55**, (*12*), 2894–2901, 2007.

[25] L. Yang, L. J. Martin, D. Staiculescu, C. P. Wong, and M. Tentzeris, "Conformal magnetic composite RFID for wearable RF and biomonitoring applications," *IEEE Trans. on Microwave Theory and Techniques*, **56**, (*12*), 3223–3230, 2008.

[26] V. Lakafosis, A. Rida, R. Vyas, *et al.*, "Towards the first wireless sensor networks consisting of inkjet-printed, paper-based RFID-enabled sensor tags," *Proceedings of the IEEE*, **98**, (*9*), 1601–1609, 2010.

[27] G. Orecchini, V. Palazzari, A. Rida, *et al.*, "Design and fabrication of ultra-low cost radio frequency identification antennas and tags exploiting paper substrates and inkjet printing technology," *IET Microwave Antennas & Propagation*, **5**, (*8*), 993–1001, (Jun.) 2011.

[28] M. Nelo, A. Sowpati, V. Palukuru, J. Juuti, and H. Jantunen, "Utilization of screen printed low curing temperature cobalt nanoparticle ink for miniaturization of patch antennas," *Progress In Electromagnetics Research*, **127**, 427–444, 2012.

[29] L. Basirico, P. Cosseddu, A. Scida, *et al.*, "Electrical characteristics of ink-jet printed, all-polymer electrochemical transistors," *Organic Electronics*, **13**, (*2*), 244–248, 2012.

[30] Stevan Preradovic and Nemai Karmakar, "Fully printable chipless RFID tag," in *Advanced Radio Frequency Identification Design and Applications,* Dr Stevan Preradovic (Ed.), ISBN: 978953–307–168–8, 2011, InTech, DOI: 10.5772/14555. Available from: http://www.intechopen.com/books/advanced-radio-frequency-identification-design-and-applications/fully-printable-chipless-rfid-tag.

[31] X. Jingtian, Y. Na, C. Wenyi, *et al.*, "Low-cost low-power UHF RFID tag with on-chip antenna," *Journal of Semiconductors*, **30**, (*7*), 075 012/1–075 012/6, 2009.

[32] M. Law, A. Bermak, and H. Luong, "A sub-μW embedded CMOS temperature sensor for RFID food monitoring application," *IEEE Journal of Solid-State Circuits*, **45**, (*6*), 1246–1255, 2010.

[33] *Alien Technology Corporation White Paper: Fluidic Self Assembly*, Alien Technology, 1999. [Online], available: http://www.alientechnology.com.

[34] F. Alimenti, M. Virili, G. Orecchini, *et al.*, "A new contactless assembly method for paper substrate antennas and UHF RFID chips," *IEEE Trans. on Microwave Theory and Techniques*, **59**, (*3*), 627–637, 2011.

[35] C. Hertleer, H. Rogier, L. Vallozzi, and L. V. Langenhove, "A textile antenna for off-body communication integrated into protective clothing for firefighters," *IEEE Trans. on Antennas and Propagation*, **57**, (*4*), 919–925, 2009.

[36] X. Li, J. Liao, Y. Yuan, and D. Yu, "Eye-shaped segmented reader antenna for near-field UHF RFID applications," *Progress In Electromagnetics Research*, **114**, 481–493, 2011.

[37] J.-J. Tiang, M. Islam, N. Misran, and J. S. Mandeep, "Circular microstrip slot antenna for dual-frequency RFID application," *Progress In Electromagnetics Research*, **120**, 499–512, 2011.

[38] Y. Amin, Q. Chen, H. Tenhunen, and L.-R. Zheng, "Performance-optimized quadrate bowtie RFID antennas for cost-effective and eco-friendly industrial applications," *Progress In Electromagnetics Research*, **126**, 49–64, 2012.

[39] Y. Amin, Q. Chen, L.-R. Zheng, and H. Tenhunen, "Development and analysis of flexible UHF RFID antennas for 'green' electronics," *Progress In Electromagnetics Research*, **130**, 1–15, 2012.

[40] F. Viani, M. Salucci, G. Olivieri, F. Robol, and A. Massa, "Design of UHF RFID/GPS fractal antenna for logistic management," *Journal of Electromagnetic Waves and Applications*, **26**, (*4*), 480–492, 2012.

[41] S. Tedjini, N. Karmakar, E. Perret, *et al.*, "Hold the chips: Chipless technology, an alternative technique for RFID," *IEEE Microwave Magazine*, **14**, (*5*), 56–65, (Aug.) 2013.

[42] S. Preradovic and N. C. Karmakar, "Chipless RFID: Bar code of the future," *IEEE Microwave Magazine*, **11**, (*7*), 87–97, (Dec.) 2010.

[43] F. Alimenti, P. Mezzanotte, M. Dionigi, M. Virili, and L. Roselli, "Microwave circuits in paper substrates exploiting conductive adhesive tapes," *IEEE Microwave and Wireless Components Letters*, **22**, (*12*), 660–662, 2012.

[44] C. Jones, "Invisible tattoo ink for chipless RFID safe, company says," *EE Times White Paper (Online)*, available: http://eetimes.eu/industrial/196900063 June, 2009.

[45] J. Collins, "RFID fibers for secure applications," *RFID Journal (Online),* available: http://www.rfidjournal.com/article/articleprint/845/-1/1/April, 2006.

[46] J. McVay, A. Hoorfar, and N. Engherta, "Space-filling curve RFID tags," *IEEE Radio and Wireless Symp. Dig*, San Diego, pp. 199–202, 17–19 Jan. 2006.

[47] Tagsense, Inc., "Chipless RFID products. Data Sheet" available: http://www.tagsense.com/ingles/products/product_chipless.html, Oct. 2006.

[48] S. Preradovic, I. Balbin, N. C. Karmakar, and G. Swiegers, "A novel chip-less RFID system based on planar multiresonators for barcode replacement," *IEEE International Conference on RFID*, pp. 289–296, 16–17 April 2008.

[49] I. Balbin and N. Karmakar, "Novel chipless RFID tag for conveyor belt tracking using multi-resonant dipole antenna," *IEEE Microwave Conference, EuMC 2009 European*, pp. 1109–1112, 29 Sept.–1 Oct. 2009.

[50] R. Vyas, V. Lakafosis, Lee Hoseon, *et al.*, "Inkjet printed, self powered, wireless sensors for environmental, gas, and authentication-based sensing," *IEEE Sensors Journal*, **11**, (*12*), 3139–3152, 2011.

[51] C. Mandel, M. Schussler, M. Maasch, and R. Jakoby, "A novel passive phase modulator based on LH delay lines for chipless microwave RFID applications," in *IEEE MTT-S*

International Microwave Workshop on Wireless Sensing, Local Positioning, and RFID, IMWS 2009, pp. 1–4, 24–25 Sept. 2009.

[52] I. Balbin and N. C. Karmakar, "Phase-encoded chipless RFID transponder for large-scale low-cost applications," *IEEE Microwave and Wireless Components Letters*, **19**, (*8*), 509–511, 2009.

[53] S. Mukherjee, "Chipless radio frequency identification by remote measurement of complex impedance," *European Microwave Conference*, pp. 1007–1010, 9–12 Oct. 2007.

[54] G. Orecchini, L. Yang, A. Rida, *et al.*, "Green technologies and RFID: Present and future," *Applied Comput. Electromagnetics Society Journal*, **25**, (*3*), 230–238, 2010.

[55] S. Steudel, S. D. Vusser, K. Myny, *et al.*, "Comparison of organic diode structures regarding high-frequency rectification behavior in radio-frequency identification tags," *J Appl. Phys.*, **99**, (*11*), 114519, (Jun.) 2006.

[56] L. Valentini and J. Kenny, "Novel approaches to developing carbon nanotube based polymer composites: Fundamental studies and nanotech applications," *Polymer*, **46**, (*17*), 6715–6718, (Aug.) 2005.

[57] V. Marinov, Y. Atanasov, A. Khan, *et al.*, "Direct-write vapor sensors on FR4 plastic substrates," *IEEE Sensor Journal*, **7**, (*6*), 937–944, 2007.

[58] T. Unander and H.-E. Nilsson, "Characterization of printed moisture sensors in packaging surveillance applications," *IEEE Sensor Journal*, **9**, (*8*), 922–928, 2009.

[59] S. Couderc, B. Kim, and T. Someya, "Cellulose-based composite as a raw material for flexible and ultra-lightweight mechanical switch devices," in *IEEE 22nd International Conference on Micro Electro Mechanical Systems*, Sorrento (Italy), pp. 646–649, Jan. 2009.

[60] M. Bozzi, A. Georgiadis, and K. Wu, "Review of substrate-integrated waveguide circuits and antennas," *IET Microwaves, Antennas & Propagation*, **5**, (*8*), 909–920, (Jun.) 2011.

[61] F. Alimenti, P. Mezzanotte, L. Roselli, and R. Sorrentino, "A revised formulation of modal absorbing and matched modal source boundary conditions for the efficient FDTD analysis of waveguide structures," *IEEE Trans. on Microwave Theory and Techniques*, **48**, (*1*), 50–59, 2000.

[62] F. Alimenti, P. Mezzanotte, G. Tasselli, *et al.*, "Development of low-cost 24-GHz circuits exploiting system-in-package (SiP) approach on commercial PCB technology," *IEEE Trans. on Components, Packaging and Manufacturing Technology*, **2**, (*8*), 1265–1274, 2012.

[63] F. Alimenti and L. Roselli, "Theory of zero-power RFID sensors based on harmonic generation and orthogonally polarized antennas," *Progress in Electromagnetic Research*, **134**, 337–357, 2013.

[64] B. Colpitts and G. Boiteau, "Harmonic radar transceiver design: Miniature tags for insect tracking," *IEEE Trans. on Antennas and Propagation*, **52**, (*11*), 2825–2832, 2004.

[65] W.-H. Tu, M.-Y. Li, and K. Chang, "Broadband microstrip-coplanar stripline-fed circularly polarized spiral antenna," in *IEEE International Antennas and Propagation Symposium Digest*, Albuquerque (USA), pp. 3669–3672, Oct. 2006.

[66] S. Maas, *Nonlinear Microwave and RF Circuits*, 2nd Edn. Artech-House, Inc., 2003.

[67] G. Monti, R. D. Paolis, and L. Tarricone, "Design of a 3-state reconfigurable CRLH transmission line based on MEMS switches," *Progress In Electromagnetics Research*, **95**, 283–297, 2009.

[68] J.-J. Lyu and T.-L. Chen, "Optimize a RFID-based turbine maintenance model – a preliminary study," in *IEEE International Conference on Industrial Engineering and Engineering Management*, Singapore, pp. 501–505, Nov. 2008.

[69] R. Collin, *Antennas and Radiowave Propagation*, McGraw-Hill, 1985.

10 Energy-efficient off-body communication using textile antennas

Luigi Vallozzi, Patrick Van Torre, and Hendrik Rogier

10.1 Introduction

Wearable RFID systems should be low-profile, lightweight, flexible, and breathable, in order not to hinder the user's body movements. Planar textile antennas are a perfect choice for integration into clothing as they meet all the above requirements. Moreover, textile antennas are composed of environmentally friendly materials.

Exploiting the large platform garments have to offer, multiple antennas are readily implemented onto the human body. The high radiation efficiency of the proposed textile antennas, combined with a multi-antenna setup, provides a solution to the "energy problem" by creating very reliable communication links at a low transmitted power. Therefore, textile antennas are an excellent choice for green RFID applications.

Multi-antenna systems increase communication reliability using several techniques. A MIMO technique such as space–time block coding provides diversity gain, which mitigates the effects of fading by the environment. The use of dual-polarized textile antennas doubles the number of exploitable signal paths for an equal number of antenna patches on the body. Beam forming techniques can concentrate the transmitted power towards the receiver, which potentially results in a performance increase for an off-body RFID system at a reduced cost, compared to space–time coding techniques for transmit diversity used in MIMO systems.

In this chapter, the basics of textile antenna design are outlined, including design examples. Using the described textile antennas, a number of applications are documented for off-body RFID transmission links in indoor environments. Further application domains are highlighted and suggestions for future work are proposed.

10.2 Basics of textile antenna design

10.2.1 Textile materials/characterization

Textile materials, easily available on the market, are used as a main component of textile wearable antennas. Since an antenna necessarily requires electrically conductive parts, the use of special types of conductive textiles (in contrast to most fabrics that are classically insulating) is required. Such materials are known as *electrotextiles* and they are constructed by coating dielectric yarns with a conductive material. Their appearance and mechanical properties are very similar to common textiles, showing electrical conductivity as an additional property. In addition to electrotextiles, low-loss insulating textile materials may be required as a component in textile antennas, for example as a substrate of a textile patch antenna: many materials are commercially available and very suitable for this purpose, such as felt, aramid, shock absorbing foams, etc.

Such materials typically exhibit a value of dielectric permittivity ε_r between 1 and 2, and a very low value of tan δ (in the order of 10^{-3}). For antenna design, an accurate dielectric characterization of textile materials is required first, in order to determine the values of ε_r and tan δ. Several techniques have been developed for this purpose, such as the one proposed by Declercq *et al.* [1]. The technique relies on two transmission lines that are implemented on the material under test, which acts as a dielectric substrate. Based on a matrix-pencil method, the dielectric properties of the textile substrate are then extracted from the measured S-parameters.

10.2.2 Literature overview of textile antennas

Textile antennas emerged as a novel research field in the last decade. The first antenna for wearable applications, though not completely made out of textiles, was introduced in 1999 by Salonen *et al.* [2], consisting of a planar inverted-F antenna for GSM applications with a copper-foil patch placed on a printed circuit board, suitable for integration into garments. Next, Massey of Philips laboratories proposed, in 2001, the first antenna completely made out of textile materials [3]. After that, several research groups started to work on the subject, proposing other novel textile antennas, as described in [4–6]. Moreover, adverse effects on antenna performance were the object of several investigations, such as the vicinity of the human body [7, 8], antenna bending [9, 10], humidity [11], etc.

10.2.3 Design examples

In this section two significant examples of textile antenna design are considered, in particular a *dual-polarized textile patch antenna* and a *textile patch antenna array*. Both are realized completely out of textile or wearable materials, can be completely integrated into protective garments and show resistance to harsh environmental conditions. The dual-polarized antenna enables the exploitation of second-order polarization diversity with a single patch antenna, whereas the array can produce a very directional radiation pattern with high gain and steering capabilities.

10.2.3.1 Dual-polarized textile patch antenna

General description: this textile antenna consists of a microstrip rectangular patch on a flexible substrate, operating in the 2.45 GHz ISM band, entirely made out of textile and wearable materials, such as a highly-conductive electrotextile, *Flectron*, produced by *LessEMF*, for the patch and ground plane, and flexible protective foam for the substrate. The main feature of such an antenna is its dual-polarization operation. This means that the antenna enables simultaneous transmission or reception of two linearly polarized

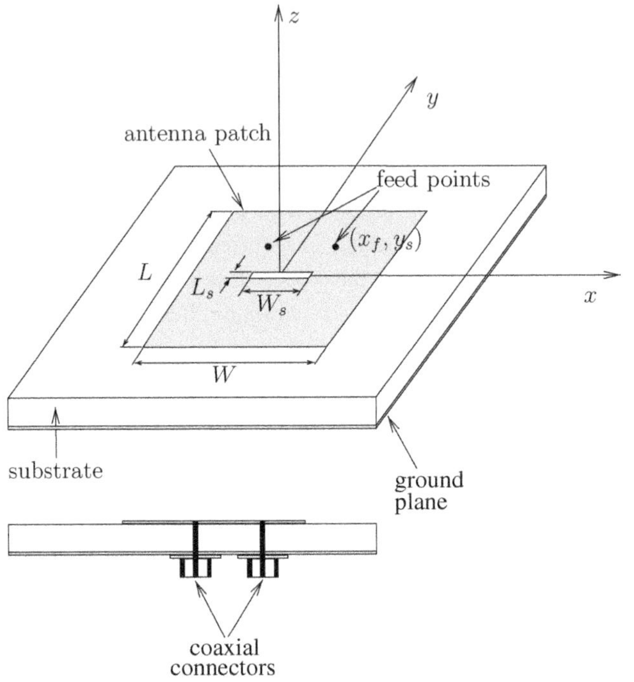

Figure 10.1 Dual-polarized patch antenna.

orthogonally polarized waves, by means of two separated and highly isolated antenna ports, as shown in the antenna scheme in Figure 10.1.

This feature is particularly interesting for communication systems implementing *diversity* at the receive and/or transmit side, as an effective means to combat fading in multipath environments. Moreover, the materials employed for this antenna are robust to harsh environmental conditions, such as high relative humidity levels and high temperatures, making it suitable for rescue workers' applications.

The antenna has been designed in open-space conditions and, after that, its performance, in terms of reflection coefficient, gain pattern, and polarization, was studied also in close proximity to a human wearer, showing that the body has only a moderate effect on the overall performance.

Topology, materials, and design: the adopted topology is a simple nearly square patch with the two coaxial feed points nearly located on the two diagonals, which ensures orthogonality of the two radiated waves. A slot is present at the center of the patch, whose dimensions have been chosen with the aim of improving the matching of the feed lines to the antenna ports. In order to achieve impedance matching (for $Z = 50\ \Omega$) at both antenna ports, and isolation between the two transmitted/received signals, two conditions must be enforced as design goals. In particular, in the frequency band $2.4 \leq f \leq 2.4835$ GHz (2.45 GHz ISM band), it is required that:

$$\begin{aligned} \text{Reflection coefficients: } |S_{11}|,\ |S_{22}| &< 10\,\text{dB}, \\ \text{Isolation: } |S_{21}|,\ |S_{12}| &< -15\,\text{dB}. \end{aligned} \tag{10.1}$$

Table 10.1 Parameters of the realized antenna.

Optimized parameters	patch length	$L = 44.46$ mm
	patch width	$W = 45.32$ mm
	slot length	$L_s = 14.88$ mm
	feed points	$(\pm xf, yf) = (\pm 5.7, 5.7)$ mm
Fixed parameters	slot width	$W_s = 1$ mm
	substrate height	$h = 3.94$ mm
	substrate permittivity	$\varepsilon_r = 1.53$, $\tan\delta = 0.0012$
	conductive materials	ShieldIt® (patch) Flectron® (ground plane)
	substrate material	protective closed-cell foam
	feed connectors	SMA

The materials chosen for the antenna realization are Flectron, a highly conductive electrotextile, for the patch and ground plane, and flexible, fire-retardant, shock absorbing foam, for the substrate. This last material is commonly used in protective rescue-workers' clothing, in order to protect vulnerable parts of the human body from shocks and high external temperatures.

The dimensions of the antenna patch and the slot were optimized in order to meet the design goals in Equation (10.1) by means of the EM simulation software ADS Momentum.

The geometrical and material parameters of the antenna, including the final values resulting from the optimization, are displayed in Table 10.1.

10.2.3.2 Textile antenna array

This novel wearable array is composed of four equilateral triangular microstrip patch antennas (ETMPA) and operates in the 2.45 GHz ISM band. The main advantage over a single textile patch antenna lies in its increased directivity and narrower beam width in the radiation pattern, which makes the antenna effective in mitigating fading in a multipath environment. In addition, by using an appropriate feed network, the array's main beam can be easily steered along the elevation plane, which allows enhancing of the reliability of indoor communication between a moving subject equipped with the array and a fixed base station. Figure 10.2 shows the realized prototype and how this was integrated into a rescue-worker's jacket worn on a human subject.

10.3 Off-body links relying on space–time coding and textile antennas

10.3.1 Introduction

Many exciting smart fabrics/interactive textiles applications require off-body RFID data communication links using textile antennas integrated into clothing. The reliability of such a communication link is compromised by shadowing and multipath fading in the indoor radio propagation environment. Multipath and shadowing effects make the signal levels fluctuate severely while the mobile users move around. The quality of the link can be greatly improved using multiple input multiple output (MIMO) techniques that

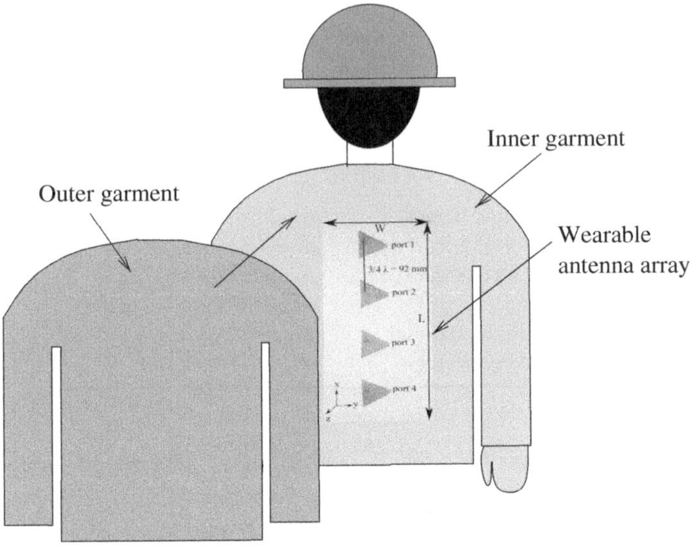

Figure 10.2 Textile patch antenna array.

generate or combine signals transmitted and/or received by multiple antennas. In the same indoor environment, SIMO systems, generating receive diversity gain only, were studied in [12, 13]. In contrast, MIMO systems provide diversity gain at both link ends and also enable beam forming and spatial multiplexing. Off-body MIMO links were recently studied in [14–17].

Wearable textile multi-antenna systems allow a convenient implementation of wireless off-body MIMO communication systems. The body area provides a large platform to deploy multiple antennas, enabling MIMO communication, allowing the full exploitation of spatial and pattern diversity. By deploying dual-polarized antennas [18, 19], two signals are handled on each single compact antenna patch, further increasing the MIMO system's degrees of freedom and the overall performance of the communication link.

The antennas in this chapter operate in the 2.45 GHz ISM (Industrial, Scientific, and Medical) band; the corresponding wavelength of 12 cm allows the design of comfortably sized antenna patches for off-body communication.

In the scenario described below, a pair of dual-polarized antennas is integrated in a firefighter jacket, such that the antennas are worn on the front and back of the body. A similar set of antennas is used at the base station to obtain a 4x4 MIMO link [20] with pairs of dual-polarized antennas at both link ends. At the base station, the antennas are separated by a distance of several wavelengths.

The performance of the MIMO link is evaluated using test persons of very different sizes [20], whose characteristics are listed in Table 10.2.

10.3.2 Measurement setup

The wireless test bed at the transmit and receiving ends of the wireless link ensures synchronous transmission and reception on four channels. The transmitted frames contain a

Table 10.2 Size and weight of the test persons.

Test person	Height [m]	Weight [kg]
Small	1.67	52
Medium	1.75	73
Large	1.85	104

sequence of BPSK (binary phase shift keying) modulated pilot symbols, non-overlapping in time, for each transmit antenna. The pilot symbols are followed by a QPSK (quadrature phase shift keying) modulated data block, transmitted using an orthogonal space–time code [21, 5]. Four signals are transmitted using both antenna feeds of both dual-polarized antennas deployed on the human body, in order to enable the evaluation of a real 4x4 MIMO system in operation [20]. At the base station, the reception is also performed on four channels. After conversion to baseband, the synchronously sampled data are recorded on a hard disk, providing full flexibility for later analysis.

In order to assess MIMO links of varying order, the SISO channels composing the MIMO links are estimated based on the received pilot symbols. As the transmission is performed in a 4x4 configuration, 16 SISO channels must be estimated. Each SISO channel represents a 1x1 link. For the 2x2 links, the SISO channels composing the MIMO link are combined to enable a performance evaluation when applying the Alamouti [22] space–time code.

Measurements are performed along a path with non-line-of-sight propagation. This path is situated at a distance of approximately 17 m from the transmitter and oriented perpendicular to the direction of the transmitter. This sideways path is obstructed by brick walls and office equipment, causing fading and shadowing.

10.3.3 Measurement results

Statistics, average bit error rate graphs, and real-time performance are assessed based on the measured channel gains for the 1x1 link as well as for 2x2 and 4x4 MIMO communication. The demodulation of the transmitted data allows real-time error performance evaluation of the practical system in operation, including the effects of channel estimation errors and Doppler shift. Since in the experimental transmission the data were transmitted using a fourth-order space–time code [21, 5], the demodulation of data is limited to MIMO configurations with fourth-order transmit diversity. In a practical situation the performance is suboptimal (worse than predicted by theoretical Rayleigh fading characteristics in ideal conditions) due to signal correlation, unequal gain, and shadowing. For off-body communication, unequal gain is partially caused by the shadowing effect of the body, combined with the directional radiation pattern of the textile patch antennas. The shadowing is different for persons of different size.

To assess the influence of a person's body size, a measurement series was performed with three persons of substantially different stature and weight. The results illustrate that the gains obtained by switching to higher-order MIMO configurations are similar for all

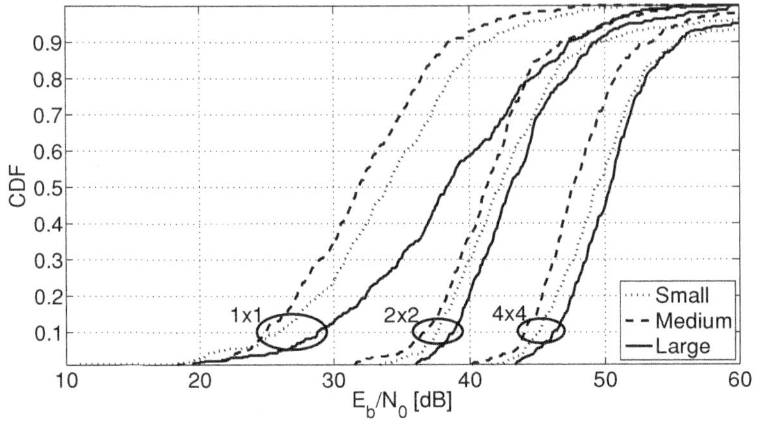

Figure 10.3 Cumulative distribution functions for the median 1x1, 2x2, and 4x4 channels, for each test person.

three persons. In this measurement the transmit power was set to a total transmit power of 100 mW.

A small remark should be made here concerning the limits of maximum power that can be absorbed by the human body, as imposed by international safety regulations: they allow a maximum transmit power of 20 mW, which corresponds to the allowed SAR limit of 2 W/kg averaged over 10 g of tissue [IEC 62209–1]. The value of 100 mW used in our measurement campaign, which is substantially higher than the above-mentioned limit, was chosen exclusively in order to obtain higher SNR and accuracy on the measurements. This is not a problem, as thanks to the directional radiation pattern of the patch antenna with ground plane near the body, the user is very much shielded from this radiation. However, in a real application, a substantially lower power, complying with the safety limits, independent of the antenna used, is sufficient to establish a reliable wireless communication link in the described situation.

10.3.3.1 CDF and outage probability

For each person a set of 325 frames is captured. The cumulative distribution functions (CDFs) displayed in Figure 10.3 represent the probability of receiving a signal with a given signal-to-noise ratio (SNR $= 2E_b/N_0$ for a QPSK transmission). A curve is presented for each of the differently sized test persons and for 1x1, 2x2, and 4x4 MIMO links. For the 1x1 and 2x2 transmissions, respectively, 16 and 36 potential configurations are assessed. From the calculated CDFs, the cases with the median 10% outage probability level are selected. This was preferred instead of the average for a set of configurations, as the median case corresponds to a real physical SISO or MIMO channel of the given order.

The 10% outage probability level (signal level corresponding to 0.1 in the CDF) illustrates the similarity of the channel fading experienced by the different persons. The CDF curves clearly form groups according to the order of the MIMO system considered. Relatively small differences result from a combination of many factors and should not

Table 10.3 10% outage probability for different test persons.

				SISO/MIMO [dB]		
Size	Height [m]	Weight [kg]	Statistic	1x1	2x2	4x4
			MINIMUM	11.1	21.7	45.2
Small	1.67	52	MEDIAN	26.3	37.7	45.2
			MAXIMUM	38.8	45.0	45.2
			MINIMUM	12.3	21.2	44.1
Medium	1.75	73	MEDIAN	24.8	36.7	44.1
			MAXIMUM	37.5	43.8	44.1
			MINIMUM	12.8	23.2	46.2
Large	1.85	104	MEDIAN	28.0	38.8	46.2
			MAXIMUM	40.0	46.1	46.2

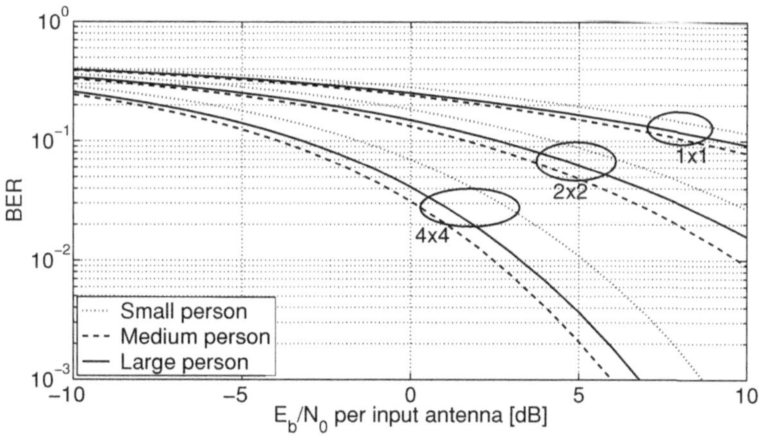

Figure 10.4 Bit error characteristics for 2x2 and 4x4 MIMO communication with QPSK modulation, for persons of different size (small, medium, large), compared to the performance of the average SISO (1x1) link.

be directly linked to body weight or height. The minimum, median, and maximum 10% outage probability values for the SISO and MIMO links are listed in Table 10.3.

Although the distribution of the signal levels is influenced by shadowing, MIMO communication offers a substantial performance gain for each increase in the degree of diversity. Based on the median values for all three persons, the 2x2 system offers 10.8–11.9 dB of diversity gain, compared to the 1x1 system. Similarly, the 4x4 system offers 15.3–18.9 dB diversity gain, compared to the 1x1 system. The gain observed by switching to a higher-order MIMO system is similar for all three persons of substantially different size.

10.3.3.2 Bit error characteristics

The BER curves in Figure 10.4 are calculated based on the set of channel estimation values, determined by using the received pilot symbols. The BER is then calculated for the different MIMO orders and the applied space–time codes following a method further detailed in [20]. The curves also display a significant improvement in performance for each increase in the link's diversity order. These results are consistent for all three test persons. A spread of up to 3 dB is present when comparing the BER characteristics of the same diversity order between different test persons. Nevertheless, the curves clearly form groups corresponding to the given MIMO orders.

When ordering the persons of different size as a function of bit error performance, the order is different when compared to the CDF in Figure 10.3. For the CDF, the absolute E_b/N_0 received by the fireman is important, but for the BER curve (as a function of E_b/N_0) it is not. Again it is more interesting to focus on the performance increase for each increment of the MIMO order, which is always very significant and comparable for the different persons.

Some difference is unavoidable as the persons can never walk exactly the same trajectory. For the performance of an off-body communication system in real life, variation in performance is to be expected due to many factors, including the environment, body posture, and the specific path followed. When the antennas are worn at a different height above the floor (due to body size) they consistently experience a different fading pattern, even for a random walk. Body mass is not expected to have a large impact as the torso of even a small adult is still much wider than the size of the antenna patch.

10.3.3.3 Real-time error performance

While the BER curves display the average performance of the link over an extended period of time, the real-time error performance is an important indication of the system's behavior over a short time-span. In order to compare 1x1, 2x2, and 4x4 links, the real-time performance is assessed by determining the SNR and BER for each transmitted frame. The BER is calculated for QPSK modulation, based on the SNR of the received pilot symbols. The SISO channel powers are combined according to the space–time code considered and the resulting MIMO channel SNRs are calculated per frame for a given average SNR per input antenna. The BERs are then computed as function of the resulting SNR.

Figure 10.5 illustrates the real-time behavior of the E_b/N_0 ($= \frac{1}{2}$ SNR) and BER for the different MIMO systems, based on the measurements for the large firefighter. Only the first 100 consecutive frames are selected out of the measurement set, to produce a more readable graph.

Note that the total transmitted power on all antennas used is kept equal to 100 mW. Hence, there is no array gain at the transmitter. However, the highest SNR nearly always corresponds to the highest-order MIMO link for each single frame. The dips in the signal are always much smaller for the higher-order MIMO systems. The graph displays the median 1x1 and 2x2 channel combinations. The average received E_b/N_0 ratio is set to 10 dB for the selected median 1x1 channel by adding noise to the measurement data. This is necessary because the actual measurement is performed at a high total transmit power, equal to 100 mW, producing average received E_b/N_0 levels over 30 dB. Adding noise lowers the average E_b/N_0 in the received samples and produces bit errors; the effect is equivalent to lowering the transmitted power. The bar graph for the BER

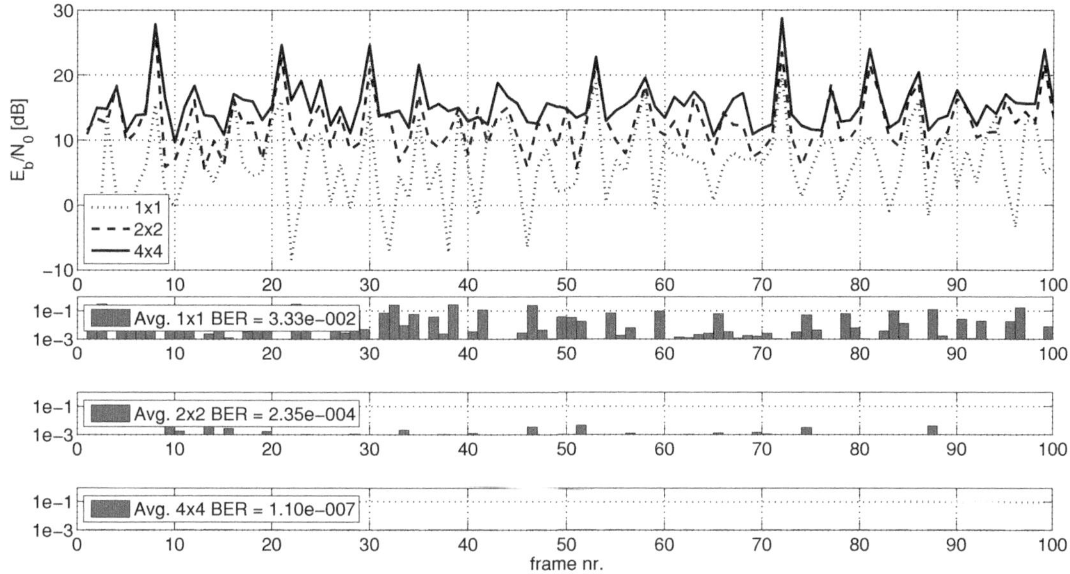

Figure 10.5 SNR and BER as a function of time for the large fireman. Median 1x1 channel with 10 dB average E_b/N_0, median 2x2 MIMO channel combination, and 4x4 MIMO link. The total transmitted power on all antennas is equal for all cases.

illustrates that frame by frame, the higher-order MIMO links perform significantly better than the lower order.

10.4 Off-body beam forming versus space–time coding

10.4.1 Experimental setup

The performance of beam forming versus space–time coding using body-worn textile antenna arrays is experimentally evaluated for an indoor environment, where a walking person transmits RFID data in the 2.45 GHz ISM band, relying on two vertical textile four-antenna arrays integrated into the front and back of his garment. The two transmission scenarios considered are static beam forming at low elevation angles and space–time code-based transmit diversity. Signals are received by a base station equipped with a horizontal array of four dipole antennas providing spatial receive diversity through maximum-ratio combining. Signal-to-noise ratio, bit error rate characteristics, and signal correlation properties are assessed for both off-body transmission scenarios.

The indoor propagation environment introduces space- and time-varying channel behavior, including fading, shadowing, and non-constant path loss. The movements of the person wearing the communication system will cause varying shadowing by the human body, as well as antenna reorientation. As the reliability of the link is of paramount importance for RFID applications, systems consisting of multiple textile antennas distributed over the body of the wearer are used to improve the robustness of the link. Such an enhancement is possible by employing transmit and/or receive diversity, using space–time coding and maximal ratio combining, respectively.

Diversity systems reduce the variance of the resulting signal at the detector, leading to better bit error characteristics. Receive diversity also results in array gain, producing a higher average signal-to-noise ratio at the detector. Transmit diversity in practice does not provide array gain.

Transmit beam forming techniques are able to produce a higher average signal-to-noise ratio at the receiver by concentrating the available transmission power towards the receiver. Dynamic beam forming techniques, using channel estimates to determine the optimal set of transmitted signals, theoretically offer the best performance. Their implementation for off-body communication with a moving person requires high-rate channel feedback, increasing the complexity, cost, and power-consumption of battery-operated systems beyond acceptable limits.

Therefore, measurements are presented for a static off-body beam forming system. This system includes two vertically oriented uniform linear arrays consisting of four textile patch antennas [23] at the front and back of the human body, greatly reducing the body's shadowing effect.

Figure 10.2 showed the textile antenna array and its position on the human body. A spacing of $\frac{3}{4}\lambda$ between the center points of the antenna elements is used. A tip-truncated equilateral triangular topology is adopted, resulting in low mutual coupling and providing sufficient gain in the broadside direction, which is oriented away from the user. The antenna patches are excited in phase to produce a zero-elevation beam, concentrating the power towards a receiver situated on the same floor. The performance of this beam forming array is compared to a space–time coding system using the same off-body four-element antenna configuration. The corresponding signal-to-noise and bit error characteristics are presented.

A similar measurement campaign with one textile antenna array has already been analyzed and documented earlier [24]. Those measurements were highly influenced by shadowing effects of the human body. Nevertheless, a higher average received signal-to-noise ratio is always obtained by means of beam forming. However, receiver diversity is essential, as the resulting reduction in signal variance is required to produce acceptable bit error characteristics in the demanding indoor propagation environment.

The system shown in Figure 10.6 realizes fourth-order receiver diversity at the access point side using a horizontal array of four vertical dipoles. The dipole antennas are connected to the measurement test bed which down-converts and synchronously samples the signals. The data are then sent to a PC and recorded on a hard disk for further processing in Matlab.

The performance of the off-body array in the indoor environment is illustrated in Figure 10.7. The received SNR is displayed for a range of beam elevation angles. With the antenna array pointed towards the receiver (line-of-sight, beam towards RX), the zero-elevation beam clearly results in the strongest signal. This is also the case when the array is directed away from the receiver (one dominant specular reflection), demonstrating the importance of low-angle reflections in the environment.

Figure 10.6 Measurement equipment at the receiving side, including horizontal four-dipole array.

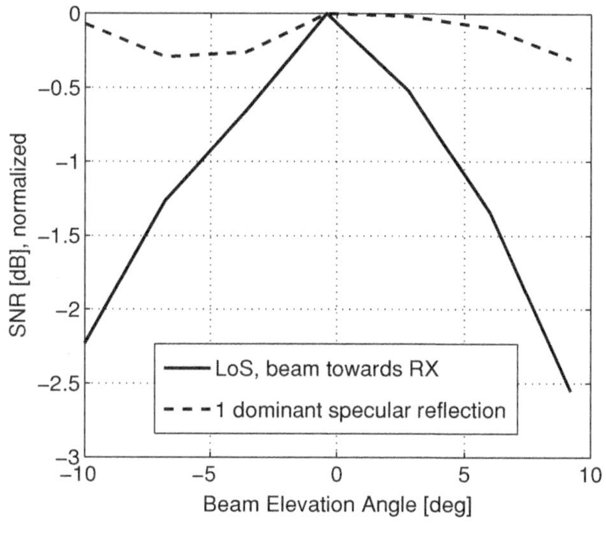

Figure 10.7 SNR for various beam elevation angles under different indoor propagation conditions.

The measurement campaign documents the performance of a system deploying both front and back textile antenna arrays on the human body in similar conditions. This configuration significantly reduces shadowing by the human body. Both arrays are driven with the same signals for equal patch numbers, resulting in a more omnidirectional radiation pattern in the azimuth plane, while still concentrating the power towards low elevation angles. Plots of the time-domain evolution of signal-to-noise ratios for textile array systems are shown in Figures 10.8 and 10.9.

Table 10.4 Received E_b/N_0 for space–time coding and beam forming.

Statistic	Space–time coding [dB]	Beam forming [dB]	Beam forming power gain [dB]
Non-line-of-sight, no receive diversity			
Min	34.1	19.3	
Avg	44.1	47.0	2.9
Max	50.8	55.4	
Non-line-of-sight, 4th order receive diversity			
Min	41.6	43.9	
Avg	51.0	54.4	3.4
Max	57.1	61.7	
Line-of-sight, no receive diversity			
Min	33.5	37.5	
Avg	53.2	56.0	2.8
Max	59.9	62.3	
Line-of-sight, 4th order receive diversity			
Min	49.7	48.5	
Avg	59.4	62.3	2.9
Max	65.6	68.9	

10.4.2 Analysis of the received signals

10.4.2.1 Beam forming power gain

E_b/N_0 refers to the ratio of the received energy per bit to the one-sided spectral noise density, a value that is equal for space–time coding and beam forming in our experiments, under equal channel attenuation conditions. An overview of the average, minimum, and maximum E_b/N_0 received for different situations is listed in Table 10.4.

Beam forming power gain is the difference between the average E_b/N_0 in dB for beam forming and space–time coding. Because of the concentration of power along low elevation angles, this factor amounts to approximately 3 dB for all measured cases. In [25], a slightly higher beam forming gain results for the line-of-sight situations with the beam directed towards the receiver because the power is more concentrated in a narrower range of azimuth angles when employing only one antenna array. When, for the dual-array system, the front array is directed towards the receiver, half of the power is radiated at the back array, away from the receiver.

10.4.2.2 Time-dependent signal behavior

The E_b/N_0 for subsequent frames received with diversity along the line-of-sight path is shown in Figure 10.8.

Clearly, the shadowing by the human body is mitigated by the dual-antenna array, installed at the front and back of the body. For the two-array measurement considered in this chapter, the behavior of E_b/N_0 is dominated by changing path lines-of-sight and fading. The beam forming transmission produces received signals that are significantly

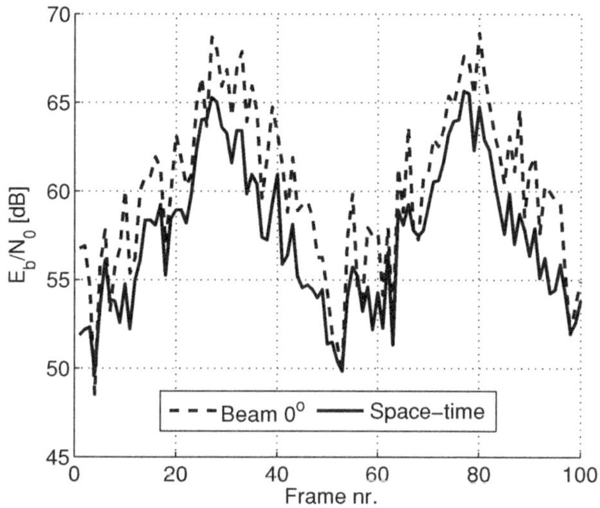

Figure 10.8 E_b/N_0 along the line-of-sight path, with receiver diversity (MRC) for beam forming and space–time coding.

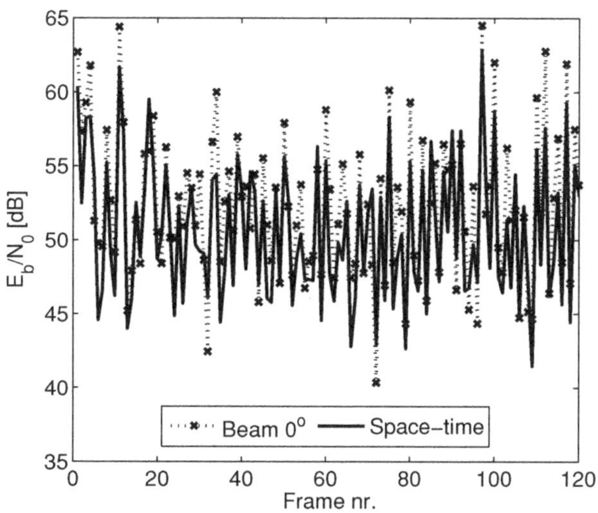

Figure 10.9. E_b/N_0 along the non-line-of-sight path, with receiver diversity (MRC) for beam forming and space–time coding. This graph displays the behavior with only the back antenna array active.

stronger than for space–time coding for nearly every received frame. The average E_b/N_0 is 2.9 dB higher for the beam forming than for the space–time coding.

The non-line-of-sight performance with 4th order receiver diversity is shown in Figure 10.9. The average E_b/N_0 is 3.4 dB higher for the beam forming than for the space–time coding. This value is even higher than for the line-of-sight situation, which indicates that the non-line-of-sight signal propagation over larger distances occurs mainly through reflections and scattering at low elevation angles.

10.5 Energy-efficient channel tracking of off-body communication links

Wearable textile RFID systems are excellent for wireless identification and remote monitoring. In these applications power efficiency is a very important matter. It is desirable for a wearable system to have very low power consumption, especially for autonomous systems for which the necessary energy for the operation of the devices is scavenged from the surrounding environment and ideally no additional power supply is present. In particular, the hardware and software that realize wireless communication between the system and the external world (i.e. wearable antennas, transceiver, and signal processing) must be designed in such a way that they ensure a required value of bit error rate (BER) using the smallest possible amount of power.

Unfortunately, several adverse factors must be overcome in order to realize power-efficient wearable communication systems. The wireless channel in which wearable textile systems are employed often represents a hostile environment, making reliable and power-efficient wireless communication a challenging task to realize. In fact, for body-worn indoor RFID applications in particular, transmitter and receiver are typically in relative motion inside a rich-scattering environment, giving rise to phenomena that are detrimental to power efficiency, such as *fading* and *shadowing*. The combination of such phenomena produces a random fluctuation of the received signal, with deep fades that produce errors in the received data stream. In such cases, to ensure a minimum bound on the error rate, it is necessary to increase the transmit power with respect to the ideal case without fading and shadowing. In order to combat fading and shadowing and to be able to use a lower transmit power for a given BER (thus realizing a power-efficient system) a wearable communication system with the following characteristics is employed:

(1) Multiple wearable antennas are integrated at different locations inside the clothing worn on a human body, implementing **diversity**. Multiple antenna systems with combining schemes represent an effective method to combat channel variability: placing four antennas at the front, back, and right/left sides of the human body mitigates shadowing by the body, and combining the four signals relying on an appropriate scheme allows the minimization of signal fading. Such a diversity scheme requires channel knowledge only at the receive side, which can be obtained by pilot-based channel estimation techniques.

(2) As a second feature for improving the power efficiency of the system, we propose robust **channel tracking** at the receive side. In time-variant channels, data transmission is often performed by transmitting short symbol packets, whose duration is smaller than the channel coherence time (i.e. the time interval during which the channel can be considered constant). Each packet is composed of a pilot symbol sequence for channel estimation (i.e. the packet *overhead*), followed by a data symbol field

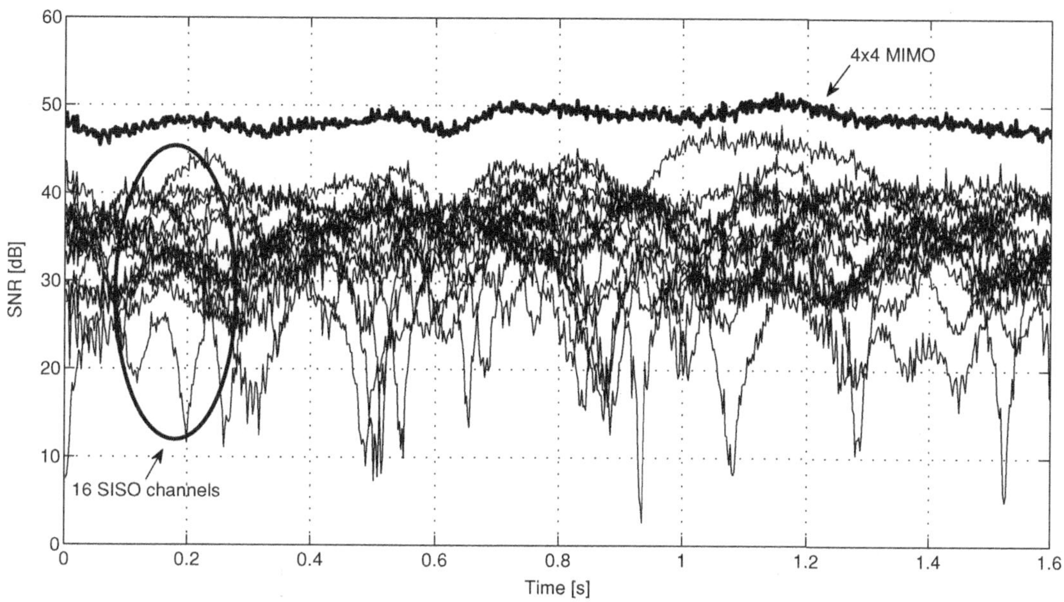

Figure 10.10 Illustrates the performance of the system for one transmitted burst of 1.6 seconds duration.

(i.e. the actual information to transmit). The use of many short packets does not represent a power-efficient solution, since it implies the use of a large relative amount of transmit power for the overhead, which does not contain useful information, but is required for channel estimation only. For this reason, we propose the use of longer packets (with a duration greater than the channel coherence time) combined with dynamic channel estimation or channel tracking, resulting in a reduced overhead with consequent decrease of the required transmit power. The channel tracking is performed by means of a simple algorithm, causing no significant additional power demands for the receiving system.

The system was realized and its performance was investigated by means of a measurement campaign conducted in an indoor office environment. A test person was equipped with the system and, by means of a measurement test bed, a data transmission link was implemented with the test person moving around in the chosen environment. Results, in particular BER curves, showed that antenna diversity in combination with channel tracking allows large savings in transmit power (for a given required BER) and validate the proposed approach as a valid scheme for power-efficient communication systems.

Signals for different receive channels are displayed in Figure 10.10. Four curves are shown for each receive channel, corresponding to the four transmit signals. In total, 16 SISO channels are present, each experiencing an enormous variation in signal-to-noise ratio (SNR). The average bit error rate experienced by a system is predominantly determined by the minima in the SNR curve, which produce the largest share of bit errors. For this reason, the transmit power has to be dramatically increased in order to still achieve an acceptable bit error rate during the signal minima.

Exploiting MIMO techniques to combat fading, as well as taking advantage of the antenna array gain at both sides of the link results in the top curve, labeled 4×4 MIMO. From the graph, two advantages of the MIMO system are clear:

- The average SNR is significantly higher for the MIMO system.
- The SNR variation is dramatically reduced by the MIMO system.

The average SNR at the detector is 12.0 dB higher for the MIMO system, compared to the average SISO channel. This corresponds to two times 6 dB array gain from the four-antenna arrays at the transmitter and the receiver. When the total transmitted power for the MIMO system is equal to the transmitted power for an SISO system, no transmitted array gain is present. In this case 6 dB lower performance increase is obtained.

The minimum SNR at the detector is 42.8 dB higher for the MIMO system, compared to the minimum for the worst SISO channel. This has a profound effect on the average bit error rate. Calculations show that in the case of QPSK modulation, to achieve an acceptable average bit error rate of 10^{-3} for the SISO channels, an average SNR of 26.0 dB is required at the detector. For the MIMO system, an SNR of only 10.4 dB results in the same bit error rate. Taking into account 6 dB receive array gain, this means that the total required transmitted power to achieve a bit error rate of 10^{-3} with the 4x4 MIMO system is 21.6 dB less than for the average SISO system.

Reducing the transmit power over more than two orders of magnitude results in significant energy savings and a longer autonomy for battery-operated communication systems or increased usefulness of energy-scavenging systems. The time-varying channel can be tracked without significant processing overhead at the receiver, as illustrated in [24].

10.6 Application domains

The number of possible applications for off-body RFID systems is virtually unlimited. Some application domains are:

- Monitoring of the location of patients in psychiatric hospitals or at homes for the elderly. Infrastructure in the building can track persons and also provide selective access.
- Selective access control in a large number of organizations, including companies, hospitals, and prisons. Personnel can open and close doors according to their access rights in a very convenient way.
- Restricted use of equipment for police or in the military. Equipment, such as vehicles, can only be operated by people carrying the proper RFID.
- Military applications might include the wireless detection of the presence of given soldiers in a team.
- In sports events, RFID can be used to register the time when people pass a finish line.

10.7 Conclusions

Based on the experimental investigations described in the previous sections, the following conclusions can be drawn:

- From the measurements on off-body links using textile antennas and space–time coding (described in Section 10.3) we conclude that, despite the different body sizes of the test persons, MIMO off-body RFID communication always greatly improves the reliability of a wireless link operating in an indoor multipath propagation environment.
  Cumulative distribution functions allow a comparison of the 10% outage probability levels, providing a quantitative indication of the gain realized by the MIMO system. Measured median gains (relative to a 1x1 system) are 10.8–11.9 dB and 15.3–18.9 dB for the 2x2 and 4x4 systems, respectively. Bit error characteristics are derived for the systems of different MIMO orders. An increase in the number of antennas, going from a 2x2 to a 4x4 MIMO link, results in a significantly better bit error characteristic for all three firefighters under test. Despite the large variation in body size, the BER curves group according to the MIMO order used. This indicates that the performance increase, which is expected by increasing the off-body system's MIMO-order, is effectively obtained for a wide range of body sizes.
- With regards to the performance comparison of off-body beam forming and space–time coding, measurements indicated that without receiver diversity the performance of space–time coding is generally better. In the case of 4th order receiver diversity, beam forming is superior in line-of-sight conditions. For non-line-of-sight propagation, the space–time codes perform better as soon as bit error rates are low enough for a reliable data link.
- Concerning the investigations on energy-efficient channel tracking for off-body communications, the experimental results obtained showed that antenna diversity in combination with channel tracking allows large savings in transmit power (for a given required BER) and validates the proposed approach as a suitable scheme for power-efficient communication systems.

10.8 Future perspectives

Potential future work can use channel knowledge to transmit RFID information with higher efficiency. When the RFID is responding to a transmission from the base station, the channel estimation information based on the signals received at the RFID tag is then used for an efficient transmission specifically towards that base station. The following techniques can be used to achieve this.

10.8.1 Dynamic beam forming

Using a multi-antenna RFID transmit system, the phase relationship between the signals is chosen to produce a radiation pattern with the energy directed mainly towards the base station. Dynamic beam forming allows a significant reduction in transmit power.

10.8.2 Dominant eigenmode transmission

Dominant eigenmode transmission uses the channel information captured during the reception of signals from the base station. Power is then distributed over the different transmit antennas to use the total transmitted power as effectively as possible. Dominant eigenmode transmission uses complex weight factors, optimizing also the phase relationship between the signals, leading to reduction in total transmitted power by generating the most power-efficient communication link for transmission towards a specific base station.

References

[1] F. Declercq, H. Rogier, and C. Hertleer, "Permittivity and loss tangent characterization for garment antennas based on a new matrix-pencil two-line method," *IEEE Transactions on Antennas and Propagation*, **56**, 2548–2554, (Aug.) 2008.

[2] P. Salonen, L. Sydänheimo, M. Keskilammi, and M. Kivikoski, "A small planar inverted-F antenna for wearable applications," *Third Int. Symposium on Wearable Computers*, pp. 95–100, 19 October 1999.

[3] P. J. Massey, "Mobile phone fabric antennas integrated within clothing," *Eleventh International Conference on Antennas and Propagation*, vol. 1, pp. 344–347, 2001.

[4] P. Salonen, Y. Rahmat-Samii, H. Hurme, and M. Kivikoski, "Dual-band wearable textile antenna," *Antennas and Propagation Society International Symposium*, vol. 1, no. 4, pp. 463–466, 20–25 June 2004.

[5] P. Salonen and M. Keskilammi, "Softwear antenna," *IEEE Military Communications Conference, MILCOM 2008*, pp. 1–6, 2008.

[6] C. Hertleer, H. Rogier, L. Vallozzi, and L. Van Langenhove, "A textile antenna for off-body communication integrated into protective clothing for firefighter," *Special issue of Trans. Antennas Propag. an Bodycentric Wireless Comm.*, **57**, (*4*), 919–925, 2009.

[7] P. Salonen, Y. Rahmat-Samii, and M. Kivikoski, "Wearable antennas in the vicinity of human body," *Proc. Antennas and Propag. Soc. Int. Symp.*, pp. 467–470, 2004.

[8] L. Vallozzi, H. Rogier, and C. Hertleer, "Dual polarized textile patch antenna for integration into protective garments," *IEEE Antennas Wireless Prop. Lett.*, **7**, 440–443, 2008.

[9] F. Boeykens, L. Vallozzi, and H. Rogier, "Cylindrical bending of deformable textile rectangular patch antennas," *Int. Journal of Antennas and Propagation*, article 170420, 11 pp., 2012.

[10] T. Kellomäki, J. Heikkinen, and M. Kivikoski, "Effects of bending GPS antennas" *Asia-Pacific Microwave Conference, APMC 2006*, pp. 1597–1600, 12–15 Dec. 2006.

[11] M. Scarpello, D. Vande Ginste, and H. Rogier, "Design of a low-cost steerable textile array operating in varying relative humidity conditions," *Microwave and Optical Technology Letters*, **54**, (*1*), 40–44, 2012.

[12] L. Vallozzi, P. Van Torre, C. Hertleer, *et al.*, "Wireless communication for firefighters using dual-polarized textile antennas integrated in their garment," *IEEE Transactions on Antennas and Propagation*, **58**, (*4*), 1357–1368, 2010.

[13] P. Van Torre, L. Vallozzi, H. Rogier, and J. Verhaevert, "Diversity textile antenna systems for firefighters," in *Proceedings of the Fourth European Conference on Antennas and Propagation (EuCAP)*, pp. 1–5, April 2010.

[14] S.L. Cotton and W.G. Scanlon, "Channel characterization for single- and multiple-antenna wearable systems used for indoor body-to-body communications," *IEEE Trans. Antennas and Propag.*, **57**, (*4*), 980–990, 2009.

[15] I. Khan, P. S. Hall, A. A. Serra, A. R. Guraliuc, and P. Nepa, "Diversity performance analysis for on-body communication channels at 2.45 GHz," *IEEE Trans. Antennas and Propag.*, **57**, (*4*), 956–963, 2009.

[16] I. Khan and P. Hall, "Experimental evaluation of MIMO capacity and correlation for narrow-band body-centric wireless channels," *IEEE Trans. Antennas and Propag.*, **58**, (*1*), 195–202, 2010.

[17] Y. Ouyang, D. J. Love, and W. J. Chappell, "Body-worn distributed MIMO system," *IEEE Trans. Vehicular Technology*, **58**, (*4*), 1752–1765, (May) 2009.

[18] J. M. Molina-Garcia-Pardo, J. V. Rodriguez, and L. Juan-Llacer, "Polarized indoor MIMO channel measurements at 2.45 GHz," *IEEE Trans. Antennas and Propag.*, **56**, (*12*), 3818–3828, 2008.

[19] A. Serra, A. Guraliuc, P. Nepa, *et al.*, "Dual-polarisation and dual-pattern planar antenna for diversity in body-centric communications," *Microwaves, Antennas Propagation, IET*, **4**, (*1*), 106–112, 2010.

[20] P. Van Torre, L. Vallozzi, C. Hertleer, *et al.*, "Indoor off-body wireless MIMO communication with dual polarized textile antennas," *IEEE Trans. Antennas and Propag.*, **59**, (*2*), (Dec.) 2010.

[21] C. Oestges and B. Clerckx, *MIMO Wireless Communications: From Real-World Propagation to Space-Time Code Design*, p. 194, Academic Press, 2007.

[22] S. M. Alamouti, "A simple transmit diversity technique for wireless communications," *IEEE Journal on Selected Areas in Communications*, **16**, (*8*), 1451–1458, 1998.

[23] M. Scarpello, D. Vande Ginste, and H. Rogier, "Design of a low-cost steerable textile array operating in varying relative humidity conditions," *Microwave and Optical Technology Letters*, **54**, (*1*), 40–44, 2012.

[24] P. Van Torre, L. Vallozzi, H. Rogier, M. Moeneclaey, and J. Verhaevert, "Channel characterization and robust tracking for diversity reception over time-variant off-body wireless communication channels," *EURASIP J. Adv. Signal Process*, **2010**, 7:1–7:13, (Feb.) 2010. [Online], available: http://dx.doi.org/10.1155/2010/978085.

[25] P. Van Torre, M. Scarpello, L. Vallozzi, *et al.*, "Indoor off-body wireless communication: Static beamforming versus space–time coding," *International Journal of Antennas and Propagation – Special Issue on MIMO Antenna Design and Channel Modeling,* (Feb.) 2012.

Index